Parameters of Biological Materials

parameter	air	water	ice	fatty tissue	bone	blood	tendon	muscle
density ρ [kg/m³]	1.29	1000	920	≈ 940	1800			1058
electrical resistivity ρ_{el} [$\Omega \cdot$m]		$2 \cdot 10^3$		$25 \cdot 10^7$		$1.5 \cdot 10^7$		$5 \cdot 10^7$
impedance Z [kg/(m²s)]	440	$1.5 \cdot 10^6$		$1.4 \cdot 10^6$	$6 \cdot 10^6$			$1.7 \cdot 10^6$
index of refraction n	1	1.33						
kinematic viscosity v [m²/s]	$1.6 \cdot 10^{-5}$	$1.0 \cdot 10^{-6}$				$4 \cdot 10^{-6}$		
specific heat C [kJ/(kg K°)]		4.18	2.0					
specific heat const. pressure C_p	723							
speed of sound v [m/s at 20°C]	-			≈ 1480	3360			1570
tensile strength V [··					$2 \cdot 10^{10}$		$1.5 \cdot 10^9$	$\approx 2 \cdot 10^5$
thermal condu								

Often Used M

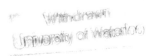

body, model anim	....me V	surface area	mass moment of inertia I
circle		πR^2	
sphere, radius R (bird)	$(4\pi/3)R^3$	$4\pi R^2$	$(2/5)\,m R^2$
long cylinder, mass m (whale)	$A L = \pi R^2 \cdot h$	$2\pi R^2 + 2\pi R h$	
cross section $A = \pi R^2$, length L			
... rotating about center of mass			$(1/12)\,m L^2$
... rotating about hip (elephant leg)			$(1/3)\,m L^2$
cone (chicken leg, mass m)	$(1/3)\,A L = (\pi/3)R^2 L$		$(1/10)\,m L^2$

Physical Constants

constant	value
acceleration of gravity	$g = 9.81$ m/s²
Avogadro's number	$N_A = 6.02 \cdot 10^{23}$ molecules/mole
Boltzmann's constant	$k_B = 1.38 \cdot 10^{-23}$ J/K
electron mass	$m_e = 9.11 \cdot 10^{-31}$ kg
fundamental charge	$e = 1.6 \cdot 10^{-19}$ Cb $= 1.6 \cdot 10^{-19}$ J/V
gas constant	$R_g = 8.31$ J/(mol K)
magnetic field of earth	$B_e = 5 \cdot 10^{-5}$ T $= 0.5$ G
Planck's constant	$h = 6.6 \cdot 10^{-34}$ Js $= 4.1 \cdot 10^{-15}$ eV $\cdot$ s
proton mass	$m_p = 1.67 \cdot 10^{-27}$ kg
solar constant	$S = 1.37$ kW/m²
speed of light	$c = 3 \cdot 10^8$ m/s
electric constant	$\varepsilon_0 = 8.85 \cdot 10^{-12}$ F/m
Stefan's constant	$\sigma = 5.67 \cdot 10^{-8}$ W/(m² K⁴)

Energy and Power Units

1 cal $= 4.18$ J $= 4.129 \cdot 10^{-2}$ liter $\cdot$ atm
1 J $= 1$ N $\cdot$ m $= 10^7$ erg
1 eV $= 1.6 \cdot 10^{-19}$ J
1 horse power $= 746$ W

Zoological Physics

Boye K. Ahlborn

Zoological Physics

Quantitative Models
of Body Design,
Actions, and Physical Limitations
of Animals

With 259 Figures and 67 Tables

Springer

2004

Professor Emeritus Boye K. Ahlborn

Department of Physics
University of British Columbia
6224 Agricultural Road
Vancouver V6T 1Z1
Canada
e-mail: ahlborn@physics.ubc.ca

Cataloging-in-Publication Data applied for
Bibliographic information published by Die Deutsche Bibliothek
Die Deutsche Bibliothek lists this publication in the Deutsche Nationalbibliografie; detailed biblio-
graphic data is available in the Internet at <http://dnb.ddb.de>.

ISBN 3-540-20846-1 Springer-Verlag Berlin Heidelberg New York

Springer-Verlag is a part of Springer Science+Business Media

springeronline.com

© Springer-Verlag Berlin Heidelberg 2004
Printed in Germany

The use of general descriptive names, registered names, trademarks, etc. in this publication does not
imply, even in the absence of a specific statement, that such names are exempt from the relevant pro-
tective laws and regulations and therefore free for general use.

Typesetting: Michael Ascheron using QuarkXPress®
Cover design: Erich Kirchner, Heidelberg

Printed on acid-free paper   SPIN 10757502   57/3141/ba   5 4 3 2 1 0

To my wife, Eike for unending love and sharing with me the fascination for animals, and to my daughters Birgid, Caren, and Dorit, who gave us a golden retriever some 15 years ago. This animal has made me aware that intelligence, individuality, and self-awareness are not restricted to the human race.

Prolog

Life is not fixed in stone as one might think,
Because it is enacted by the genes,
Which guide a being like a book of laws.
Life gradually yields to new designs
That outperform the old shape in some way.

The laws of evolution are not fixed
They change according to new needs.

Life is much wiser than some laws of men
That tie the offspring to some ancient truths,
Which may no longer meet the present needs,
And contradict the moral of the time.

But all behind the dictum of the genes
Loom physics laws that hold –
As we believe – to all eternity, and rule
The universe in silent elegance.

We catch a glimpse in splendor of design
When we inspect the microbe and the whale,
When we observe the jumping of a cat,
Or listen to the honking of the goose.
The forceful drummers in the back
Are physics laws that set the rhyme and rhythm
That dictate what must be and what cannot.
There is no good or bad in physics laws
Just consequence: A leads to B, and B demands the C.
And all this comes together in a being.

Take what you need, and care not for your loss,
But leave enough for your environment
That nurtures you in turn
In interacting harmony, called love.
Just carry on and pass the torch of life
To your descendant who is next in line,
To take his share of energy and time
From the eternal universe
That is so bare of feeling.

Life uses *energy* and *information* to flourish and evolve. Physical and chemical principles lurk behind the life functions and body design of animals. Genes, cell dynamics, metabolism, photo-synthesis, and other cell activities, and microscopic structure questions are studied in the traditional field of *Biophysics*.

There is however more, and other physics in larger living structures: *Energy* plays a role in metabolism, thermal control, and locomotion. These activities must conform to the laws of thermodynamics, fluid flow, and mechanics. *Information* collection by the long distance senses relies on the principles of optics, acoustics, and electricity, and magnetism. Physics laws enable and limit the body design, and control the activities of animals, and the gradual acquisition of new physics tricks, like running, flying, seeing, and hearing, are the stepping stones of evolution.

Such topics are generally not included in *biophysics* studies. Yet they are of interest to people of all ages and many walks of life. Such topics also provide an easy entry into a new arena of applied physics at the crack lines between Physics and Zoology. The main aim of this text is to weave the tenuous webb of unrelated zoological observations, and physical laws into a coherent text on body design, actions, and physical limits of animals within the new field *Zoological Physics*.

Since 1995 colleagues from Zoology and myself have dealt with such topics in a course, *Physics of Animals*, given to senior undergraduate students at the University of British Columbia. The themes of this book were tried out as lecture material, and issued to the students as annually upgraded lecture notes. The topics were introduced through naïve questions about animals, and life functions. Physics was brought in only on the basis of "need to know". However, in the attempt to look at various life-functions the whole spectrum of classical physics: mechanics, thermodynamics fluid flow, optics, acoustics, and electricity and magnetism comes into focus. Initially articles from Scientific American, Physics Today, Nature, and various popular books inspired the course material. These topics were gradually supplemented by problems that seemed to fit, by anecdotal knowledge from colleagues, by quantitative material from books, by articles from over 30 different journals with Physics-, or Biology content, and by model calculations of my own.

A rich specialized zoological literature exists for many topics described here. But to my knowledge this book is the first attempt to define a new teaching and research field that brings macroscopic Physics to overlap with Zoology. Many other topics could have been included. In the attempt to develop the models from first physics principles, often in ignorance of ongoing work in specialized areas, I may have occasionally come to conclusions that differ from the accepted views. Also quite likely am I unaware of many relevant studies. No doubt will these pages contain some errors and misrepresentations. For these shortcomings I apologize. Further, the reader should not take the derivations or conclusions as gospel, but rather treat the text as a set of methods and backup facts to interpret the actions and body designs of animals within the framework of physics laws. And thus I hope that the principal aims of this book are recognized: First, to show how simple physics models can give a deductive perspective that adds to the inductive method of observational biology augmenting the repertoire of applied physics; and second, to provide the tools for the interested reader to develop Zoological Physics models of his or her own.

By stressing how concepts lead to formulas, and by avoiding calculus, whenever possible, I attempted to make the material accessible to readers, who only possess some basic training in Physics and Zoology. Over the years the audience of this course has included students from Biology, Chemistry, Engineering Physics, Human Kinetics, Integrated Science, Mathematics, Physics, Plant Science, and Zoology. These diverse students found challenge in different aspects of the course material. Everyone advanced to the level of independent project work, on a question of his or her choice, which combines physics and zoology. Some of this student research is incorporated here. The mathematics-physics inclined students detected new applications for their analytical skills as they learned to model the design, and the actions of animals. The life science students discovered that physics is actually useful for finding quantitative answers to some zoological questions.

The topics have lent themselves to popular talks given to lay audiences, to first year general science students, and to special seminars in different Science and Engineering departments in Canada and abroad, always evoking lively debates. Therefore, I would not be surprised if this book will find use in other undergraduate courses, and it might even fit onto the book-shelves of private homes.

This book is an extensions of concepts and calculations from my lecture notes, now in the 7th edition, used at the University of British Columbia [Ahlborn 2002]. It owes its existence to an intense interaction between the author, and colleagues from the Department of Zoology: Robert W. Blake, John Gosline, Margo Lillie, and William Megill. They helped to select the topics, and contributed material from their own fields of specialization: Biomechanics and Biomaterials. They made sure that the biological perspective did not suffer from oversimplified physics models. First John, and William, and lately Margo have shared in the teaching, and the development of course strategy. Robert Blake has collaborated with myself on several bio-mechanics and bio-thermodynamics research topics which form part of the text. The close interaction of physicists and zoologists leads to a quite comprehensive but novel view of Zoological topics, which enriches the understanding of biologists and physicists alike. I sincerely thank these wonderful colleagues, and these great students for all their help and inspiring questions. I am also much indebted to Eckhard Rebhan for many discussions during the last 40 years about physics, and how to find, in each case, the core of the puzzle, so that solutions can be uncovered.

I would also like to acknowledge valuable comments from David Measday who read the acoustics chapter, and from Franck Curzon, who read the electricity and magnetism chapter. Special thanks is due to Jeff Gordon who has developed a similar course at the Ben Gurion University of the Negev in Bersheeba. We have talked about zoological physics topics on numerous occasions, and Jeff helped me with some optics sections. Finally I am very grateful to Margo Lillie, Mia Love, and my daughter Dorit Mason, for help with the proof reading, and I would like to thank Cherry Wu for writing the Chinese poem at the beginning of Chap. 3 and Don Witt for his patient help with software, and computer problems.

Vancouver,   *Boye Ahlborn*
February 2004

Contents

1. Life: Information, Matter, and Energy

> Leben ist,   *Life's secret is*
> *daß im Wandel der Materie*   *to remember its form*
> *die Form erhalten bleibt.*   *as matter passes through it.*
>
> Thomas Mann

Necessities of Life

All organisms are made up from matter drawn directly or indirectly from their inanimate surroundings. They consume energy that ultimately comes from the sun or from thermal vents. All organisms rely on information, which is passed on in their genes or gleaned from their senses. Genes contain the words of instructions to build the body and to carry out certain actions. For every organism life begins with its genetic words. Therefore in, a true sense, as St. John says in the gospel *In the beginning was the word*.

The major topics of this book are outlined in Sect. 1.1. Life is action and response. However, the transition from inanimate material to living structures is gentle. Even inanimate matter exhibits actions if there are differences of certain physical quantities like pressures, forces, number densities, or charges. This is described in Sect. 1.2. Every organism must eat. It takes in materials and chemical energy at a well-defined daily rate, called the metabolic rate Γ, which is a certain function of the body mass M. The resting metabolic rates Γ_0 of inactive warm blooded animals scale with their body masses as $\Gamma_0 = \text{const } M^\alpha$, where $\alpha \approx \frac{3}{4}$. This type of equation, where the body mass is raised to some power α is called an allometric relation. Many other empirical allometric relations have been discovered. Metabolic rates and allometry are discussed in Sect. 1.3. The examples in this book are developed as quantitative models of body design, actions, and physical limitations of animals. Some words of caution about the accuracy of such calculations are given in Sect. 1.4. It is further explained how Physics dovetails with Zoology, and how *Zoological Physics* fits within the framework of other fields of science.

1.1 Physics and Life

The transition from inanimate world to living organisms is blurred. At the microscopic level "dead" lipids spontaneously assemble into lipid bilayer vacuoles that look like cell walls and may deform like living cells. On the other hand some virus structures have the appearance of inorganic crystals.

At the macroscopic level inanimate materials are involved in *automatic* actions and changes. These events are driven by differences of some physical parameters such as temperatures, concentrations, pressures, forces, and electric charges. Differences of temperature induce heat to flow, differences of particle densities cause diffusion flows. Differences in pressures create winds or push objects around. Differences of static electrical charges let lightning strike. These differences set the arrow of time; they drive the hourglass of change. Such automatic actions are the precursors to similar events occurring in all organisms, and are responsible for certain action of animals.

What sets living organisms apart from inanimate matter is the purpose and organization of form, features, and activities arising from the genetic code and other information. Based on the inherited information animals assemble their body mass M, and consume a certain daily measure of energy, at the metabolic rate Γ. Both, matter and energy are constantly flowing through each organism. They are temporarily "on loan" from the surrounding. However, truly owned by each organism for its entire life span is its genetic code, and the external data that are collected by the senses, processes by the brain and stored as "information", "experience" or "knowledge" in the memory of each individuum. Evolution is accompanied by an explosive growth of the information, which becomes available to each succeeding generation.

As Shermer [2001] observes "Science is an exquisite blend of data and theory... The facts never just speak for themselves. They must be interpreted through ideas". Zoology is traditionally an empirical science. *Zoological Physics* attempts to derive quantitative models of body design, actions, and physical limits of animals, which are then compared to data from living organisms. Such an endeavor is possible because physical models are very simplified images of the complex structures of life. They use just a few well defined physical quantities such as mass time, length, speed, temperature, energy, power, force, frequency, wavelength. The most important parameters of life forms are energy, matter, and information.

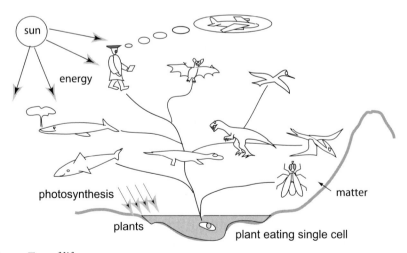

Fig. 1.1. Tree of life

Genetic information, energy from the sun and material from the earth all contribute to evolution. This is sketched as a tree of life in Fig. 1.1, from single cell organisms at the bottom to Homo Sapiens at the top.

In a very simplified way the evolution can be seen as an increase of body mass with time, Fig. 1.2. How life began has been a fundamental question for philosophers of all ages. The German poet Goethe struggles with it in Dr. Faustus' translation of the first line of genesis:

Geschrieben steht:	The scripture reads:
"Im Anfang war das Wort."	"Initially was the word."
Hier stock ich schon. Wer hilft mir weiter fort?	Here I get stuck who gives me help of sort?
Ich kann das Wort so hoch unmöglich schätzen	For sure the word is overrated
Und muß es anders übersetzen	It must be differently translated
Wenn ich vom Geiste recht erleuchtet bin	If inspiration clears my thinking's mist
Geschrieben steht:	The scripture means:
"Im Anfang war der Sinn."	"Initially was the gist."
Bedenke wohl die erste Zeile	No, it means "sense!" Don't write it down too fast
Daß deine Feder sich nicht übereile	Consider that your work should stand to last.
Ist es der Sinn der alles wirkt und schafft?	Is it the sense that sets occurrence's course?
Es sollte stehn: "Im Anfang war die Kraft."	It should translate: "Initially was the force."
Doch auch indem ich dieses niederschreibe	But even as I write this in the annal
So warnt mich was das ich dabei nicht bleibe	I feel a warning that this is not really final
Mir hilft der Geist. Auf einmal seh ich Rat	All of a sudden I find satisfaction
Gerschrieben steht: "Im Anfang war die Tat."	The scripture says: "Initially was the action."

(translated by the author)

These lines encompass four fundamental concepts that are important for all organisms: *Action* and *Force*, which are connected to the energetics of life, and *Sense* and *Word*, which are connected to information. Living and dead matter is subject to the action of irreversible flux processes or forces: diffusion, fluid flow, heat-energy transfer, and charge flow. Mechanical forces create motion of macroscopic structures. But life is more than motion. As Thomas Mann observes "Life is remembering the form as matter flows through it." The form is maintained by the internal information, the connection of the molecular letters ACGT into words of the genetic code of each cell. Life is also the continuous evolution that brought sin-

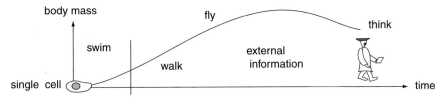

Fig. 1.2. Evolution of body mass. Early animals only swam, in the ocean later the they learned to walk and fly to occupy the land and the air

gle cells to the level of octopus and Homo sapiens. And life is the ongoing evolution that we witness around us. This evolution would not proceed without input of information from the surroundings gathered by the senses.

This book addresses these four components of life and related material from a physics perspective in order to derive quantitative models of body design, action and physical limitations of animals. The work generally contains the following steps: discuss the underlying physical principles, present the relevant fundamental equations, construct simplified models, collect and insert the pertinent material constants, derive simplified equations, and calculate numerical values.

The topics are arranged, as they might have become important for the young life on earth. Chapter 1 begins with the metabolic rate Γ, an empirical relation for daily ration of food, which is needed to maintain the body functions. Also discussed are some other empirical allometric scaling relations, which are later used to compare animals of different sizes. The metabolic rate affects many life functions. Its dimension is energy divided by time.

The first steps of physical modeling must involve relations between energy and matter. It is the realm of thermodynamics, presented in Chap. 2, where temperature T, and the first law of thermodynamic rules supreme over animals of any size. Since higher animals operate at constant body temperature, questions of heating and cooling are included in this chapter. Evolution leads from small to larger animals, and from water to land and into the air. This brings up questions about body forms, and materials that can withstand all the forces that animals may encounter, Chap. 3. Single cells can feed themselves by diffusion alone. More involved fluid flow system become necessary for larger animals. They are described in Chap. 4. Big animals must move around to gather their food. Locomotion, Chaps. 5 and 6, is governed by the physics topic dynamics with Newton's equation, at the center. Swimming, flying, and running are the ultimate mechanical achievements of animals.

Organisms rely on internal – and external information. Genes store the fixed internal information, which is passed on from generation to generation. Animals have developed various senses to collect external data, and brains to interpret the information. External information helps in many ways in the struggle for existence. The second half of the book describes the physical principles that enable the collection of external information. The primary long-distance senses rely on waves to receive messages. A short review of the different types of collected information, and properties of the carrying medium, waves, is given in Chap. 7. Different forms of eyes and limitations of the imaging process are described in Chap. 8. Sound for hearing and voice communication is discussed in Chap. 9. Some electrical effects, important to animals, are outlined in Chap. 10. Chapter 11 is a summary of general findings, amplifying Darwin's prophetic words "in the struggle for existence... the merest trifle (of difference) would give the victory to one organic being over another". Darwin continues "but probably in no one case could we precisely say why one species has been victorious over another". It is hoped that after going through the derivations presented here the reader will disagree with Darwin. Instead it appears that modeling performed within the framework of *Zoolog-*

ical Physics can help to lift the veil of mystery. It becomes apparent that animals at every level of evolution have used physics laws to their advantage, millions of years before these principles were discovered by scientists and engineers.

The purpose of the first chapter is threefold. First, we describe how some actions and processes that living organisms utilize are already present as physical phenomena in the inanimate world. Second, we report the empirical quantitative relation between body mass M, and the daily energy consumption Γ, of warm blooded animals. It is known as the metabolic rate $\Gamma \approx 4 M^{3/4}$, and we describe some other body parameters f which can be given as allometric relations, namely functions of the form $f = a M^{\alpha}$, where $\alpha \neq 1$. We believe that all these allometric functions ultimately express some physics limitation imposed onto life. Indeed some of these functions will be derived from first principles or traced back to the metabolic rate in later chapters. Third, we delineate how the new subject of *Zoological Physics* differs from traditional Biophysics, and how it fills in holes between the disciplines of Biology, and Physics. Further, we comment on the strength and shortcomings of Physical models of biological systems.

1.2 The Hourglass of Change

The inanimate world is full of action and motion. Much of it seems to go on "by itself" or in the language of physicists due to irreversible processes. Continental drifts, volcanism, climatic variations, shape the continents. Rain, wind, ice, and the diurnal and seasonal variations of temperature, pressure, and solar radiation contribute their part to change the local terrain. Winds blow, lightning strikes. Hurricanes whip up the oceans and uproot trees, rain falls, lakes freeze over, rivers tumble down the mountains. Waves pound the shores, and within this scenery life flourishes and evolves. What is cause and what is effect of these automatic reactions? We will see later that these reactions are actually used by living organisms.

1.2.1 Differences and Chemistry Make Things Happen

It seems that many things happen by themselves. However, in reality the differences, generically called potentials Δf, make things happen. The quantity f can stand for temperature, T, pressure, p, number density, n, force F, and charge, e.

We measure time by the events that happen: years by the seasons, days by the rotation of the earth, minutes by the dropping of sand in an hour glass, or the dripping of water in a water clock, microseconds by the ringing of some quartz crystals in a computer chip, nano seconds by the return time of a light beam in a laser cavity. Time has meaning for us if we can relate it to some action. Figure 1.3 indicates some familiar events.

A foot on ice gets cold, balloons loose their shape after a while. A sugar cube in water gradually dissolves and after a few days all the water tastes equally sweet, an iron nail in water will start to rust. Batteries loose their charge, objects perched on

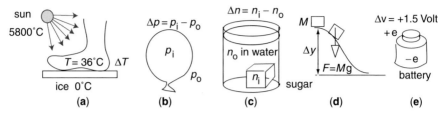

Fig. 1.3. Quantities that can induce change. (a) Temperature differences ΔT, (b) pressure differences Δp, (c) concentration differences Δn, (d) gravitational potential differences Δy, and (e) electrical potential differences Δv

a cliff will slide down. Such automatic actions, are the physical background against which life unfolds. We intuitively know that all these automatic actions – these predictable events – reduce the differences that caused them. They are "just chemistry". We know that the actions stop when the differences have disappeared, when "equilibrium" is reached, when stable, that is less energetic, chemical compounds have been formed. No differences, no impromptu actions,… and no life on earth. Time seems to stand still. These automatic actions have been harnessed by life in many ways and they are incorporated in the actions of every organism on earth. These automatic effects of the inanimate world are indispensable for many activities of animals. The rusting iron nail illustrates a chemical reaction. Eating and digestion involves chemical reactions that supply the body with energy.

The very close link between the daily energy consumption and body mass will be discussed later in this chapter. The warm foot standing on the ice initiates heat conduction. As time goes on heat will flow, the foot loses energy and gets cold. Simultaneously some ice gains the same amount of heat energy and melts. Warm-blooded animals must avoid such loss of heat. The temperature difference ΔT drives the heat flow: the larger ΔT, the faster the heat will flow. Actions and processes associated with heat flow will be discussed quantitatively in Chap. 2.

The balloon in Fig. 1.3b keeps its shape by the excess pressure $\Delta p = p_i - p_o$. Similarly a caterpillar gets its shape from the excess pressure Δp of its body fluids which are in balance with the tension forces in its skin. Such processes are discussed in Chap. 3. As time goes on some air will diffuse through the rubber membrane, and the balloon will shrivel up, the higher Δp, the faster the air escapes through minute holes in the balloon. When an animal dies the skin becomes a limp bag. This is diffusion through a membrane, a process basic to the absorption of oxygen in the lung. A sugar cube in water gradually dissolves. First the sugar molecules stay close by, but then they gradually diffuse into every corner of the fluid. The sugar molecules "diffuse" away from the crowd. Diffusion wipes out the difference and gradually fills every segment of the fluid; however, the farther away from the sugar cube the lower the diffusion rate. Diffusion flow proceeds the faster the larger the difference $\Delta n = n_i - n_o$. Diffusion processes limit the maximum size of single cell organisms. When a spider sucks out some of the juicy interior of its prey a convection pipe flow is set up. Fluid flow and diffusion are discussed in Chap. 4.

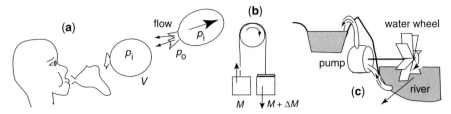

Fig. 1.4. (a) Inflating a balloon to make it fly. (b) Lifting a weight by lowering a larger one. (c) Pumping water with the help of a water wheel that extracts energy from a river

If the orifice of the balloon is not tightly sealed, a jet stream of air will be expelled by the pressure difference and the balloon will lurch forward, Fig. 1.4a. The pressure difference generates an air-flow that propels the balloon. Such jet propulsion is actively used by squid, and octopus, see Chap. 5.

A rock perched precariously on a cliff will slide down sooner or later. The force of gravity pulls it down. As it slides down the slope it loses potential energy, converting most of it into friction work. Often more than one force, is at work. Think of two dogs fighting over a sock. The outcome is quite predictable. In the case of two equally strong dogs there will be a stalemate, Fig. 1.5a. The forces are equal and opposite $\Delta F = F - F = 0$. No motion occurs. This is static equilibrium, a topic of Chap. 3. However if the dogs have different size, Fig. 1.5b, the stronger dog will have the upper hand, because it can muster the force F_a, which is larger than F_b. As we will see in Chap. 5, the resultant $\Delta F = F_a - F_b$ generates an acceleration towards the left, which in turn causes a motion at increasing speed. The big dog does work as he pulls, and work is a form of energy. All living cells accumulate useful atoms or molecules and keep the unwanted ones out. This is the basis of life. It so happens that many atoms preferentially occur in the form of negatively or positively charged ions. Therefore living cells separate charges as they select certain ions. They are little batteries continuously charging themselves to some electric potential ΔV. Charging and discharging of cells is a fundamental action of living organisms that enables nerve conduction and the gathering of information by the senses. Sight and sound are discussed in Chaps. 7, 8, and 9. When a battery is discharged, an electric current j flows, and the electric power $P = j \Delta V$ is dissipated. This is the basis of some electrical effects discussed in Chap. 10.

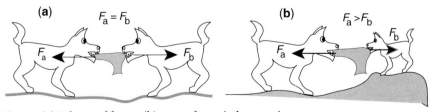

Fig. 1.5. (a) Balance of forces, (b) excess forces induce motion

In each one of these irreversible processes energy changes hands. The energy is lost from the object with the higher potential.[1] However, in total no energy disappears. Part of this lost energy accumulates in the object with the lower potential. The rest dissipates in, friction, or in Ohmic loss, and reemerges as heat which warms up the conducting medium, and finally ends up in the environment. Think of a cable that heats up when an electric current passes through, or think of your hands that heat up when a rough rope slips through your fingers. The common link between these processes is the gradual disappearance of differences, nature's perfect socialism.

One can accurately measure how much energy changes hands, by determining how much energy is needed to repair the damage, and reestablish the old conditions. For instance to re-inflate the limp balloon, Fig. 1.4a one must invest the energy $E = (p_I - p_o)V$, where V is the balloon volume. Given enough energy all these *irreversible processes*, described above, can be reversed. A weight can be raised by lowering a slightly larger one, Fig. 1.4b. Water can be raised with a pump that extracts energy out of the flowing river. Figure 1.4c. Therefore, these processes should rather have been called "ill-reversible".[2] The approach towards total equilibrium can be described with a scaled quantity: energy divided by temperature, called entropy.

1.2.2 Life and Entropy

All these irreversible processes have a common feature: they erase differences, they have a time direction, and they transfer energy to the lower potential. Consider the heat transfer from a hot body T_h, to a cold body T_c. An amount of heat energy $\Delta Q_h = -\Delta Q$ is removed from the upper potential, and is added as $\Delta Q_c = +\Delta Q$ to the lower potential. The sum $\Delta Q_h + \Delta Q_c$ vanishes, because no energy is destroyed or created in the transfer process. However something happens to the scaled quantity.[3] $\Delta S_h = -\Delta q/T_h$ and $\Delta S_c = +\Delta q/T_c$, called entropy S. If the amount of heat Δq passes from the hot to the cold body, there is the change of total entropy $\Sigma S = \Delta S_h + \Delta S_c$, but the transferred heat $|\Delta q|$ is the same. However, since the temperatures are different, the gain ΔS_c is larger than the loss ΔS_h. Therefore the total change of entropy of the universe due to this irreversible process is positive.

$$\Sigma S = \Delta S_c + \Delta S_h \geq 0 \tag{1.1}$$

---

[1] Lost energy Δq is counted negative; energy gained is counted positive

[2] Different processes could be happening at the same time. An electrostatically charged balloon, filled with hot argon gas that is heavier than air, could be perched on a ledge. This balloon would cool, loose its fill gas, and its electrostatic charge, and roll down the preci-pice

[3] Scaled quantities are useful. For instance the stock value of a company when scaled by the annual earning yields the yardstick *price per earning*

For each one of the irreversible flow processes mentioned in the previous section the entropy increase ΣS can be calculated. In every case the total entropy increases as the differences disappear. The price for the automatic action is an increase of the entropy of the universe.

Living organisms distinguish themselves from inanimate structures by the ability to control and utilize potential differences for starting and maintaining actions. Control arises from information provided by the genes or gathered by senses and processed by the brain. Life's secret is the ability to make and maintain potential differences against the natural flow of events, namely to accumulate, and maintain concentrated regions of materials, of pressure, of temperature, of charge. This requires energy, which living organisms extract from metabolic processes that consume more external energy. Animals exist because they can create the concentrations of *good* molecules inside their cells, because they can maintain pressure gradients to make their blood flow, and because they can hold constant body temperatures even in blazing heat of the tropics or in the frigid waters of the Arctic. Survival means that all living organisms are able to assemble and maintain their operating conditions, which are optimized for their position in the biosphere against the trend of natural events. In short, all living organisms continuously work against the disappearances of differences. There is no real socialism in nature only "marriages of convenience", in which each organism gains something in return for giving up some of its resources. In life functions like in many physical processes of dead matter the potential differences are used up in the most gentle way, to increase entropy at the least possible rate.

Entropy has the unit energy/temperature. One does not have an intuitive feeling for entropy. Energy is easier to perceive. One is aware how much energy it takes to climb a mountain. A mountain climber knows how a granola bar, eaten after the ascent, reactivates the body. One knows how far the energy in one liter of gasoline will move one's car. One is afraid of the energy in an avalanche, or the energy in a charging bull. A convenient energy scale for atomic process is temperature T multiplied with Boltzmann's constant $k_B = 1.38 \cdot 10^{-23}$ J/K. The quantity $k_B T$ is the thermal energy per unit mass of an object. Therefore the k_B-scaled entropy

$$\sigma = \Delta S/k_B = \Delta Q/k_B T \qquad (1.2)$$

is a dimensionless number, namely the ratio of the energy (gained or lost) to the average thermal energy per unit mass of the object. Hence $\sigma = \Delta S/k_B$ can be viewed as the percentage-change of heat contents of the object. A percentage change is easy to perceive. We deal with it in everyday life. A much-talked about percentage change is the increase of take home pay in salary negotiations. The most notorious percentage change is the increase in the cost of living. It is somewhat akin to the increase of scaled entropy in the universe.

Entropy is also related to information. An increase of entropy is like an increase of the general uniformity. Less stands out. The information content is lowered when the total entropy $\Sigma \Delta S$ increases. Inversely when information is accumulated there must be a local decrease in entropy, bought most likely at the expense of increasing the entropy in another part of the system.

New treasures of information yield an elevated understanding, a decrease of the general confusion. Clarity of thinking, Genius, lowers the information entropy, and exciting events happen at the crack lines between disciplines like the subject of Zoological Physics.

1.3 Energy, Metabolic Rate, and Allometry

Matter is the substance from which bodies are built, energy is the fuel that drives all processes of life. Both matter and energy are constantly flowing through each animal. They are only temporarily borrowed.

Energy can exist in many different forms, it can move from one object to another, but it is never destroyed. All actions are driven by the conversion of energy from one form to another. The laws that govern the flow of energy and information rule not only the inanimate world and technical devices, but they also govern the structure and action of living organisms. The overall energy consumption is reduced when animals substitute information for body motion; it takes less energy to detect and identify something from a distance by sight and sound, or by electric or magnetic fields instead of going there to taste, or touch. Sharp senses assist in communication, and they can make the difference between life and death. The physical laws directly or indirectly associated with the use of energy are discussed under the headlines heat transfer, statics, dynamics, acoustics, optics and electricity, and magnetism. These laws provide many opportunities and set some limits, which animals have exploited to the fullest.

All animals must continuously support their metabolism. Their energy consumption is called the metabolic rate, Γ with the unit Watt. It came as a great surprise when biologists realized that there is an analytic relation between the body mass and the resting metabolic rate of animals.

1.3.1 Empirical Determination of Metabolic Rates

All animals eat. How much food do big animals need to exist? It is well known that small animals eat less than big animals. Therefore there should be some relationship between the body mass M, and the average (food) energy consumption Γ. Ingesting food gives animals the ability to do physical work in their daily routines. In order to generate energy from a certain mass Δm of food the animal must also inhale a quantity of oxygen (O_2). The reaction yields a certain amount of heat of reaction $\Delta H = \Delta m \cdot h$, and reaction products. The process can be expressed as an energy balance equation:

$$\text{Fuel} + \text{oxygen} = \Delta H + \text{reaction products (waste)} . \tag{1.3}$$

This relation holds for animals as well as for combustion engines. Combustible substances are characterized by their specific heat of the reaction h J/kg. The total

Table 1.1. Specific heat of combustion h of food, and fuel. Data adopted from Tennekes [1997]

	h MJ/kg	\$/kg	\$/MJ		h MJ/kg	\$/kg	\$/MJ
prime beef	4	20	5	starch	21	3.6	0.18
whole milk	2.8	8	2.86	bacon	29	8	0.28
sugar	14–15	0.8	0.06	butter	32	6	0.19
cheese	15	12	0.80	gasoline	42	0.60	0.014
corn flakes	15	6	0.4	natural gas	45	0.30	0.007

heat of reaction ΔH may be given in Joule, J, in calories, cal, or in large calories, Cal, which are converted as 1 cal = 4.18 J, 1 Cal = 1 kcal = 4 180 J. Values for various substances are given in Table 1.1.

The heat of reaction is used to support all life functions. Heat of reaction *pays* for the energy costs of motion, growth, supporting the sensory organs, transmitting nerve signals, and transporting nutrients and wastes as well as for the replacement of tissue and organs. Since oxygen is a necessary component of the reaction, one can measure the energy release rate $\Delta H/\Delta t$ by measuring the oxygen consumption (J = mass of oxygen atoms consumed per second) as indicated in Fig. 1.6. Such measurements have been made for many animals.

The metabolic power consumption Γ of an animal of the body mass M, namely the daily rate of intake of chemical energy ΔH, can then be determined.

$$\Gamma = \frac{\Delta H \text{ Joule}}{24 \text{ h/day} \cdot 60 \text{ min/h} \cdot 60 \text{ s/min}} \tag{1.4}$$

Kleiber [1932] plotted such values for various resting warm-blooded animals on log-log paper, and found a unique function, Fig. 1.7 which has the form

$$\Gamma_0 \approx a M^\alpha \approx 4 M^{3/4}, \tag{1.5}$$

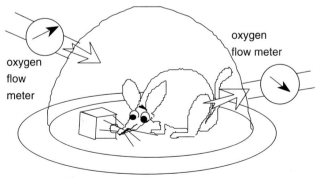

Fig. 1.6. Oxygen consumption

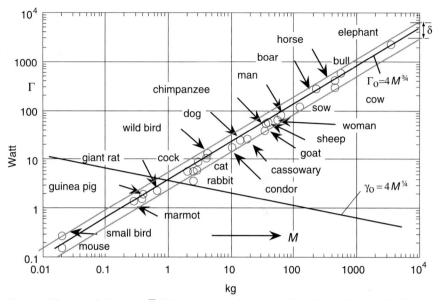

Fig. 1.7. The metabolic rate Γ, "Mouse to elephant curve", and specific metabolic rate $\gamma_0 = \Gamma_0/M$. Data adopted from Schmidt Nielsen [1984]

where M is measured in kg, and the constant $a = 3.6 \approx 4$ has the dimension Watt/kg$^{3/4}$ to give Γ_0 the unit Watt. Active animals exceed the resting metabolic rate Γ_0 by a factor b in the range $2 \leq b \leq 15$. One would expect that the metabolic rate is a linear function of the body mass ($\alpha = 1$), however the numerical value of the metabolic exponent α is not easy to explain.

Initially it was thought [see for instance K. Schmidt-Nielsen 1984] that Γ_0 should scale with the exponent $\alpha = 2/3$, which arises if the metabolic power production just compensates the heat losses of the warm-blooded body conducting heat through its surface. The exponent 0.75 can be derived from physical models that look at the mechanical power loss of the fluid flow in the blood system due to the branching of arteries into smaller and smaller vessels [see Spatz, 1991, West et al. 1997, and Bejan 2000]. However, it seems that neither of these models is complete, because the metabolic rate must account for *all* irreversible losses. It has to include the mechanical power losses of fluid flow as well as the reaction heat losses ($\delta Q = TS$) to the surrounding. Furthermore the scaling $\Gamma = \text{const } M^{3/4}$ with the metabolic exponent $\alpha \approx 3/4$ even holds to the size of bacteria [see Hochatchka, and Somero 2002, and Darveau et al. 2002] where none of the macroscopic heat and convection transfer processes are involved, but where diffusion dominates. The relation (1.5) contains the second law of thermodynamics for living organisms. Quite likely (1.5) implies a net entropy production at the minimum rate.

The metabolic rate has far reaching consequences for the body design and performance, as we will see in many examples. At this place we only draw one conclusion from this experimental result, namely bigger is better in many respects: The specific metabolic rate $\gamma_0 = \Gamma_0/M$, namely the metabolic rate per unit body mass, decreases with M.

$$\gamma_0 = \Gamma_0/M = 4\,M^{-\frac{1}{4}} \tag{1.6}$$

This function is drawn into Fig. 1.7. Big animals have a smaller specific metabolic rate than smaller ones. Pound by pound they don't need so much food and oxygen. That can certainly be an advantage when food is in short supply. As we will discuss in detail later, big animals have less surface area per body mass, and therefore don't lose heat so rapidly as small animals. Big marine mammals can dive longer than small ones, because they can store more oxygen in their muscles.

The animal "human" consumes food containing about 3 000 kcal = 12 540 kJ per day. This is an average power of 12 540 kJ per day/(24 h/day · 3 600 sec/h) = 145 W. Part of the power maintains the body functions, part generates mechanical muscle power, about ¼ supports the brain functions. The rest appears as heat.[4] With this intake a person can deliver about 80 Watt of mechanical power on a continued basis, and can produce short power spurts of 200–1 000 Watt.

Metabolic rates scale with body masses M raised the power $\alpha \approx \frac{3}{4}$. Many other body parameters also scale with the body mass M raised to some power $\alpha \neq 1$. Such functions are called allometric.

1.3.2 Allometry

Body functions that do not scale linearly with body mass M are generally called *allometric*. Quite a number of such functions are empirically known for living and inanimate objects. For instance the power $P = \mathrm{const}\,M^{3.5}$ consumed (emitted) by main sequence stars can be approximated as allometric relation of the star mass M. A short survey of such relations is given here. In later chapters some of them will be derived from first principles, in order to show how physics rules behind the scenes.

Animals have honed their body plans for eons, taking advantage of all the laws of physics, and the limitations of the material constants. It is therefore to be expected that each animal has found – for its niche in the biosphere – the right proportions of muscles, skeleton, lung, brain, heart. That it has found an appropriate lifetime, locomotion- and heart frequency, and has reduced its energetic cost of transport to an absolute minimum. It turns out that some body parameters, labeled f, can be represented as allometric functions of the body mass M.

$$f[\mathrm{u}] = a\,M^\alpha \tag{1.7}$$

The quantity u in the square brackets behind f indicates the unit. For instance if f represents the lifetime τ of an animals the unit of τ could be given in seconds, minutes, hours, days, or years. Relations like (1.7), in which the exponent α differs

---

[4] Today the average power consumption per person in North America is about 80 times as large.

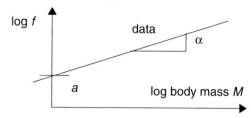

Fig. 1.8. Allometric function. The parameter f with slope $\alpha \neq 1$ and intercept a is shown as function of mass M

from 1 are called "allometric". Such relation can be displayed as a straight lines with the slope α in a log f versus log M plot, Fig. 1.8.

If for instance the life times of animals are displayed as an allometric function each data point describes the average life time of one particular animal. Elephants show up as a point on the right edge of the graph, dogs as a point somewhere in the middle, and mice as a point on the left edge of the graph. Table 1.2 gives the parameters a, and α for various organs or body functions adapted from the literature. Although these parameters are quoted with 2 significant figures these numbers should be rounded off to only one significant figure in recognition of the fact that they describe statistical averages of animals of all ages.

In allometric relations the quantity f cannot be expressed as a fixed percentage of the body mass for small and large animals. Compare for instance the brain mass of a small rodent of $M = 0.1$ kg and a water buffalo at $M = 1\,000$ kg. The rodent

Table 1.2. Allometric parameters for mammals. Most data are data adapted from Vogel [1988], more extensive allometric data are given by Schmidt-Nielsen [1984]

parameter	factor a	exponent α
body surface in m²	0.11	0.65
brain mass (man) in kg	0.085	0.66
brain mass (non primates) in kg	0.01	0.7
breathing frequency in Hz	0.892	-0.26
cost of transport (running) in J/m · k	7	-0.33
cost of transport(swimming) in J/m · kg	0.6	-0.33
effective lung volume in m³	$5.67 \cdot 10^{-5}$	1.03
frequence of heartbeat in Hz	4.02	-0.25
heart mass in kg	$5.8 \cdot 10^{-3}$	0.97
life time in years	11.89	0.20
metabolic rate in W	4.1	0.75
muscle mass in kg	0.45	1.0
skeletal mass (cetaceans) in kg	0.137	1.02
skeletal mass (terrestrial) in kg	0.068	1.08
speed of flying in m/s	15	1/6
speed of walking in m/s	0.5	1/6

has the brain mass $M_{br} = 0.01 \cdot 0.1^{0.7} = 0.002$ kg, which is 2% of its body weight. The water buffalo has the brain mass $M_{bw} = 0.01 \cdot 1000^{0.7} = 1.26$ kg. This represents only 0.12% of its body weight. Relatively speaking the rodent is much brainier. It is quite likely that each one of these relations expresses some optimization involving the laws of physics or limits of phases of materials, and physical constants, such as diffusion constants, strength of materials, latent heats, drag coefficients, electrical properties, etc. One can speculate that similar allometric relations should exist for many more body functions and design parameters.

The most important of these allometric relations is the metabolic rate, which governs many of the body functions such as the mechanical power production of animals. Some of the empirical constants, a and α will be derived from first principles in later chapters. Some of them can be traced back to the still poorly understood metabolic rate scaling.

The allometric relations indicate that body mass is a key parameter for animals. Often bigger is better, but, as will be shown later, there are ranges and limits on the body mass due to various physics principles.

1.3.3 The Benefits of Large Bodies

Large size has many advantages for an animal. Big bodies can store energy in fatty tissue to carry the organism through lean periods. As a quantitative measure one can define a body mass decay time, $\Delta t_{0.1}$ namely the time which it takes for a starving animal to lose 10% of its body mass M.

The metabolism of a warm blooded animal utilizes chemical energy at the rate $\Delta H / \Delta t = \Gamma_0 \approx 4 M^{3/4}$. Assume that the energy is derived from burning body fat with a specific enthalpy $h \approx 2.5 \cdot 10^7$ J/kg. A loss of 10% of body tissue represents the mass $\Delta m = 0.1 M$ with the energy content $\Delta H = h \cdot \Delta m = 2.5 \cdot 10^7 \cdot 0.1 M = 2.5 \cdot 10^6 M$. The 10% body mass decay time is

$$\Delta t_{0.1} = \Delta H / \Gamma = 2.5 \cdot 10^6 \cdot MJ / (4 M^{3/4} J/s) = 6.25 \cdot 10^5 M^{1/4} \text{ s} . \qquad (1.8)$$

This energy decay time increases with $M^{1/4}$. Large animals can stand starvation better than small ones.

The surface area A of an object of mass M scales as $A = \text{const } M^{2/3}$. The surface area per unit mass therefore scales as $A/M = \text{const } M^{-1/3}$. Big bodies have less surface area per unit mass. Therefore they will lose less heat when it is cold. Deep diving whales make frequent excursions into very cold water. Their skin may get cold but the inner organs stay warm. When they dive they have to hold their breath. But oxygen may be stored in tissues, so again large size is an advantage. Large land animals having large legs can roam farther and run faster to find food. Big animals generally have less to worry about enemies. Big size is definitely desirable if food is plenty.

However, the burden of a large body falls onto bones, joints, ligaments, tendons, and muscles to keep the structure up and moving. As we will show later, bodies

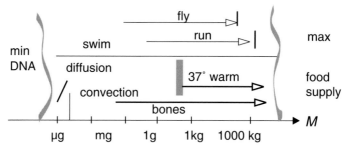

Fig. 1.9. Ranges of body mass of animals

cannot be scaled up like photographic enlargements. Such geometric scaling is ruled out (i) for reasons of statics design, (ii) for limitations in the thermal control, (iii) for restrictions imposed by inertia forces, and (iv) for fluid drag problems associated with laminar and turbulent flow. Geometric scaling generally does not apply, yet some parameters of different animals like the *masses* of their organs, *body functions* like their breathing parameters, and heart frequency, and their *propagation velocities* in water, air and on land, can be described systematically as simple functions of their body mass M.

Each new achievement in the evolutionary chain meets without fail new physical limits, set by the laws of physics and the magnitude of material constants. Evolution has lead from very small organisms that only contained the genetic code to animals with ever-larger masses; very small animals get by with diffusion alone. Animals with body masses larger than about 10^{-9} kg need a convection system to transport nutrients and wastes in and out of the body. Land animals larger than about 0.001 kg must have exoskeletons or bones. Lower mass limits exist for such achievements as active flying and a constant elevated body temperature. Warmbloodedness enables animals to inhabit large tracts of land in colder climates. Active flying gives them a chance to migrate annually across the globe.

Bigger is better because big animals have a smaller specific metabolic rate $\gamma_0 = \Gamma_0/M = \text{const } M^{-\frac{1}{4}}$ and a lower energetic cost of transport. However, being too big is not an asset either: birds larger than $M \approx 10$ kg cannot fly, and huge animals will starve when food becomes scarce. In fact there is a trend in the last 200 million years to replace size by intelligence, or to substitute matter, and energy by information collection-, and processing abilities. Model calculations that support these statements are reported in later chapters, and are summarized in Fig. 1.9.

1.4 Zoological Physics Modeling

Scientific understanding of the world progresses through observation and theory, which mutually reinforce each other. Darwin [1861] described this scientific method with some astonishment: "How odd is it that anyone should not see that all observation must be for or against some view if it is to be of any use." Inherent in this statement is the notion that everything we know is governed by a sense of order, that the operating principle of life is not chaos but plan.

Physicists build models, they practice the art of reducing real problems to simpler situations. Thereby they attempt to bridge the vacuum between empirical facts. Models are often non-exclusive. They are just one possible approach to describe a particular problem amongst others that would yield similar results.

Zoological Physics attempts to derive quantitative models of body design and actions of animals for the purpose of explaining or interpreting zoological data. Therefore this new field of study presents a Physicist's viewpoint of life functions. Physics deals with just a few parameters such as mass, temperature, length, time, and with derived quantities such as frequency, speed, energy, intensity, pressure, and force. Zoological Physics uses these parameters to describe the construction of animals and their various activities, using principles from all fields of classical physics: Thermodynamics, Statics, Dynamics, Acoustics, Optics, Electricity, and Magnetism.

Zoological Physics deals with large-scale effects but occasionally one cannot fail to see how the microscopic events reveal themselves in the macroscopic phenomena. Life rests softly but firmly on the canopy of molecular interactions, utilizing chemical energy sparingly and without haste.

A word of caution of the numerical accuracy of the results is in order. Physics equations generally have a high degree of accuracy. However, in these models one encounters parameters of quite different character: (i) Group behavior parameters like the factors a and α in the allometric equation (1.7), (ii) material constants like the thermal conductivity of blubber, or the tensile strength of tendons, and (iii) physical constants, like Planck's constant or the specific heat of water. Physical constants are known to many decimal places. Material constants of a particular specimen can be measured precisely. But another sample might yield a slightly different value differing perhaps by a few percent. Group behavior parameters may vary significantly: scaling laws that sumarize the behavior of different animals have an inherent scatter of data. This may be illustrated by metabolic rate function Fig. 1.7, where upper and lower bounds have been drawn as lines parallel to the Γ_0 function. These lines are separated by the factor $\delta \approx 2$ indicated on the right border of the figure. Therefore the numerical answers derived here from model calculations should be viewed with some caution. If material constants are involved the answers might be good to $\pm 10\%$. However, group-scaling calculations may easily involve uncertainties of a factor 2. They are however still better than dimensional analysis, and they often allow determining whether the results are upper or lower bounds.

In order to remember that the model calculations leave a certain leeway in the interpretation it is attempted here to avoid "tierischen Ernst" (animal-like seriousness) by sprinkling the text from time to time with light hearted comments and sketches.

1.4.1 Where Zoological Physics Fits in

The building blocks of all organisms are individual cells, which typically measure one micrometer $1\mu = 10^{-9}$ m across, one thousandths of a millimeter. Countless organisms only consist of a single cell. Animals that we can see with the naked eye are often made up from millions of cells. The properties of these atoms of life set the stage for all aspects of life. Therefore an enormous research effort, in *biophysics*, is concentrated on the study of single cells and the design and the workings of their components, particularly the genes. Genes control the reactions and actions that rule the lives of all animals. Genes are the alphabet of the code, which each organism strictly obeys. However, the script of the genetic code in turn is slave to the rules and laws of physics, which are the supreme constitution by which dead matter, from dust to stars interacts, and by which all animals must live.

Single cell organisms deal with many physical effects related to energetics: statistics, light and dark, hot and cold. However, larger animals must heed additional constraints, and enjoy additional opportunities given by the physics of macroscopic scales. For instance single cell organisms have no voice, they cannot see, they are unable to jump or to actively fly, and they cannot maintain a constant body temperature. Such effects become important only for larger organisms. These topics of Zoology are modeled here with physics concepts and equations.

To set these topics apart from the traditional field of Biophysics, we have called the subject *Zoological Physics*.

Of interest are the design, evolution, and actions of animals and their organs. This involves temperature control, allometric scaling, internal transport, limb motion, locomotion, senses, and communication; in short all the effects that are connected to large assemblies of cells, and that are used to manipulate the environment of the organism. This line of interdisciplinary inquiry is fairly new, and much can be explained, as will be shown, with the concepts of classical physics.

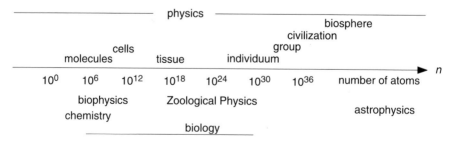

Fig. 1.10. Where Zoological Physics fits in

The distinction between Biophysics and Zoological Physics becomes very apparent when one classifies the objects of study by the number of atoms of which they are composed, see Fig. 1.10. Zoological physics, with subjects ranging from tissue, and organs to individua, and groups of animals occupies the biology of large scales.

1.4.2 The Warp and Weft of Zoological Physics

Life has an extraordinary ability to maintain successful features. What worked once is not easily abandoned, even if it proves to be a dead end given new external circumstances. At the same time life possesses a vital originality and opportunistic inventiveness.

Physics and Zoology have been traditionally treated as separate subjects. However, new insight is often gained at the crack lines between disciplines. *Zoological Physics* intertwines Physics and Zoology like the warp and the weft of a fabric. When studying this new subject one recognizes first that biological and physical principles together, enable present day's animals to function, and second that the evolution of life on earth proceeded because animals "learned" new physics.

The corner stones of life as seen by zoologists are energy and information. Energy is central to the metabolism, to heat budget, and locomotion. The biological processes in the senses and the nervous system together with all aspects of genes and inheritance fall under the headline of information. Consider these topics as the warp of Zoological Physics.

Table 1.3. The warp and the weft of Zoological Physics. The numbers in the table refer to the chapters

zoological principles	zoological topics	thermodynamics	statitics	fluids, waves	dynamics	optics	acoustics	electr. & magn.
energy	metabolism	1, 2	4					
	materials		3					
	locomotion			4	5, 6			
	reproduction							
infor-mation	communication			7		7, 8	7, 9	10
	senses			7		8	9	10
	eyes					8		
	voice & hearing						9	
	nerves							10

Physics generally deals with the inanimate matter. Physics topics may be divided into Modern Physics, and Classical Physics. The first addresses properties of very small and very large objects: atoms and nuclear particles, quantum mechanics, relativity, and the universe. Classical Physics is divided into thermodynamics, mechanics, fluid dynamics, acoustics, optics, electricity and magnetism. These classical fields of physics are the wefts of *Zoological Physics*.

1.4.3 Strength and Limits of Physical Models

Physicists start by defining the properties of their objects such as length, mass, force, time, velocity, temperature, electrical current, thermal conductivity, color etc. They then look for simple laws that relate these quantities, and thereby show the inherent possibilities and limitations of the device. For instance the period of pendulum only depends on its length; a short pendulum must have a short period. A short arm modeled as a pendulum hence must make rapid oscillations.

The main advantage of the physics perspective is tunnel vision: Physicists distinguish primary and secondary effects. First they concentrate onto the primary aspects of a problem. Then they deal with the less important parts, and they often ignore entirely what they consider minor effects. Physicists call this method modeling. Zoologists apply such methods for instance when describing parameters related to body size, and call some results *allometry* and other results *scaling*. Often their allometric relations are found empirically, and they become subjects of debate because Biologists are keenly aware of the multitude of biological effects that might confuse the interpretation. In contrast, physicists try to derive scaling and allometric relations from first principles, neglecting what they deem to be minor effects.

Think for instance of an object moving through the air. If the object is large and heavy say a bison jumping off a cliff, physicists ignore the air resistance entirely. If the object is small and light like a spider jumping down from a branch, the air resistance becomes important. It slows down the motion, and yields a terminal velocity. If the object is shaped like the wing of a bird, the generated lift becomes the main effect. By this approach of distinguishing principal effects, secondary contributions, and negligible phenomena, physicists possess a divide and conquer strategy, which facilitates a simplified description of the processes, called first approximation. In most cases one can improve the accuracy of the description, by including less important effects in a "second" approximation.

Physical models of organs and animal actions have their strengths and weaknesses. Three nice aspect of physics modeling are: (i) the ability to fill in gaps between existing data, (ii) the power to extrapolate and make predictions about the actions of organisms under extreme conditions, and (iii) to reveal ultimate limits. Three examples may illustrate this predictive power: (i) What is the maximum size of organisms that entirely rely on diffusion to nourish every corner of their body? We will see that critters with body dimensions below typically 1 μm can get by with diffusion, but larger animals need an internal convection system. (ii) Suppose Astronomers discovered somewhere in our milky way the planet Gaia, with a mass four times the mass of earth and with an atmosphere, possibly harboring life, as on earth. With methods described here we can predict the body mass of the largest flying creatures there, and the minimum size of warm-blooded animals living in Gaia's oceans. (iii) Physical scaling laws allow us to predict with certainty that no animal eyes can *see* a virus or a CO_2 molecule.

Physical laws enable and constrain the design and the action of animals. Animals that use these laws judiciously, and find optima for material strength, body dimensions, process control, or modes of collecting and processing information will benefit in a given situation. Therefore one could say that Darwin's rule of "survival of the fittest" is really survival of the best physics; the fitness landscape of animals is sculptured by the laws of physics.

Zoological Physics modeling has both an obvious and a hidden weakness. The clearly apparent weakness is the uncertainty of numerical results when extrapolating from one set of data to similar actions of another specimen. The hidden weakness is the lack of exclusive proof for the uniqueness of a particular result. When a model, which is based one particular biological function, yields agreement with some experimental data, one cannot rightfully claim that this and only this function is responsible for the observed action. There may very well be other phenomena of even more importance to the organisms that lead to the same body design or activity. The animal just has learned to optimize its activities simultaneously to satisfy quite diverse requirements, as if it had managed to dance gracefully to the simultaneous tunes of different drummers. On the physics side of modeling there is also ambiguity. Different starting assumptions and physics concepts could also be invoked, and might in fact yield a better understanding under slightly different circumstances. The diversity of approaches is well illustrated in the context of co-evolution. Aims can often be reached by different means; birds fly on feathers, bats fly on skin. Mammals hear with ears on the head, some insects hear with their legs. Frogs achieve high image resolution with lens eyes, while flies achieve it with the shaped light pipes in their facet eyes.

Hence when interpreting the results of model calculations, one must keep an open mind. The physical model is like the shadow of a multi-dimensional structure, it never reveals the full view of the object. Physics is somewhat like Poetry; the essence lies not in the symbols and words but in the message, which may be conveyed with different allegories. This knowledge should help to take one's own results not as the exclusive truth, but as an approximation of reality.

Table 1.4. Frequently used variables of Chap. 1

variable	name	units
a	allometric constant	
b	activity factor	
E	energy	J
F	force	N
h	heat of reaction	J/kg
H	chemical energy	J
j	electrical current	A
J	Mass flow rate	kg/s
M	body mass	kg
n	number of molecules	
N_m	number of mols	
p	pressure	$N/m^2 = Pa$
P	power	W
Q	heat	J
Q'	heat flux	W
S	entropy	J/K
T	temperature	C°, or K
v	voltage	V
V	volume	m^3
y	vertical height	m
Γ	metabolic rate	W
α	allometric exponent	
$\gamma = \Gamma/M$	specific metabolic rate	W/kg
$\sigma = S/k_B$	scaled entropy	

Problems and Hints for Solutions

P 1.1 Skeletons
a) Determine the skeletal masses of a cat with the body mass $M_c = 10$ kg, and an elephant ($M_e = 5\,000$ kg). b) Give the skeletal masses M_{sc}, and $M_{s,e}$ as percentages of the body masses of each animal. What mechanical problems can you envisage?

P 1.2 How Many Heartbeats in a Lifetime?
Calculate the number of heartbeats, N, during the full lifetime of warm-blooded animals. A typical technical device, like a spark plug is built for about $N_t = 10^7$ cycles. Comment on the differences of the number of heartbeats for big and for small animals with $M_b = 1\,000$ kg and $M_s = 0.01$ kg.

P 1.3 Lifetime Energy Consumption by the Pound
Calculate the specific lifetime energy $e_L = (\Gamma/M)T$, using allometric relations. T is the lifetime. Then calculate e_L for a mouse and an elephant. Which animal makes better use of the energy from the environment?

P 1.4 So Sweet so Mean

A humming bird weighing $M = 3.9$ g visits 1000 flowers daily and thereby collects nectar with an energy content of $\Delta H = 7 - 12$ kcal. It's breathing frequency is $f_{br} = 4$ Hz, the heart frequency is $f_H = 20$ Hz, and the wing beat frequency is typically $f_{wb} = 40$ Hz. In contrast to bees, who only see blue-green light, humming birds can see red like humans [see R. Conniff 2000]. a) Take an average value of $\Delta H = 9$ kcal/day. Determine the metabolic rate of the little bird, and the constant a for the metabolic rate function $\Gamma = aM^{3/4}$. b) Compare your calculated value a to the constant $a_0 \approx 4$ of the mouse to elephant curve, equation (1.5), and determine the factor $b = a/a_0$. c) What problems can you foresee for such a high metabolic rate? d) Do heart beat- and breathing frequency agree with the predicted allometric scaling relations of Table 1.2?

P 1.5 On a Diet

How many days does it take a 30 kg wolf to loose 8% of its body mass if it cannot find any food? Assume an average body mass enthalpy $\Delta h = 25$ MJ/kg.

P 1.6 New Caramel Almond Crunch

The company Häager Dazs has brought out a new ice cream bar with the ingredients 5.1 g protein, 25 g fat, and 32 g carbohydrate. The package claims that this snack contains 1560 kJ of energy. a) Is this claim correct? b) For how long would this bar support an active person (activity factor $b = 5$) of body weight 65 kg?

Hints and Comments for Solutions

S 1.1 Use the relation for skeletal mass $M_{s,terrs} = 0.068\,M^{1.08}$ kg from Table 1.2. Big animals carry a lot of bones, and need to develop different motion strategies compared to small animals.

S 1.2 Big and small animals have nearly the same number N of heartbeats. This number is about a factor 100 larger than N_t for technical devices. Animal organs have self-repairing mechanisms. All atoms in a human body are typically replaced within 7 years.

S 1.3 This problem is investigated in depth by Andresen [2002].

S 1.5 Mass loss $\Delta M = 0.08\,M$, energy loss $\Delta E = \Delta M \Delta h$. For $\Delta h = 25$ MJ/kg. The 8% body mass decay time is $\Delta t_{0.08} = \Delta E/\Gamma_0 = 0.08\,M \cdot 25 \cdot 10^6$ J/kg/$(3.6\,M^{3/4}$J/s$) = 5 \cdot 10^5$ $M^{1/4}$ s. For a wolf with $M = 30$ kg one finds $\Delta t_{0.08} = 5 \cdot 10^5\,30^{1/4} = 1.17 \cdot 10^6$ s. A day has 86 400 s. Therefore the wolf looses 8% of its body weight in $\Delta t_{0.10} = 1.46 \cdot 10^6$ s/86 400 s/d $= 13.5$ days.

2. Energy and Temperature

Epirrhema

Müsset im Naturbetrachten	*As you see an organ's role*
immer eins wie alles achten.	*look at details and the whole.*
Nichts ist drinnen nichts ist draußen	*None is inside none is out*
denn was innen ist ist außen.	*since within, too, is without.*
So ergreifet ohne Säumnis	*So perceive without delays*
heilig offenes Geheimnis.	*nature's open secret ways.*
Freuet euch des wahren Scheins,	*Do enjoy the physics clues*
euch des ernsten Spieles:	*playful seen in any*
kein Lebendiges ist eins,	*no life form has single use*
immer ist's ein vieles.	*always it has many.*

J. W. Goethe   (translated by the author)

Life – a Station in the Energy Chain

Radiation energy flows in the universe from its origin in the sun to its cold grave in deep space. On the way some rays strike the earth and warm the earth to moderate temperatures in the range between 270 K and 380 K where water exists and life can flourish.

Myriad of plants on earth convert water, air, and sunlight or geothermal energy into living tissue on which, in turn, countless animal species thrive. All living beings are entwined and mutually interdependent in the biosphere of which they are a part. Each animal receives energy and materials by ingesting food and absorbing sunlight, and then passes on energy and through excrements and activities. Every animal finally relinquishes its entire body mass M, either when it is eaten, or when it decays after death.

Energy is the currency that drives events. Energy can appear in many different disguises, but is never destroyed. Energy arrives from the sun as radiation at an intensity $S \approx 1.37 \cdot 10^3$ W/m². This power density is equivalent to operating one electric teakettle on each square meter of ground. Plants at the bottom of the food chain convert solar energy into chemical energy that is stored in starches, fats, and proteins. Plants are the food of herbivores, which are the sustenance of carnivores, which may in turn be digested by stronger predators or consumed by microbes after death. Thus, the sun's energy is passed along the food chain, appearing in many different forms such as chemical energy, internal energy, mechanical work, latent heat, kinetic energy, potential energy, or heat.

Animals that want to move need energy. They extract energy as the heat of reaction or chemical energy ΔH, out of their food. Chemical energy can be converted partly into mechanical energy ΔW, which can appear as work, as potential energy, as elastic energy, or as kinetic energy. As a consequence of the first law of

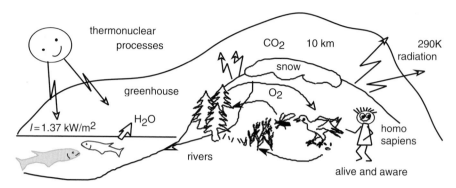

thermonuclear processes

CO2   10 km   290K radiation

snow

greenhouse O_2

$I=1.37$ kW/m2 H_2O

homo sapiens

rivers   alive and aware

Fig. 2.1. The food chain

thermodynamics only a fraction $\eta = \Delta W / \Delta H$ of the heat of reaction ΔH, can be converted into mechanical energy. The remaining energy appears as internal energy U, which will raise the body temperature unless it is removed by conduction, convection, or radiation. Warm blooded (homeotherm) animals that live in cold climates have learned to use this heat as an asset.

The first topic of this chapter, Sect. 2.1, reviews the parameters of the energy chain. Next, in Sect. 2.2 the conservation of energy and the compound efficiency are discussed. Section 2.3 describes thermal problems of warm-blooded animals, how they control their temperature by dealing with heat losses, and overheating. The last Section of this chapter outlines limitations inherent in the principles of thermodynamics.

2.1 Parameters of the Energy Chain

All animals eat to support their metabolic rate. Animals are part of the food chain and therefore part of the energy chain. The basic metabolic rate Γ_0 supports the functioning of the vital organs and enables animals to grow, keep warm, reproduce, see, hear, and think, and maintain the body functions at an optimal level. Active animals increase their basic metabolic rate by some factor b to the level $\Gamma = b \Gamma_0$ in order to move around. Parameters of the energy chain include the ingested chemical energy H, the mechanical energy, or work W that is drawn out of the food, the efficiency η at which this mechanical energy is extracted, the internal energy U that appears as by product of work production, and the body temperature T. These parameters are linked by principles of thermodynamics.

2.1.1 Temperature

All biological processes only occur in *liquid water*. Therefore life can only flourish within the narrow temperature limit between the freezing point T_{ice} also called triple point $T_{tr} = 273.16$ K (because liquid, gas, and solid can be found at this temperature), and the boiling point $T_b = 373.16$ K.

Actually some life forms push these limits. Some critters have antifreeze in their body fluids so that their body fluids stay liquid a few degrees below T_{tr}. Some other organisms live near thermal vents at great depths in the ocean were the water temperature is a few degrees above T_b.

Generally all objects acquire the temperature of their surroundings unless they posses an internal heating of cooling system. The temperature increases when heat energy is released. The temperature decreases when heat flows out of the object. Heat is a form of energy.

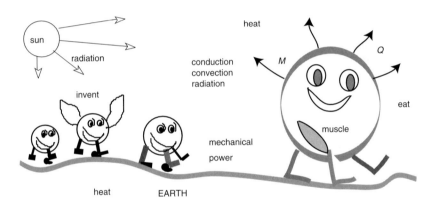

Fig. 2.2. Heat, work, and inventions

Many biological processes have an optimum temperature. Organisms try to obtain and maintain their optimum temperatures. Primitive animals have to live at the temperature of their environment. They become active only when the temperature is right. Advanced animals control their temperature: lizards move to a warm or cold spot, crocodiles wait till the sun is up. Mud wasps fan their brood with their wings. Some desert animals wait till the sun is down. Warm-blooded critters, called homeotherms, regulate their body temperature continuously. They have fur or feather insulation to reduce heat losses in winter, or alternately they may expose their tongue in order to remove heat. Migratory animals annually move to the most favorable climatic zones. Butterflies and birds leave the northern latitudes in the fall when the weather gets cold.

2.1.2 Forms of Energy and Power

All life forms extract chemical energy out of their environment. Plants generate carbohydrates from water and carbon dioxide with the help of sun light. Plant eaters generate chemical energy when these carbohydrates react with oxygen. Specific chemical or *heat of reaction range* up to $h = 50$ kJ/g, see Table 1.1. A sugar cube weighing about 1 gram contains approximately 20 kJ.

Energy may appear in many forms, such as heat, work, internal energy, kinetic energy, potential energy, see Table 1.2. Energy can be converted from one form to another with but it can never be destroyed. This fundamental fact is called the first law of thermodynamics. Energy is measured in Joule J, or in calories 1 cal = 4.18 J. The food industry uses large calories 1 Cal = 1 000 cal = 4.18 kJ. Internal energy can appear as heat that passes from hotter to colder bodies.

Mechanical energy can appear in three forms, kinetic, potential, or elastic. An object of mass M raised to the height y has the potential energy $E_{pot} = M g y$. If this object moves at the speed u it has the kinetic energy $E_{kin} = \frac{1}{2} M u^2$. If the object is elastic like a spring, with the spring constant k, and it is stretched by the distance s, it will contain the elastic potential energy $E_{el} = \frac{1}{2} k \cdot s^2$. All mechanical energy can be converted into heat. However, in the reverse process only a fraction η of heat energy can be converted into mechanical energy, and even this requires special machinery. One can warm the fingers by rubbing the hand. Yet, if one has hot fingers and holds the hands together, they do not suddenly start to perform rubbing motion.

Table 2.1. Forms of energy and power

form	energy [Joule J]	power [Watt W]
light	intensity $I \cdot$ area $\cdot \Delta t$	$I \cdot$ area
chemical	specific enthalpy of reaction h,	h times mass flow rate
internal	heat of fusion L_f,	ΔU times mass flow rate
(thermal)	heat of vaporization L_v:	C_v = specific heat of gas
	gas; $U = (3/2) R_g T$, $\Delta U = C_v \Delta T$	R_g = gas const mechanical
mechanical	work: $\Delta W = F \cdot \Delta s$,	$P_{mech} = F \cdot u$
	kinetic: $E_{kin} = \frac{1}{2} M u^2$	force F,
	potential: $E_{pot} = M g \Delta y$,	velocity u
	elastic potential: $E_{el} = \frac{1}{2} k \cdot \Delta x^2$	height difference Δy
	$\Delta W = \eta \Delta Q$, energy efficiency η	Δx extension of spring
heat	$\Delta Q = c M \Delta T$; c specific heat,	$Q' = dQ/dt$

In many processes one must consider the *flow* of energy. This quantity is measured in Watt (1 W = 1 J/s). For instance when an animal metabolizes a certain nutritious energy ΔH in the time Δt, the animal has the metabolic rate $\Gamma = \Delta H / \Delta t$. When a certain amount of work ΔW is delivered, or some quantity of heat ΔQ is removed in a time interval Δt one deals with the average physical quantities mechanical power $P_{av} = \Delta W / \Delta t$, and heat flux $Q'_{av} = \Delta Q / \Delta t$. Mechanical power and heat flux also have instantaneous values: $P = dW/dt$ and $Q' = dQ/dt$.

When an animal is resting, its metabolic energy production is at the minimum, Γ_0. Its *engine idles*. The metabolic rate of warm-blooded animals of different body masses M can be described by the unique function (1.5) approximated as $\Gamma_0 \approx 4\,M^{3/4}$. When an animal is active, its metabolic rate goes up to the value $\Gamma = b\,\Gamma_0$. For instance birds have an activity factor $b = 10$ to 15. The resting metabolic rate of a 70 kg person is approximately 80 W, the power consumed by of a light bulb. The mechanical power $P = dW/dt$ of an animal is only a fraction η of its metabolic power Γ. The efficiency η has typical values $\eta \approx 25\%$.

2.1.3 How Animals Do Work and Generate Mechanical Energy

Animals need mechanical energy in order to do work. Work W is done when a force, described by the vector F generates a displacement s. The displacement is a vector. If $\emptyset$ is the angle between the force and the displacement, the quantity $s \cdot \cos \emptyset$ is the component of the displacement moved in the direction of the force. The work done is $W = F \cdot s = F \cdot s \cdot \cos \emptyset$. When F, and s point in the same direction like a contracting muscle, the angle is $\emptyset = 0$ and $\cos \emptyset = 1$. Hence a muscle of force F that contracts by the incremental length ΔL delivers the work

$$W = F\,\Delta L. \tag{2.1}$$

Example: How much <u>work</u> does a coyote produce when is drags a 15 kg dog carcass through $\Delta s = 50$ m of back lane? Work is generated if an object moves against a force F through some distance Δs. The coyote must overcome the friction force $F = \mu M g = 0.3 \cdot 15\,\text{kg} \cdot 9.81\,\text{m/s}^2 = 44.1$ N. Here we have assumed a friction coefficient $\mu = 0.3$. The work done by the coyote is $W = 44.1\,\text{N} \cdot 50\,\text{m} = 2.2$ kJ.

The muscle force F depends on the muscle's cross section area A_m, and the specific muscle stress $f \approx 2 \cdot 10^5\,\text{N/m}^2$, which is about the same for muscle tissue of all animals. F does not depend on the length L of the muscle. Details about muscle contraction are discussed in Sect. 3.2.

$$F = A_m\,f\,\text{N} \tag{2.2}$$

A biceps of cross section $A_m = 30\,\text{cm}^2 = 3 \cdot 10^{-4}\,\text{m}^2$ generates the force $F_b = 3 \cdot 10^{-4}\,\text{m}^2 \cdot 2 \cdot 10^5\,\text{N/m}^2 = 600$ N as it contracts isomerically. Muscles, like all tissue, consist mainly of water, and water is practically incompressible. Therefore the muscle volume $V = A_m L$ must remain constant so that the muscle cross-section area A_m increases as the muscle length L shrinks. One can see the biceps bulge when one lifts the hand. Muscles push laterally as they pull at the connecting tendons.

The weight-force of an object $F_w = g M$ is the product of its mass M, and the gravitational acceleration $g = 9.81$ m/s^2 of the earth. When we use our biceps to lift a weight the biceps might shorten its length by $\Delta L = 2$ cm. With the force F_b calculated above the muscle delivers the work $W = F_b \cdot \Delta L = 600$ N $\cdot$ 0.02 m $= 12$ J. The biceps is attached close to the elbow. However, the hand is at a much larger distance from the elbow joint so that the hand sweeps through a large arc and raises the weight by the distance Δh, which is much larger than ΔL. The work of the muscle is converted without loss into the potential energy E_{pot} of *all* the masses that are lifted, namely weight, hand and lower arm. The change of the *potential energy* ΔE_{pot} of the weight equals the work W done, when lifting the weight $\Delta E_{pot} = W = F_w$ $\Delta h = M g \Delta h$. If the work W is done in Δt seconds the mechanical power $P = W/\Delta t$ is delivered.

Potential energy is encountered by animals when they climb trees. A cat of mass $M = 5$ kg climbing up a tree of height $\Delta y = 12$ m must apply the force $F = M g = 5 \cdot 9.81$ kg m/s²s $= 4.9$ N. The cat gains the potential energy $E_{pot} = M g \Delta y = 588$ J. This potential energy is converted in to kinetic energy $E_{pot} = E_{kin} = \frac{1}{2} M u^2$ if the cat jumps down from its perch. Halfway down the cat has still $\frac{1}{2}$ of the potential energy, and some kinetic energy. Close to the bottom the cat has lost all the potential energy, and now has the velocity $u = \sqrt{(2g\Delta y)} = 15.3$ m/s. Kinetic energy depends on the velocity u of an object.

> It is instructive to measure your own mechanical power output while running up a flight of stairs. Consider this example: The flight of stairs up to my office has 23 steps, each 0.17 m high, yielding $\Delta y = 3.91$ m. When I am in a hurry it takes me $\Delta t = 6$ s to quickly run up this flight of stairs. My body mass is $M = 83$ kg. Thus my leg muscles have produced the work $\Delta W = \Delta y \cdot M g = 3.91$ m $\cdot$ 83 kg $\cdot$ 9.81 m/s² $= 3.18 \cdot 10^3$ J in 6 s, and they generated the power $P = W/\Delta t = 2.68 \cdot 10^3$ J/6 s $= 531$ W. This power is considerably larger than my resting metabolic rate $\Gamma_0 \approx 100$ W.

If a dolphin of mass $M = 150$ kg cruises at its top speed $u = 10$ m/s it has the *kinetic energy* $E_{kin} = 0.5 \cdot 150$ kg $\cdot$ (10 m/s)² $= 7.5$ kJ. This kinetic energy would be liberated if the dolphin crashed into the hull of a ship. A historic record of a similar encounter is reported by Philbrink [2001]. A huge sperm whale (estimated at $M \approx 80\,000$ kg) attacked the whaler ship Essex west of the Galapagos Islands in 1820. It swam towards the ship at about $u = 9$ knots $= 9$ s m/h $= 9 \cdot$ (1852 m/h) $\cdot$ (1 h/3 600 s) $= 4.6$ m/s coming to a complete stop as it crashed into the hull. The impact, delivering the energy $E_{kin} \approx 0.5 \cdot 8 \cdot 10^4\ 4.6^2 = 846$ kJ, broke the hull and the ship sank within less than an hour.

2.1.4 Internal Energy, and Heat

Muscles heat up when they perform work. The temperature rises in the muscles. When the muscles are hotter than the blood, they can heat up the blood (by thermal conduction) thereby cooling the muscles down. The blood carries the heat away by convection, leading to a general temperature increase of the body subsequently. Some of the heat subsequently may be removed by sweating.

Warm-blooded animals try to maintain a constant temperature. The temperature rises in a contracting muscle because not all of the chemical energy ΔH consumed in the contraction turns into work W. The remainder appears as heat that increases the internal energy U, and may flow away to colder body parts or to the environment. Internal energy U is related to the molecular motion (translation, rotation, and vibration) of all the molecules of an object. This relationship holds for dead matter as well as living organisms. The molecules are never at rest. Each carries a kinetic energy that is proportional to $k_B T$, where T is the molecule's temperature T, and $k_B = 1.38 \cdot 10^{-23}$ J/K is Boltzmann's constant. One finds the internal energy U of an object by adding up the energies of all its molecules. Since the energy of each molecule is proportional to $k_B T$ the total internal energy can be written as

$$U = \text{const } k_B T . \tag{2.3}$$

Generally one is more interested in the change of the internal energy ΔU than in its absolute value. If a muscle of mass M suddenly converts some chemical energy ΔH in order to generate work W the muscle will simultaneously release a certain amount of heat energy ΔQ. Unless the released heat is immediately carried away, it will increase the local temperature by

$$\Delta T = \Delta Q / CM \tag{2.4}$$

where C is called the specific heat. Since the body consists mainly of water the specific heat of tissue is approximately equal to that of water, namely $C_w = 4.18 \cdot 10^3$ J/kg°. By using (2.4) one can determine the heat ΔQ by measuring the temperature rise. For instance if a muscle of $M = 0.6$ kg heats up by $\Delta T = 2°$ it has released the heat $\Delta Q = C M \Delta T = 4.18 \cdot 10^3$ J/kg $\cdot 0.6$ kg $\cdot 2° = 5$ kJ.

Heat Q always flows from hotter to colder objects. Quite generally, when one encounters heat flow only the hot and the cold objects need be considered. The thermodynamicist calls the hot and the cold object a *system*. In such a system heat always flows from the hotter to the colder part. For instance, a system could consist of a sailor who falls overboard in a cold ocean. His body loses heat to the water, and he dies quickly of hypothermia, unless he is pulled out within a short time. Another thermodynamic system with heat flow is a Thanksgiving turkey in the oven. Here the heat flows from the heating elements to the meat. The flow of heat occurs in three different ways, by conduction, convection and radiation, see Table 2.2. More details will be discussed in Sect. 2.3.

Most often the direction of heat flow is intuitively clear: from the hot to the cold. Hot objects share their energy with cold things, like eternal justice in a welfare state. Occasionally it is not obvious in which direction the heat is flowing. One then uses a more general statement describing the same fact. Heat will flow in a direction so that the total *scaled heat exchange* $Z = \sum \Delta Q_i / (k_B T_c)$ is positive.

To quantify this statement consider a system with a hot and a cold object in contact. The heat gained by the cold object is $+\Delta Q_c$, and the heat lost by the hot object is $-\Delta Q_h$. Since energy is conserved one part loses what the other gains or $\Delta Q_c + \Delta Q_h = 0$. The heat changes may be compared to the internal energies $U_h = \text{const } k_B T_h$, and

U_c= const $k_B T_c$ of the hot and the cold bodies defining the scaled heat exchanges $\Delta Z_h = -\Delta Q_h/(k_B T)$, and $\Delta Z_c = -\Delta Q_c/(k_B T_c)$. These quantities might be compared to the relative value changes of two bank accounts when a sum of money has been transferred from one to the other. Since the internal energy of the hot object U_h is larger than the internal energy of the cold object U_c, but the exchanged energy is the same, the scaled loss of the hot object is always smaller than the scaled gain of the cold object. Therefore the sum of the percentage changes $Z = \Delta Z_h + \Delta Z_c = \Sigma \Delta Q_i/(k_B T_c)$ is always larger than zero. The flow of heat reduces the differences between hot and cold part of a system, and increases $\Sigma \Delta Z$. The quantity $\Delta Z = \Delta S/k_B$ is proportional to the entropy ΔS, and the relation $\Delta k_B \Delta Z = \Sigma \Delta S \geq 0$ is the second law of thermodynamics, which states that the total entropy of a system always increases.

Energy can not be destroyed; however, energy can be exchanged , and can appear in many different disguises. Internal energy, heat, and work are all forms of energy, which are linked by the principles of thermodynamics, see Table 2.2.

Table 2.2. Parameters and relations of thermodynamics. The symbol Δ is used to emphasise which quantities undergo a change

	thermodynamics	mechanics
parameter	temperature T [K, °C, °F, °R]	mass M [kg]
	volume V [m³]	length L [m]
	density ρ [kg/m³]	time t [s]
	pressure p [N/m² = Pa]	velocity u [m/s]
	energy 1 Joule [J = 1 N·m]	acceleration a [m/s²]
		force $F = M \cdot a$ [N; 1 N = 1 kg · m/s²]
energy	heat & internal	work ΔW = force · distance [J = Nm]
	ΔQ J, 1 cal = 4.18 J	gravity $F = M \cdot g$ [g $\approx$ 9.81 m/s², Δy = height]
	$\Delta Q = CM(T_f - T_{in})$ [caloric equation]	potential energy $E_{pot} = M$ g Δy
	$\Delta Q = L_f \Delta M$ [melting]	kinetic energy $E_{kin} = \frac{1}{2} M u^2$
	i $= L_v \Delta M$ [vaporization]	elastic energy $E_{el} = \frac{1}{2}$ k ΔL^2
	$\Delta U = C_v n \Delta T$ [internal energy]	k = spring constant, ΔL = change of length
	n = number of moles	typical chemical energy h $\approx$ 30 kJ/g
	C_v specific heat/mole	
1. law	$\Delta Q = \Delta U + \Delta W$	P = work/time $= F(\Delta L/\Delta t) = F \cdot u = p \cdot V$,
power	heat flow $Q' = \Delta Q/\Delta t$	unit W = J/s
units W	conduction $Q'_{con} = kA(T_h - T_c)/L$	if work W arises from pressure $p = F/A$
	convection $Q'_{conv} = JC \Delta T$	J = mass flow rate in pipe of radius R
	radiation $P_{rad} = Q'_{rad} = A \varepsilon \sigma T^4$	$J = \rho \pi R^2 u$
	$\sigma = 5.67 \cdot 10^{-8}$ W/m² K⁴,	
	ε = emissivity	
	heating, cooling $Q' = CM(dT/dt)$	efficiency $\eta = \Delta W/\Delta Q$
intensity	heat: heat flow/area	power flow/area
equation of state	$pV = nR_g T$	$R_g = 8.31$ J/mole K gas constant
		$k_B = 1.38 \cdot 10^{-23}$ J/K Boltzmann's const.

2.2 The Conservation of Energy

Animals encounter energy in various forms: as chemical energy, as radiation, as heat, as internal energy, as kinetic energy, as gravitational potential energy, and as elastic potential energy. Since energy cannot be destroyed, one can describe quantitatively some motions and activities of animals using the law of conservation of energy.

Animals need mechanical energy to walk around, to pump blood, and to expel air for making sounds, and for many other activities. Only a certain fraction of the chemical energy can be converted into mechanical energy. The key word here is efficiency η.

Homeotherm animals must promote heat flow to cool their muscles when they are moving. They must insulate their body so that excessive heat can not enter when it is hot outside, and can not leave the body in a cold environment. Internal energy, heat, and work are connected by the *conservation of energy*. Conservation of energy rules the metabolic processes, and the muscle activities of animals. The fundamental law of conservation of energy was recognized at about the same time by two very different individuals: an engineer, Joule, and a medical doctor, Mayer. This illustrates that some physical laws can appear under very different circumstances. Joule saw the connection between heat and work as he was drilling the holes into gun barrels, and Mayer recognized the conservation of energy as he observed that venous blood of sailors in the tropics looked as red as the arterial blood: The oxygen had not been used up. The fundamental relation between heat, work, and internal energy, discovered by Joule, and Mayer around 1840, is known as the first law of thermodynamics.

So töricht ist der Mensch... Er stutzt	*How stupid can this person get?*
Schaut dämlich drein und ist verdutzt,	*Looks puzzled, and is all upset*
Anstatt sich erstmal solche Sachen	*Instead of seeing clear and plain*
In aller Ruhe klarzumachen.	*The energy conversion chain.*
Hier strotzt die Backe voller Saft:	*Here glows the cheek, the beauty's source.*
Da hängt die Hand gefüllt mit Kraft.	*There rests the hand, teaming with force.*
Die Kraft, infolge der Erregung,	*The force, when freed by animation,*
Verwandelt sich in Schwungbewegung.	*Provides the quick acceleration.*
Bewegung, die in schnellem Blitze	*This moves the hand with rapid speed.*
Zur Backe eilt, wird hier zu Hitze.	*The cheek receives the impact heat.*
Die Hitze aber durch Entzündung	*The heat excites the nerve again*
Der Nerven, brennt als Schmerzempfindung	*Which then is recognized as pain.*
Bis in den tiefsten Seelenkern,	*The pain sinks deep into the soul.*
Und dies Gefühl hat keiner gern.	*This is resented by us all.*
Ohrfeige heißt man diese Handlung.	*Slap on the face is the assertion.*
Der Forscher nennt es Kraftverwandlung.	*In Physics terms just Joule's conversion.*

Wilhelm Busch   (translated by the author)

Fig. 2.3. Conversion of energy as described in Balduin Bählam by Wilhelm Busch [1883]

2.2.1 The First Law of Thermodynamics

Any system, biological organism, or technical engine, converts a given amount ΔQ of chemical energy or heat into work W and into some addition ΔU to the internal energy U. Internal energy is related to the kinetic energy of the molecules of the material. A change of the internal energy shows up as a variation of the temperature. The first law of thermodynamics states that a given (chemical) energy ΔQ must show up either as a change of the internal energy ΔU or as mechanical energy in its various forms, summarily called work W:

$$\Delta Q = \Delta U + W . \tag{2.5}$$

Since ΔU is always larger than 0, the energy fraction $\eta = W/\Delta Q$, called efficiency, is always smaller than 1. Typically animals have an overall mechanical efficiency of 25%. Then with a given heat energy ΔQ they can produce the work

$$W = \eta \, \Delta Q . \tag{2.6}$$

The three quantities ΔQ, ΔU, and W are all forms of energy. Energy is measured with the units calorie, or Joule, (1 cal = 4.18 J). A convenient practical measure is the chemical energy released by the burning of one matchstick: $\Delta Q \approx 1$ kJ.

Mechanical energy can be fully converted into internal energy, but the reverse is not true. The first law of thermodynamics has one important message: it is impossible to convert all the heat energy into work. Some energy without fail appears as change of the internal energy, which can increase the temperature of the body. The change presents a problem but it also provides an opportunity, which animals have made use of; because warm-blooded animals use the increase in their internal energy to raise their body temperature.

The flow of (thermal) energy per unit time is called heat flux $Q' = dQ/dt$. Heat flux per unit area is called *intensity*. By taking the time derivative of (2.5) one obtains a rate equation with the units J/s = W. The time derivative of the internal energy U is written as $dU/dt =: U'$, and the time derivative of work dW/dt is called mechanical power P_{mech}

$$Q' = U' + P_{mech}. \tag{2.7}$$

The mechanical power is related to Q' by the efficiency η.

$$P_{mech} = \eta\, Q' \tag{2.8}$$

When the flow of energy Q' is identified with the metabolic rate $\Gamma = b \cdot 4\,M^{3/4}$ the first law of thermodynamics in the form (2.8) describes the production rate of mechanical power.

$$P_{mech} = \eta\, b \cdot 4\,M^{3/4} \tag{2.9}$$

Equation (2.9) can be used to derive the performance of animals of all sizes.

Example: A dog of $M = 15$ kg with an efficiency of $\eta = 25\%$ and a metabolic activity factor $b = 5$ generates the mechanical power $P_{mech} = 0.25 \cdot 4 \cdot 15^{0.75} = 34$ W.

2.2.2 Energy Analysis

The conservation of energy is a fundamental law that can be applied in different circumstances. We will practice its use in many examples, often in combination with Newton's second law, the conservation of momentum. In some situations one can neglect the internal energy, and the heat, and only deal with mechanical energy in its various forms: kinetic energy of straight motion $\frac{1}{2}Mu^2$, rotational kinetic energy $\frac{1}{2}I\omega^2$, potential energy $Mg\Delta y$, elastic energy, $\frac{1}{2}\,k\,\Delta L^2$, and friction energy $F_{fr}\Delta s$. The sum of these energies remains constant. Since the total amount of energy cannot change one can write the conservation of mechanical energy as an equation linking different states before and after some event.

Think for instance of a cat, jumping down the distance Δy from a tree. We look at the event in snapshots, called *states*. The first state is the cat, of mass M, on the tree with a potential energy $E_{pot} = Mg\Delta y$. The second state could be the instant just before the cat reaches the ground. At this point the cat has lost the potential energy and gained the kinetic energy $E_k = \frac{1}{2}Mu^2$. Since these energies are equal one can write $Mg\Delta y = \frac{1}{2}Mu^2$, and solve this equation to find the speed of the cat just before impact $u = \sqrt{(2g\Delta y)}$. In general one can write the conservation of energy for a system at two different positions labeled with the subscripts 1 and 2,

$$(E_{kin} + E_{pot} + E_{el})_1 = (E_{kin} + E_{pot} + E_{el})_2 + E_{fr} \tag{2.10}$$

where: $E_{kin} = 0.5\,M \cdot u^2 + 0.5\,I\omega^2$, $E_{pot} = Mg\Delta y$, $E_{el} = 0.5 \cdot k\,\Delta L^2$, and $E_{fr} = F_{fr} \cdot \Delta s$ is the friction energy lost while moving through the distance Δs.

This energy analysis can be applied in a biological example described by Tennekes [1997]. The wagtail, a small bird of average body mass $M = 0.02$ kg (of which 0.005 kg is body fat) crosses the Sahara desert ($\Delta s = 1600$ km) on its yearly migration, by flying non stop for 2.5 days. Before the flight the bird "gases up" by putting on extra weight, namely 10g of body fat, so that its "gas tank" contains 15 g with an energy content of $h = 32$ kJ/g. We assume that the bird has a mechanical efficiency of $\eta = 25\%$, and that it must overcome an air resistance of 0.06N equivalent to an energy expenditure of 0.06 Joule/m. Energy analysis allows to answer the following questions:

(i) What is the average speed of the bird?
Answer: $u = \Delta s / \Delta t = 1.6 \cdot 10^6$ m$/(2.5 \cdot 24 \cdot 3600$ s$) = 7.4$ m/s.

(ii) What is the total mechanical energy W that the bird must generate when traveling the distance 1 600 km?
Answer Mechanical work $\Delta W = F \cdot \Delta s = 0.06$ N $\cdot$ 1 600 000 m $= 9.6 \cdot 10^4$ J.

(iii) In order to generate this mechanical work how much "fuel" energy does the bird have to use up?
Answer: First find the expended fuel energy $H = \Delta M h = \Delta Q = \Delta W / \eta = 9.6 \cdot 10^4$ J/ 0.25 $= 3.84 \cdot 10^5$ J, where ΔM is the consumed mass of the fuel and h the specific energy content per gram. Then we find the used up fuel mass $\Delta M = \Delta h / H = 3.84 \cdot 10^5/32$ kJ/g $= 12$ g of fat. This leaves a reserve of 15 g $-$ 12 g $=$ 3 g of fat, not very much indeed. The mileage is quite impressive, namely $1.6 \cdot 10^6$ m/12 g $= 133.3$ km/g.

The reserve of 3 grams would allow the bird to fly an extra $\Delta s = 3 \cdot 133.3 = 400$ km. There is not much margin for error if the bird does not forecast correctly the winds over the Sahara desert. The bird needs the mechanical power to get where it wants to go. High energy efficiency is at a premium when a large distance has to be covered.

In all their actions organisms have to pay in energy for the tasks they need to accomplish. It is instructive to see how energy flows from primary sources to the accomplishment of a certain task.

2.2.3 Compound Efficiency of the Energy Conversion Chain

There is a long chain of events from an original source of energy to its final use. Mechanical energy is produced in muscles as well as in automobile engines. However, both the muscle tissue and the combustion engine are only one part of the energy conversion chain. Mechanical force is generated in muscle fibers or in engine pistons. However, pistons and muscle fibers are not generally the locations where the force is needed, nor is the chemical energy in gasoline or muscle cells the original source of energy. One can therefore either look at the compound efficiency of the system like a car or a cat, or take the broader view and trace the energy back to its original source. This is done now for biological and technical systems.

First consider a system's efficiency η, namely the fraction $E = \eta Q$ of a given energy Q that is available for the intended purpose E. In order to lift an object with your hand the biceps force must be transmitted to the hand. The hand is directed by the brain, which must be supplied with oxygen and nutrients. The heart must be operating to supply the arm with blood and all the machinery of the body must be functioning, so that the hand will lift the weight. Similar conditions prevail in

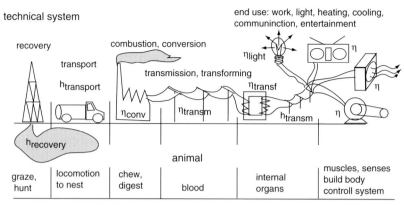

Fig. 2.4. Conversion of fuel in the ground into useful work

an automobile. Much of the mechanical energy goes into auxiliary devices. Consider a technical device: A well-tuned engine automobile engine running at city traffic speeds may have an efficiency of 36%. One half of this (18%) may be lost in the transmission, and one quarter (9%) for braking and accessories. Thus only 9% of the energy burned in the engine is left to actually push the car forward. Similarly, animals also have energy overheads associated with the operation of the intestines, the heart, the brain, and other indispensable organs. Therefore higher animals must survive with efficiencies well below 50%.

Efficiencies in the range of 25% still look fairly respectable. The picture changes when one compares the recovered work with the amount of energy taken out of the environment, Fig. 2.4. Energy is extracted with considerable cost in energy, and converted in several stages before it reaches its end use. This holds for animals as well as for technical devices. For example in the case of a car engine one can distinguish six stages from source to final application: 1) extraction from the ground, 2) transfer to the heat engine, 3) combustion, 4) conversion from heat to work, 5) intermediate transfer and storage of work, 6) end use at the wheels.

Each stage has its efficiency η_i defined as $\eta_i =$ energy recovered/(energy in +energy cost of process). Each stage has an energy loss factor $1-\eta_i$, and some stages have waste products. Similarly every activity and body function of an organism, such as food gathering, eating, idling, thinking, growth and replacement of body cells, has an associated energy cost that finds its way into the metabolic rate $\Gamma = ba\,M^\alpha$. Some stages have waste products. Note the similarities: grazing or hunting is energy extraction, and chewing is refining. Digestion is combustion. Blood is the vehicle of *fuel* transport: possibly first to an intermediate storage deposit in a fatty tissue. Ultimately the fuel is transported to muscle tissue for mechanical power generation with an efficiency of not more than 50%. Only a very small fraction of the energy taken from the environment actually goes into the wanted mechanical activity, because the losses at each stage are compounded.

The conversion inefficiencies and generation of waste products are easily understood with reference to mechanical devices. A measure for the fraction of energy reaching the end stage is the compound efficiency

$$\eta_c = \eta_1 \cdot \eta_2 \cdot \eta_3, \tag{2.11}$$

namely the product of all the partial efficiencies η_i of each step, defined as $\eta_i =$ energy recovered / (energy in + energy cost of process).

The conversion efficiency is a useful concept to recognize the great difference between technical devices and biological systems. The utility companies optimize the production of their end product: electricity, from which work, light, and heat can be derived conveniently. Technical processes contain many wasteful steps. For instance, the radiation energy from an electrical light bulb energized by a motor generator represents only a small fraction (say a few percent) of the chemical energy released from the gasoline that is driving the motor generator. Byproducts like CO_2, ashes and waste heat are discarded.

In contrast very little energy is wasted in the biosphere, because each organism is imbedded into its biological surrounding. The inefficient harvesting of one animal allows for food for other animals. The organism itself reuses much of the waste heat of its muscles, and most waste products of one organism become sustenance or building materials for the rest of the biosphere. Table 2.3 shows an attempt to identify some of the biological processes that make the biosphere energy chain so much more effective than the technical energy chain.

Table 2.3. Energy efficiency chain for engines and animals

lost fraction of energy	technical system	*energy flow arrow of time* ↓	biological system	other use in system	lost fraction
mine tailings dumped	mining	*primary recovery* ↓	harvest/kill		maintains the food supply
fumes sulfur,...	fuel extraction	*refining* ↓	chew		
ashes dumped	heat engine	*extraction, conversion* ↓	digest	raise body temp. to ideal value	feces used by other animals and plants
heat the universe	work – 220 kV	*transmission* ↓	blood system	"	
heat the universe	batteries	*intermediate storage* ↓	fatty tissue	"	
	home grid	*end distribution* ↓	blood system		
heat the universe	mech. power transportation heating, cooling communication electronics	*end use*	muscles heat evaporation nerves, brain senses, voice	"	sweat, heat the universe
garbage dump	replacement		replacement		used by other organisms

2.2.4 Optimum Rates

In Darwin's words "in the struggle for existence... the merest trifle (of difference) would give the victory to one organic being over another... but probably in no one case could we precisely say why one species has been victorious over another". We begin to lift a bit of this veil of mystery when we notice that animals save substantial amounts of energy by operating their life processes at optimum rates.

It turns out that most periodic processes have a natural frequency f or period $T = 1/f$. An arm swings like a pendulum at the frequency $f = \text{const}\,\sqrt{(g/L)}$, where L is its length, and the constant depends on the mass distribution. The arm can be made to swing faster or slower, however the power required to keep it going is at a minimum when the driving force is applied at the frequency f. This is called resonance. Driving periodic processes at their resonance saves power. Animals use resonance whenever they can. See also Sects. 3.2.4, 5.1.7, and 6.1.

The flow of heat is an energy transfer process which can also be optimized. Optimum heat transfer processes are studied in the new field of *finite time thermodynamics* [see Bejan 1997] in which technological as well as biological questions are discussed. One must distinguish the energy efficiency η_e, the ratio of work out and heat in, and the power efficiency η_p, the ratio of mechanical power and heat production. These quantities are defined as

$$\eta_e = W / \Delta Q \tag{2.12}$$

and

$$\eta_p = P_{\text{mech}} / Q' = (W/\Delta t)/(\Delta Q/\Delta t) . \tag{2.13}$$

Generally the energy efficiency η_e approaches a maximum if the process advances infinitely slowly. However, a slow event implies a long process duration Δt. Therefore the power $P_{\text{mech}} = W/\Delta t$ must approach zero as the energy efficiency reaches its maximum. In contrast, the power efficiency peaks at certain process conditions[1], where the energy efficiency is not a maximum. Power- and energy efficiencies differ because process energy is required to cause energy to flow, see Fig. 2.5, and the process energy is unavailable for conversion into work.

In all their actions animals try to optimize their efficiencies, in order to improve their survival chances. Optimum rates can be found in muscle contraction, see Sect. 3.2.4, in fluid flow, for instance in breathing, (Sect. 4.3.8) and in swimming (Sect. 5.5.4). In fact one could speculate that most evolutionary changes are just a better use of the available energy resources, and that successful species have managed to find optimum rates for their important life processes. When nothing can be gained, like in the Arctic winter, herbivores lapse into hibernation, ap-

---

[1] For instance the Carnot efficiency of a heat engine operating between two temperatures T_{hot} and T_{cold} has the maximum value $\eta_{\text{Car}} = 1 - T_{\text{cold}}/T_{\text{hot}}$. In contrast the efficiency maximum power efficiency of a Carnot engine has the slightly smaller value $\eta_{\text{max power}} = 1 - (T_{\text{cold}}/T_{\text{hot}})^{1/2}$; [see Curzon and Ahlborn 1975]

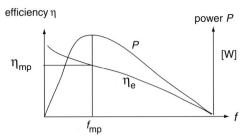

Fig. 2.5. Energy Efficiency η_e, and power efficiency η_p as function of frequency f. In all cyclic processes there is an optimum frequency f_{mp} at which the power efficiency has a maximum. For muscle efficiency see also Fig. 3.8

proaching the energy efficiency optimum of very slow processes. On the other hand when danger looms, fish can dart off in "overdrive" using their white muscles in fantastic power bursts at great expenditure of energy.

In contrast technical devices are generally designed to maximize the profit. This could imply to minimize the cost of production, running expenses, and to maximize the durability at great costs to the environment. Energy efficiency only plays a minor role – as long as energy is cheap.

2.3 Thermal Problems of Warm Blooded Animals

The first challenge facing an animal is to extract energy out of the environment – the second challenge is to obtain the best working temperature for its body. Biological processes proceed faster at elevated temperatures. Speed can become an asset for the hunters as well as for the hunted. For the biochemistry of tissue $T = 37\,°C$ is an ideal operating temperature. Hence warm-blooded animals evolved, who maintain the ideal temperature regardless of night and day, summer and winter, in the Tropics or in Arctic latitudes. In order to reach and maintain the optimum body temperature the thermocritter must use all the tools of heat transfer and heat management. The critter must therefore master conduction, convection, and radiation, and utilize the principles of phase changes.

2.3.1 Temperature Control and Heat Fluxes

The First Law of Thermodynamics presents an opportunity for internal temperature control, because the generation of mechanical work implies the generation of internal energy ΔU, appearing as heat, which is an asset in a cold environment.

In technical applications ΔU is an unwanted by-product of power generation. In contrast, animals have taken advantage of the situation: warm blooded animals use some of the *waste* heat to maintain their body temperature. This is an example of life's opportunism, wherein natural logic has turned a handicap into an advantage. A constant temperature can only be maintained if the animals have effective temperature controls. The heat must be kept in the body by good insulation, if the outside is colder, whereas excess metabolic heat must be removed, if the outside is warmer than the body temperature. Warm-blooded animals with effective insulation were able to extend the living space into the Polar- and Sub Arctic regions, which are not accessible to cold-blooded creatures. Note that not only mammals and birds are homeotherm but also some fish (for instance tuna) have warm-blooded body parts.

An object hotter than its surrounding will loose heat until all parts of the system have the same temperature. The time for this thermal equilibrium to be established can be very different. It may happen in a fraction of a second (a moth in the campfire), minutes (milk in coffee), or it can take a month (spring breakup of the ice in the Arctic). Thermal equilibrium comes about because heat always flows from the hotter to the colder region.

There are three mechanisms by which heat can flow or temperature can be controlled: conduction, convection, and radiation. In addition the temperature can also be controlled by phase changes. Relations for these processes are given in Table 2.4. Often these processes occur at the same time. *Conduction* is the most obvious heat flow mechanism. Conduction moves heat between objects that are in contact with each other. Heat losses or excessive heat gains can be counteracted by insulation. *Convection* requires a moving medium, fluid, or gas that can absorb heat and carry it to another place. *Radiation* moves energy through air or empty space without contact. These three processes are always present. However, animals have a certain influence on the physical conditions or the material constants of heat flow processes. For instance a little bird may ruffle its feathers in order to increases the thickness of its insulation thereby reducing thermal conduction. Dolphins, and tuna may shunt off or increase some of the blood flow to their extremities, thereby manipulating their convective heat losses. Insects may move into the sunlight in order to warm up in the morning.

As soon as animals had invented the principle of maintaining constant elevated body temperature – homeothermicity – efficient insulation became a highly prized commodity. Insulation allowed animals to inhabit colder climates on the earth. Organisms found insulation solutions for this problem twice and at different locations in the tree of life: fur and feathers. Both arrangements represent a way to trap air, and use the air as insulating material. Methods of using the best possible insulation likely doubled the living space for warm-blooded animals on earth.

Table 2.4. The time derivative of the first law $Q' = U' = P$ links different parameters of living organisms

energy in Q'	heating effects $U' = dU/dt$	mechanical power P_{mech}
metabolic rate Γ	conduction	start, stop
	convection	steady locomotion, internal power
	radiation	(heart, breathing bowl movements)

2.3.2 Heat Losses by Conduction

Each section of surface area of an object that is warmer than its surrounding loses heat by conduction and radiation. Conduction works only between bodies in contact, for example between an object and the air surrounding it. Conduction always moves heat from the warm body to the colder body.

Consider a very simple system: a rod of length L that is in contact on one end with a cold object of temperature T_c, and on the other end with a hot object of temperature T_h. The conductive heat flux Q' though the rod is then found to be

$$Q' = A\kappa \frac{T_h - T_c}{L} = A\frac{T_h - T_c}{L/\kappa}, \tag{2.14}$$

where κ is the thermal conductivity, and A is the contact area. Insulators have very small values for κ and good conductors have large κ-values, see Table 2.5. Air is the best thermal insulator. Objects in contact with cold surfaces loose heat quickly.

Table 2.5. Thermal conductivity κ, and specific heat C for various materials

material	water	ice	wood[a]	muscle, bone	body fat	wool	goose down	air
κ W/m°	0.6	2.2	0.16	0.42	0.2	0.04	0.026	0.0256[b]
C kJ/kg°	4.18	2.0						

[a] this is a typical number for wood. Balsa wood has a smaller value, green lumber a larger value.

[b] Specific heat per mol for air at constant volume $C_v = (5/2)R_g$, where $R_g = 8.31$ J/mol K, and specific heat per mol of air at constant pressure $C_p = (7/2)R_g$.

Consider standing barefoot on ice, Fig. 2.6. How much heat does the foot loose in 15 min? Take a thermal conductivity somewhere between muscle and body fat $\kappa = 0.3$ W/m°. Use Celsius or Kelvin degrees, not Fahrenheit degrees in (2.14). Assume a contact area A slightly smaller than the foot outline area $0.1 \cdot 0.3$ m², say $A = 0.02$ m². Let the thickness of the skin be $L = 5$ mm $= 0.005$ m. The lost heat flux is $Q' = A \kappa (T_h - T_c)/L = 0.02$ m² $\cdot 0.3$ W/m°$\cdot (34°- 0°)/0.003$ m $= 68$ W. In $\Delta t = 15$ min $= 900$ s this energy amounts to a loss of $\Delta Q = Q'\Delta t = 68$ W $\cdot 900$ s $= 61.2$ kJ, equivalent to the burning of about 61 matches.

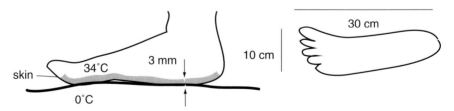

Fig. 2.6. Foot on the ice

Aquatic animals mainly use blubber as insulation. A typical value for blubber is $\kappa_{bl} = 0.1$ W/m°. Air has the smallest value of the thermal conductivity of any material: $\kappa_{air} = 0.026$ W/mK. The air must be still to act as an insulator. If the air is moving it can carry heat by convection, a phenomenon sometimes described as *chill factor*.

Equation (2.14) holds for rods as well as for any thin slab of material and for sections that have only slight curvatures. For stronger curved surfaces the radius of curvature comes into play, because a volume element on the inside has more access to surface area when the radius is small.

Table 2.6. Heat fluxes from cylindrical and spherical objects

cylindrical body	spherical body
$$Q' = 2\pi L \frac{\kappa \cdot \Delta T}{\ln(1 + \Delta R / R_i)} \quad (2.15)$$	$$Q' = 4\pi(R_i + \Delta R) \frac{\kappa \cdot \Delta T}{\Delta R / R_i} \quad (2.16)$$

Equations (2.15) and (2.16) are obtained by considering that the entire heat flux Q' at one section of the "skin" must pass through every layer dr of insulation, across which the temperature changes by dT. The flux is: $Q' = A \kappa dT/dr$. For a cylindrical body section of length Δz, and surface area $A = 2\pi r \Delta z$ one has $dT = (Q'/\kappa 2\pi r \Delta z)dr$. The infinitesimal temperature difference dT is integrated from the inside where $r = R_i$ and $T = T_h$ to the outside where $r = R_i + \Delta R$ and $T = T_c$ namely $T_h - T_c = \Delta T = -Q'/(\kappa 2\pi \Delta z) \cdot \ln(R_i + \Delta R/R_i)$. Solving for $Q' = -2\pi \kappa \Delta z \Delta T / \ln(1 + \Delta R/R_i)$, and integrating in the z-direction over the length L of the cylinder yields (2.15). Equation (2.16) is found using a similar method. Note that in the limit $\Delta R/R_i \ll 1$ the logarithm can be expanded as $\ln(1 + \Delta R/R_i) \approx \Delta R/R_i$ so that the heat flux from a cylindrical body can be approximated as the flux from a flat body of surface area $A = 2\pi R_i L$. The cylindrical geometry can be used when modeling the heat loss from a limb:

How much heat does little Joe lose when he puts his hand and forearm for 2 minutes into a pond of cold water? Suppose Joe has clenched his fist so that his arm can be approximated as a cylinder of $L = 30$ cm length and $d = 2R = 6$ cm average diameter. Assume the arm has a 3 mm thick layer of body fat insulation, ($\kappa = 0.2$ W/m°) and the water is a chilly 4°C. Hence $\Delta T = 37 - 4 = 33°$, $L = 0.3$ m, $\Delta R = 0.003$ m, $R_i = 0.027$ m, and $Q' = \Delta Q/120$ s. Then $Q' = 2\pi L \kappa \Delta T / \ln(1 + \Delta R/R_i) = 2\pi \, 0.30$m $\cdot 0.2$ W/m° $\cdot 33°/ \ln(1 + 0.003$ m/0.027 m$) = 118$ W $= \Delta Q/120$ s. This heat flux can be expressed as the loss of heat ΔQ in 2 min $= 120$ s, so that $\Delta Q = 118$ W $\cdot 120$ s $= 14.16$ kJ, or the chemical energy content equivalent to burning 14 match-sticks.

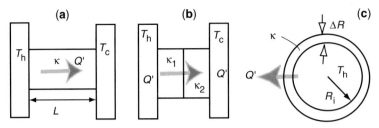

Fig. 2.7. Thermal conduction, plane-, cylindrical and spherical geometry

When the heat flows through two layers of width L_1 and L_2, and thermal conductivities κ_1 and κ_2, as illustrated in Fig. 2.7b, one does not know initially the temperature T at the interface. However, one can apply Eq. (2.14) to each of the layer. Since the same heat Q' must flow through both of the layers one has the equality

$$Q' = A \cdot \kappa_1 \cdot \frac{T_h - T}{L_1} = A \cdot \kappa_2 \cdot \frac{T - T_c}{L_2}, \qquad (2.17)$$

which only contains the unknown temperature T. With the abbreviation $B = \kappa_1 L_2 / \kappa_2 L_1$ one finds the intermediate temperature $T = (B T_h + T_c)/(B+1)$. Then T can be inserted on the left side of Eq. (2.17) to determine the unknown heat flow. $Q' = A \kappa_1 (T_h - T)/L_1$.

A simple example illustrates the use of Eq. (2.17): Is it safe to touch a $T_h = 120°C$ stove plate with an insulating glove of $L_1 = 2$ mm thickness? Assume the insulating glove has $\kappa_1 = 0.04$, and the insulating tissue under the skin has a thickness of $L_2 = 1.5$ mm and a thermal conductivity of $\kappa_2 = 0.2$. Let the body temperature under the skin be 34°C. With these values $B = \kappa_1 L_2 / \kappa_2 L_1 = 0.04 \cdot 1.5 / 0.2 \cdot 2 = 0.15$; and the intermediate temperature where the glove touches the skin is $T = (B T_h + T_c)/(B+1) = (0.15 \cdot 120 + 32)/1.15 = 43.5°$. This temperature is quite tolerable for a short time.

In general each part of an object may be insulated by different layers of material. Skin and hair, covering various parts of the body, Fig. 2.8, lose heat at different rates. Heat leaving the body flows first through the fatty tissue, than through the skin and then through the wet suit. How does one account for heat flow through various layers of insulation? The total heat flux is made up of the components

$$Q' = Q'_1 + Q'_2 + Q'_3 + Q'_4 + \dots, \qquad (2.18)$$

which must first be calculated separately, because the low temperature T_c may be different at each site if the body makes contact with objects of different temperature. We may neglect the metabolic gain Γ and consider only the heat losses 1) from the head, 2) from the dressed body, 3) and 4) from the hands, 5) and 6) from the feet. To simplify the calculations one can treat the thermal flux Q' like the flow

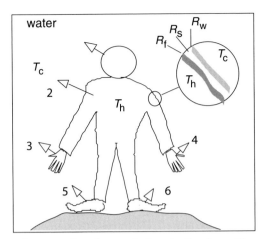

Fig. 2.8. Sum of heat fluxes. Neglect the metabolic gain Γ and consider only the heat losses 1) from the head, 2) from the dressed body, 3) and 4) from the hands, and 5) and 6) from the feet

of an electrical current j which is driven by a voltage differences V across an electrical resistance R. If one defines the thermal resistance $R_{th}=L/\kappa A$, the conduction equation becomes $Q'=\Delta T/R_{th}$. This looks exactly like Ohms law $j=V/R$, Eq. (10.6), Sect. 10.1.3. Therefore everything one knows about networks of electrical resistances in parallel and series can be applied to networks of thermal resistances in parallel and series.[2]

2.3.3 Surface Heat Transfer

Any surface of area A and temperature T_s that is exposed to open air with an outside temperature T_o will exchange heat. A thermal boundary layer is quickly established in which the temperature changes from T_s to T_o. Since the thickness ΔL of this thermal boundary layer it is not really known, one determines the heat flow with the help of a surface heat transfer coefficient h_s using the equation

$$Q'=A\,h_s\,\Delta T, \text{ where } \Delta T=(T_o-T_s).\tag{2.19}$$

The heat transfer coefficient for a vertical surface like the wall of a house is $h_s=1.8\,\Delta T^{1/4}\,W/(°\,m^2)$. If there is a chilly wind h_s will be much larger. This is *forced* convection. Values for h_s are shown in Table 2.7.

---

[2] The similarity between heat fluxes and electrical currents can be further extended to mass flow in pipes, and diffusion processes. Each flow has a force or potential that drives it. Each has a resistance that incurs a loss of potential, which is the price for maintaining the flow in the first place. Associated is a power expenditure (entropy production), representing the "energy cost per unit time" associated with the flow.

Table 2.7. Surface heat transfer coefficients h_s adopted from eported by White [1984]

	free convection	forced convection (windchill)
gases	$h_s = 5 - 25$ W/(° m²)	$h_s = 10 - 300$ W/(° m²)
oils	$10 - 60$	$50 - 2000$
light liquids (water)	$100 - 1000$	$100 - 20,000$

2.3.4 Convection

Convection is the heat flow associated with mass fluxes J kg/s. Think of a flow-through heater. A stream of water with the specific heat C (units J/°kg) enters at the flow rate J and the temperature T_c. Heat is then added at a steady rate Q' and the water leaves at the temperature T_h. The heat flux required to produce this temperature change is

$$Q' = CJ(T_h - T_c). \tag{2.20}$$

Consider the foot on the ice, Fig. 2.9, in the example discussed above. The foot suffers a power loss of about 18 W. If this heat is not constantly replaced the foot will cool quickly. If the foot receives the same amount of thermal energy it can stay at the same temperature. The blood system of the body moves heat around by convection. How much blood must flow into the foot in order to maintain it at about 34°C?

Assume $C_{blood} \approx 1$ cal/g° = 4.18J/g°. $\Delta T = 36° - 32° = 4°$, $Q' = 68$ W. The required mass flow rate for the blood is now determined from $J = Q'/\{(T_h - T_c)C\} = 18$ J/s/$\{(36° - 32°) \cdot 4.18$ J/g°$\} = 4.08$ g/s. Can this blood flow be supplied? Assume that the artery going into the foot has a diameter of $2R = 3$ mm. The artery's area is therefore $A = \pi R^2 = \pi(0.15$ cm$)^2 = 0.0707$ cm². The mass flow rate J of a fluid of density ρ and average velocity u though a pipe of cross section area A is $J = \rho u A$. Assuming a typical blood density of 1 g/cm³ (or 1 000 kg/m³), one can find the average velocity in the artery as $u = J/\rho A = (4.08$ g/s$)/\{1$ g/cm³ $\cdot 0.126$ cm²$\} = 32.3$ cm/s. This value is a typical flow velocity in arteries, see Fig 4.14.

Convective heat transfer by the blood system is used by most higher animals to maintain their body temperature at the best operating level. The blood flow can be used for heating and cooling. If a leg gets too cold additional blood can warm the

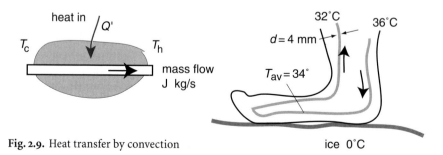

Fig. 2.9. Heat transfer by convection

limb. However, if the whole body would lose too much heat altogether the blood flow into the exposed limb may be reduced. When the body overheats blood can move heat into the tissue under the skin in order to cool the body. In an environment of arbitrary external temperature warm-blooded animals must have an efficient internal heat transfer system as well, which distributes the temperature evenly throughout the body or heats or cools organ parts on demand. This temperature control is one of the functions of the blood system, which will be described in Chap. 4.

2.3.5 How to Live with Permanently Cold Feet

A heat exchanger transfers heat from one stream J_a, into another stream J_b of fluid, without mixing the two liquids. It combines the principles of conduction and convection. Two arrangements are possible: co- and counter stream heat exchangers as shown in Fig. 2.10.

A co-stream heat exchanger acts like a mixing chamber. The hot and the cold stream flow parallel to each other. The final temperature T_{out} at the exit is approximately equal to the mixing temperature of the two streams. The final temperature is found from the energy balance $J_a C_a (T_{a,in} - T_{out}) \approx J_b C_b (T_{out} - T_{b,in})$, where C_a and C_b are the specific heats of the fluids.

In a counter stream heat exchanger the temperature can be "exchanged" except for some difference ΔT that gets smaller the longer the two fluid streams are close to each other.

Consider the effects of co- and counter flow heat exchangers: counter flow exchangers normally have a lower rate of heat transfer, but they can exchange the temperature, whereas co- flow heat exchangers only allow to reach the mixing temperature, see for instance Schmidt Nielsen [1981].

Numerous animals live permanently in cold climates. For instance sea gulls, and penguins can perch on ice or stand in cold water, yet they are warm-blooded. Cold water is a very effective thermal contact material, and the birds would quick-

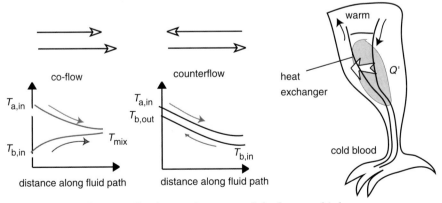

Fig. 2.10. Co- and counter flow heat exchangers, and the foot a seabird

ly lose all their heat energy if the feet were supplied with warm blood. These birds developed a nearly 100% efficient heat exchanger system in their legs. It is based on counter-flow. The feet are cold because the arterial blood, which is pumped by the heart into the legs, warms up the venous blood, which returns from the feet. The heat transfer between veins and arteries is accomplished by thermal conduction. For efficient heat transfer veins and arteries have to be in very close contact. Furthermore, the diameter of the blood vessels ought to be small, so that most of the blood flows near the vessel wall. This calls for many parallel vessels with the smallest possible vessel radius. Small vessels however have a large mechanical flow resistance, and the flow resistance increases the smaller the vessel gets. The tradeoff between saving body heat by good thermal resistance and requiring higher blood pressures to drive the blood through the capillaries probably has some optimum radius, that animals with permanently cold feet have found.[3]

An additional means to avoid heat losses is as *shunt*, namely a vessel with variable cross section connecting the outgoing and the incoming blood before the blood enters the extremity. The shunt acts as a switch to control the blood flow into the leg, and keeps the flow at the minimal rate required to supply the foot with oxygen and nutrients.

2.3.6 Radiation

Radiation is the "non-contact sport" of heat transfer. Radiation moves energy through free space. On Echnaton's monuments the rays of the sun are depicted as the sun god Aton's arms that touch the earth with warming hands. This is only one half of the story: the earth in turn sends radiation away, some reaching back to the sun. However, by the principle that hot things get colder much more radiation goes from the sun to the earth than from the earth to the sun.

Concepts and parameters associated with radiation are the body temperature T, the intensity of the emitted radiation I, the emissivity ε and the absorptivity α_r, which are both material constants for any surface. A body of temperature T and surface area A emits the radiation power Q' given by Stefan's law of black body radiation.

$$Q' = A\,\varepsilon\,\sigma\,T^4 \qquad\qquad (2.21)$$

In this relation T must be given in Kelvin, and A in m². The natural constant $\sigma = 5.67 \cdot 10^{-8}\,\mathrm{W/m^2\,K^4}$ is called Stefan's constant.

---

[3] Such efficient heat exchangers must also play an important role in the body design of tuna, who are known to have warm muscles but cold gills. Since the gills are in very close thermal contact to the cold water the blood flowing through them and in the nearby heart, must be cold. However, the muscles generate much heat, and they can stay permanently warm if there are heat exchangers between the muscles and the heart.

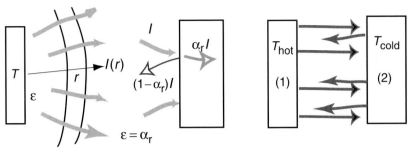

Fig. 2.11. Emission and absorption of radiation Absorption of radiation

As the radiation spreads out in space it carries the energy until the radiation is absorbed. The radiation energy passing through the unit area per second is called intensity I. It is measured in W/m².

$$I = Q'/A \tag{2.22}$$

When radiation spreads out into all directions, from a point source emitting the power P, the intensity decreases with the square of the distance r, according to the inverse square law

$$I = \frac{P}{4\pi R^2} . \tag{2.23}$$

Radiation is emitted <u>and</u> absorbed. An object exposed to radiation of the intensity I gains the power

$$Q'_{abs} = \alpha_r A I , \tag{2.24}$$

where the material constant α_r is called the absorption coefficient. In many experiments the emissivity ε and the absorptivity α_r of materials have been measured, and it has been observed that emissivity and absorptivity are exactly equal and fall into the range

$$0 \leq \alpha_r = \varepsilon \leq 1 . \tag{2.25}$$

An object which looks black has $\alpha_r = \varepsilon = 1$. The object absorbs all the light that strikes it. An object that looks silvery, shiny or white has as small emissivity, for instance $\alpha_r = \varepsilon \approx 0.1$. A body may simultaneously gain energy $Q'_{abs} = \alpha_r A I$ by absorption of radiation and lose energy $Q'_{em} = A \varepsilon \sigma T^4$ by emission of radiation. The net radiation gain is then

$$Q' = Q'_{abs} - Q'_{em} . \tag{2.26}$$

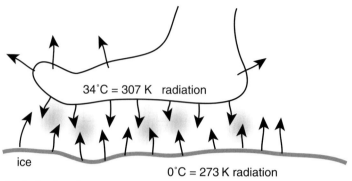

Fig. 2.12. Radiation heat loss

How does radiation contribute to the heat loss from a body part, for instance a foot? If the foot is lifted off the ice it will lose heat by *radiation*. We assume that the gap between the foot and the ice is very small, so that the emission and the absorption areas are equal. The foot emits $A\, I_f = A\, \sigma\, \varepsilon_f\, T_f^4 = Q'_{em,f}$. The ice emits the power $Q'_{em,i} = A\, I_i = +A\sigma\, \varepsilon_i\, T_i^4$, of which the foot gains the fraction $Q'_{abs,f} = A\, \alpha_f\, I_i$ of the black body radiation from the ice. The radiation exchange results in a net power loss for the foot: $Q' = +\alpha_f\, A\,\sigma\, \varepsilon_i\, T_i^4 - A\sigma\varepsilon_f T_f$. Since $\varepsilon_f = \alpha_f$ the net loss becomes $Q' = +\alpha_f\, A\,\sigma\, (\varepsilon_i\, T_i^4 - T_{f4})$.

The heat loss from the foot is calculated for the example shown in Fig. 2.12. Taking $A = 0.02$ m², $\Delta T = 34°$, and guessing $\varepsilon_f \approx 0.3$; $\varepsilon_i = 0.1$, one finds $Q' = 0.3 \cdot 0.02 \cdot 5.67 \cdot 10^{-8}$ $(0.1 \cdot 273^4 - 307^4) = -2.83$ W. This number is significantly smaller that the heat loss by thermal conduction, $Q'_{con} = 68$ W, calculated in Sect. 2.3.2.

More details about black body radiation will be given in Sect. 8.1.3.

2.3.7 Phase Changes and Evaporation Cooling

The fourth method used to transfer heat involves the phase change evaporation/condensation. Animals lose heat when they sweat, and they gain heat when steam condenses on their skin. These heat transfer processes always require a transfer of mass as well. For that reason heat transfer with phase changes may be classified as a special case of convective heat transfer. However, since the latent heat of fusion L_f, and latent heat of evaporation L_v are large numbers, one sometimes does not recognize the mass transfer. For water these values are $L_v = 2257$ kJ/kg, and $L_f = 333$ kJ/kg.

Evaporation is the process whereby ΔM kg of a liquid is turned into its gas. This process requires the heat

$$\Delta Q_v = L_v\, \Delta M\,. \tag{2.27}$$

Sublimation is the process whereby a material goes from the solid to the gaseous state. Examples are dry ice turning into CO_2 gas, or frozen clothes drying on a cloth line on cold prairie winter day.

$$\Delta Q_f = L_f \, \Delta M. \qquad (2.28)$$

In a steady process where a constant flow of mass J is evaporated or condensed a steady flow of power is consumed (negative sign) or released (positive sign)

$$P_v = L_v \, J, \qquad (2.29)$$

and

$$P_f = L_f \, J. \qquad (2.30)$$

Sweating/perspiration cooling is convective heat transfer combined with a phase change. We experience this cooling after sudden physical exercise. Dogs use sweating all the time to get rid of excessive internal energy. The reverse process of condensation may lead to severe burns, if the released heat cannot be absorbed fast enough – the temperature will stabilize at the boiling point.

Example of perspiration cooling: A dog of mass $M = 30$kg races up a steep hill raising his center of mass by $\Delta y = 120$ m in 2.5 minutes. In order to prevent overheating the dog sticks out his tongue and cools himself by evaporation. How much water will the dog loose if he does not lose any heat by conduction or surface heat transfer?

Answer: The dog produces the mechanical work $\Delta W = Mg \, \Delta y = 30 \cdot 9.81 \cdot 120$ m $= 3.5 \cdot 10^4$ J. Since it takes him $\Delta t = 2.5 \cdot 60$ s $= 150$ s the dog has generated the average mechanical power $P = 3.5 \cdot 10^4$ J/150 s $= 235$ W. At an overall efficiency of $\eta = 25\%$ his metabolic power is $\Gamma = P/0.25 = 942$W. The fraction $(1 - \eta) = 75\%$ of this power appears as heat. Set it equal to the evaporation power given in (2.29) and solve for the mass loss rate $J = 0.75 \, \Gamma / L_v = 0.75 \cdot 942W/(2\,257$ J/g$) = 0.31$ g/s. In 2.5 minutes the dog must lose $\Delta M = J \cdot \Delta t = 0.31$ g/s $\cdot 150$ s $= 46$ g of water, or heat up. The moral of the story is: have a water bottle along when you chase your dog up a steep hill!

2.3.8 Managing the Flow of Heat

The body's chemistry works best at certain temperatures. Warm-blooded animals (homeotherm) maintain body temperatures of about 37°C employing various strategies, see Table 2.8.

Animals living in extreme locations in the biosphere have learned to adapt to the temperature around them. Habitats range from the ice in Antarctica to the extremely hot thermal vents in the ocean or in hot pools in locations such as the Yellowstone National Park. Organisms living there have body temperatures that are equal to their surroundings. In cool environments their body fluids act like antifreeze, so that their body fluids will not freeze up. On the other end of the tem-

Table 2.8. Body temperature stabilization prinziples

working hard in temperate climates	cold climates	hot, dry climates
• let heat flow out by - evaporation - surface heat transfer - conduction in contact to cold objects outside • make thermal conductivity large by - wet fur - expose skin - reduce thickness of insulation layers	• make surfaced area small - roll into a sphere • thick insulation layers - ruffle feathers • minimize thermal conductivity - dry fur • avoid evaporation • keep extremities cold by - counter-flow heat exchangers • reduce T_o temporarily - torpor - hibernation • black skin to absorb sunlight	• thick insulation layers • avoid evaporation • avoid hot times of day • avoid sun light - deflect radiation - stay in the shade • use external coolants - water for evaporation - cool with sand

perature scale, the organisms in hot environments must have chemical compositions that do not coagulate like egg yolk in a frying pan.

Animals with intermediate body temperatures who make their homes in hot deserts or Arctic climates must reduce heat transfer into and out of their bodies, by insulation, by avoidance of evaporation or condensation, and by carefully choosing to be active, when the thermal challenges are less extreme. One trick employed to keep warm – by bees and penguins – is to form groups, where each individual is partly protected by its peers. Beehives maintain temperatures well above freezing through the metabolism of all the animals in the swarm. Bees can actually increase the ambient temperature in their hive by beating their wings, an ability they also use as a defense against invading insects: The bees surround the invader and *cook* their enemies to death. Male king penguins in Antarctica stand silently in tight colonies in the middle of the Arctic winter incubating a single egg on their feet. The tight formation provides some protection against the cold. The birds that face the wind stand guard for a short while and then walk slowly and graciously to the lee.

2.4 How Thermodynamics Sets Limits for Life

The laws of thermodynamics yield limits for life on different scales. On a large scale thermodynamics defines the volumes of space around stars where water based life can occur in the universe. On a small scale thermodynamics determines how small individual warm-blooded animals can make their home in a cold environment.

2.4.1 A Place Called Home

Most body functions require the presence of water; neither ice nor steam would do. Water has to be the liquid phase. Therefore life happens in a range of temperatures where liquid water can exist, namely between the ice point $T = 0°$ and the boiling point $T = 100°C$, see Fig. 2.13.

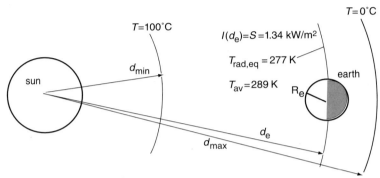

Fig. 2.13. The life belt around the sun

The temperature limits are not exactly fixed because some animals have managed to put antifreeze into their bodies so that they can operate below 0°C. Some other animals, called Extremotherms, actually live in the hot thermal vents in the ocean in geysers above 100°C. For calculations involving the radiation from hot bodies one must use the Kelvin temperature scale which has the ice point at $T = 273$ K and the boiling point at $T = 373$ K. This range of temperatures fixes a life belt around the sun, where life can prosper. Apart from geothermal sources we get all of our energy from the sun. The sun with a *body mass* $M_{sun} = 2 \cdot 10^{30}$ kg emits a power[4] of about $P_{sun} = 6 \cdot 10^{26}$ W.

For any object like the earth or a space ship, located at some distance d from the sun there is a particular *equilibrium* temperature $T(d)$ at which the radiation energy received from the sun exactly equals the black body radiation given off by the object.

If we assume that life is tied to the existence of water, life can thus only happen at places in the universe where the average temperature is in the range $T_{min} \approx 270$ K to $T_{max} \approx 370$ K. It is easy to determine at which distance from the sun such life can exist. For this calculation we use the solar constant $S = 1.37$ kW/m², namely the intensity of the sun's radiation at the distance $d_e \approx 1.5 \cdot 10^{11}$ m of the earth. With the help of the inverse square law (2.23) we find the intensity drops off with distance d from the sun as $I(d) = S(d_e/d)^2$. The earth intercepts the total amount $P = \alpha_r A I$, where $A = \pi R_e^2$ is the cross section area of the earth's shadow, and R_e is the radius of the earth. Hence the earth receives the power (energy per second)

---

[4] Similar to the metabolic rate of animals the power production $P(M)$ of main sequence stars can be approximated by the allometric function of the star mass M, namely $P(M) = P_{sun}(M/M_{sun})^{3.5}$.

$$P = \alpha A_c I = \alpha_r \pi R_e^2 S (d_e/d)^2 \text{ W} . \qquad (2.31)$$

At the same time the earth loses power like any black body, at the rate

$$Q'_L = A_s \varepsilon \sigma T^4 \text{ W} \qquad (2.32)$$

where Stefan's constant is $\sigma \approx 5.67 \cdot 10^{-8}$ W/(m^{2}K^4), $A_s = 4\pi R_e^2$ is the surface area of the earth. In equilibrium P must be equal to the lost heat flux Q'_L.

$$\pi R_e^2 S (d_e/d)^2 = 4 \pi R_e^2 \sigma T^4 \qquad (2.33)$$

The emissivity ε and the absorptivity α_r drop out of the calculation because they are equal for any object. Equation (2.33) may be solved with $d_e = d = 150 \cdot 10^6$ km to find the average temperature $T = (S/4\sigma)^{1/4} = 277$ K which a rock-planet would acquire at the earth's distance from the sun.

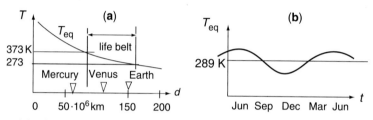

Fig. 2.14. (a) Where water based life is possible near the sun, (b) seasonal temperature variations

To determine the width of the life belt we solve (2.33) for $d = d_e (1/T)^2 \sqrt{(S/4\sigma)}$ leaving T as a free parameter. The maximum life belt distance d_{max} is found for the lowest temperature $T \approx 270$ K, where liquid water still exists, namely $d_{max} = d_e (1/270)^2 \sqrt{(S/4\sigma)} = 1.05\, d_e = 1.58 \cdot 10^8$ km. For the minimum radius of the habitable space we set T equal to the the boiling point of water, $T \approx 373$ K. This yields the distance $d_{min} = d_e (1/373)^2 \sqrt{(S/4\sigma)} = 0.53\, d_e = 8 \cdot 10^7$ km. The calculation reaffirms that we happen to live in the habitable belt of our sun. This calculation also implies that conditions for water based life must exist in a ring shaped volume around other stars in the universe.

Actually, the average surface temperature of the earth is about 16 °C = 289 K. Internal heat sources exist as well: both, radioactive decay and some heat from the latent heat of fusion of the earth's iron core, which gradually grows inside the earth. These internal heat sources might create habitable zones at larger distances from the sun, perhaps on Mars, and on the moons of Jupiter.

The earth's rotation generates seasonal and diurnal variations in temperature and light intensity, which is illustrated in Fig. 2.14b. Animals have adjusted to these temperature changes (i) by turning body functions on and off, (ii) by hiding when it is too hot or too cold, (iii) by growing big bodies that neither heat up nor cool down too fast, (iv) by maintaining a constant elevated body temperature (homeotherm), and (v) by migrating to warmer climates.

2.4.2 Why Bigger is Better in a Cold Ocean

The laws of physics enable and restrict the performance and dimensions of animate organisms: animals who learned new physics tricks may invade new niches of the biosphere, but physical properties often set limits as to what can realistically be achieved.

A case in point is the group of warm-blooded aquatic animals who have more mobility and certain more highly developed senses than do fish. The price for the increased abilities advantage is a lower limit in size, which is linked to the physics of thermal insulation. The metabolic heat production Γ of an animal must overcome the conductive heat losses Q'. A detailed analysis [Ahlborn and Blake 1999] reveals that aquatic mammals which are protected against the cold by an insulating blubber layer of thickness ΔR cannot have a diameter smaller than about $2R_{min} \approx 15$ cm. In general ΔR would be a function of R. However, for metabolic rates $Q' = \Gamma = a\, M^\alpha$ with the empirical constants a = 3.6, and $\alpha \approx 0.73$, the insulation thickness is found to be $\Delta R \approx 4$ cm, independent of the body mass M.

Warm-blooded animals lose heat through their skin from the entire surface area of their body, Fig. 2.15. Thermal conduction is easily understood for the heat transfer across plane surfaces, see Eq. (2.15). However, if an organism is so small that the thickness of its insulation ΔR becomes a substantial fraction of its radius R, the curvature of the body comes into play. For a very small organism the surface on the outside of a section of insulation is larger than the surface on its inside and therefore the thermal resistance drops. As a consequence the heat transfer from cylindrical bodies, (2.16), or from spherical objects (2.17) differs from the plane surface relation (2.15). Here we model the aquatic mammal as a cylindrical body, with the loss function.

$$Q'_{cyl} = 2\pi L \frac{\kappa\,\Delta T}{\ln(1 + \Delta R/R_i)}. \qquad (2.34)$$

The length L of the animal is still arbitrary. However, most aquatic animals have acquired over time an aspect ratio $2R/L$ in the range 0.2–0.3. The sum of form drag and friction drag happens to be a minimum for this aspect ration range

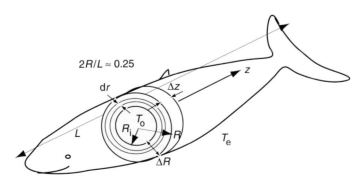

Fig. 2.15. A whale of a fish

[see MacMahon, and Bonner 1983]. This minimum will be derived from first principles in Sect. 3.4.4. For simplification we use $2R/L=0.25$ or $L=8R=8(R_i+\Delta R)$. This yields the heat loss through the cylindrical surface as function of R_i and ΔR[5].

$$Q' = 16\pi R \frac{\kappa \Delta T}{\ln(1 + \Delta R / R_i)} \, . \tag{2.35}$$

Note that the total heat flux depends linearly on the total radius R and is a logarithmic function of the insulation aspect ratio $\Delta R/R_i$. The heat losses thus become very large when the insulation thickness ΔR is small compared to R_i. The empirical metabolic rate function (1.5) was quoted for land animals, and it also appears to apply for birds. Since all aquatic mammals have terrestrial forefathers we can use the empirical metabolic rate function for the warm blooded aquatic animals as well. In equilibrium the metabolic heat production $\Gamma = aM^\alpha = a(\rho\pi R^2 8R)^\alpha$ must be in balance with the heat losses Q'_{cyl}. This condition leads to

$$a(\rho 8\pi R^3)^\alpha = \frac{16\pi R\kappa\Delta T}{\ln(1+\Delta R/R_i)}, \tag{2.36}$$

where the density ρ is given in kg/m³, the insulation conductivity in W/m°, and the radius R in m. The equilibrium equation (2.36) can be solved for the total radius R of the model animal $R = R_i + \Delta R = B\{\ln(1+\Delta R/R_i)\}^{-1/(3\alpha-1)}$, where $B=\{(2\kappa\Delta T a^{-1} \rho^{-\alpha}(8\pi)^{1-\alpha}\}^{1/(3\alpha-1)}$. According to (2.36) there is only one *correct* insulation thickness ΔR for any given core radius R_i, where the heat production and the heat losses are in balance. As an animal must produce its insulating layer the biological cost of the heat shield is better represented by the function $\Delta R(R)$. To extract this relationship one can first find R as function of the aspect ratio $x=\Delta R/R$, and then calculate $R_i=(R/(1+x))$, and $\Delta R = xR_i$. The thermal conductivity of blubber was set at $\kappa=0.1$ W/(m·K), and a cold ocean environment was selected with $\Delta T=37°C-4°C=33°C$.

Initially the Eq. (2.36) was examined for the experimentally observed exponent $\alpha \approx 0.75$, but it is easy to show the effect of varying the exponent α of the mouse to elephant curve. At the position $M_1=0.1$ kg we took $Q'_1=0.61$ W from the mouse to elephant curve to calculate the factor a for any given exponent α at the body weight of about 0.1 kg. Note that each point on these curves belongs to a particular value of the aspect ratio $x=\Delta R/R$. For a smaller R, larger $\Delta R/R$ must be chosen to maintain the thermal balance. Points for $x=1$ and 0.5 are marked on all curves. When x gets too large it becomes too expensive for the organism to manufacture its insulation. Therefore we have arbitrarily terminated at $x=\Delta R/R=1$. This point then represents a lower limit for the size of the warm-blooded animals. Note that this lower limit gets smaller as the exponent α increases.

---

[5] More precisely one should model the animals as a cylindrical body of length L_i and radius $R=R_i+\Delta R$ with 2 spherical end caps of radius R for a total length of $L=L_i+2R$, which loses heat through its entire surface area. However, the losses through the end caps only amount to a few percent. Therefore the end cap heat flux is neglected here.

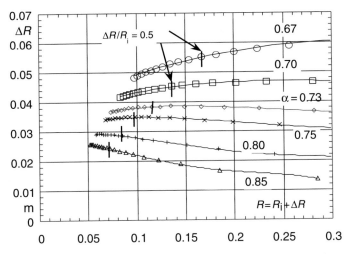

Fig. 2.16. Insulation thickness ΔR function of the total body radius $R + \Delta R$ for various exponents α

More surprising than the existence of a lower limit in size is the fact that ΔR varies only very little with the body radius for the exponent $\alpha = \frac{3}{4}$, while the required insulation thickness grows with R for $\alpha = 2/3$. To study the importance of α for the heat transfer in more detail similar curves are shown for a range of exponents, Fig. 2.16. One can see that there is a systematic change of the dependence of the insulation thickness ΔR as function of R_i: When α is below 0.73 the insulation thickness ΔR rises with radius, while for $\alpha > 0.75$ it decreases.

Neither of these relations would be biologically wanted. However, precisely at the experimentally observed metabolic exponent ≈ 0.73 the insulation thickness ΔR is independent of the radius R. The calculated insulation thickness $\Delta R \approx 0.037$ m agrees well with measurements by Kvadsheim [1997] who found average values of $\Delta R = 0.03$ m. Obviously the exponent $\alpha \approx 0.73$ chosen by Nature has advantages that go beyond the arguments presented in Sect. 1.3.1. In the context of heat conduction one could describe the experimental mouse to elephant exponent $\alpha = 0.73$ as the response of Natures to the essential cylindrical geometry of warm-blooded animals. This exponent is matched to the concept behind the heat transfer relation, namely that the thermal resistance $R_{th} = \Delta R / \kappa$ is the same for organisms of different sizes and thus the curvature effects on the thermal resistance are canceled out. The family of curves was calculated for a specific example of ΔT and κ, however, the intriguing result that the insulation thickness ΔR is independent of the body radius for $\alpha = 0.73$ holds for any ΔT and κ.

2.4.3 Why Birds Can't Be Smaller Than Bees

Evolution has produced an immense variety of shapes and designs. However, certain limits are set by the laws of physics. A case in point is the minimum size of homeotherm (warm blooded) animals, such as birds [Ahlborn and Blake 2001]. The metabolic heat production Γ must be able to overcome the conductive heat losses Q'. In the resting state one has $\Gamma_0 = 3.6\,M^{3/4}$ W. The best geometry to make the heat losses small is a sphere. Heat is generated in the volume $V = (4\pi/3)R^3$, but is lost through the surface $A = 4\pi R^2$. For a sphere the surface to volume ratio $A/V = 3/R$, is smaller than for any other geometrical shape. However small spheres have larger A/V ratios than larger spheres, so that there must be a minimum radius for any metabolic heat production where an elevated body temperature can no longer be maintained. This size limit is now determined for birds.

We model birds as spheres and assume that the skin under the feathers, Fig. 2.17b, at the radius R_i is maintained at the temperature T_I while the outside temperature is T_e. The heat loss from a sphere is

$$Q' = 4\pi (R_i + \Delta R)\frac{\kappa \Delta T}{\Delta R / R_i}. \tag{2.37}$$

The body mass M is related to the radius. Since the feathers of thickness ΔR contribute practically no mass, the body mass is $M \approx (4\pi/3)\rho R_i^3$. The metabolic heat production $\Gamma = 3.6\,M^{0.73}$ must at least equal the conduction losses. This condition leads to an equation for ΔR as function of R_i.

$$\Delta R = \frac{R_i}{(C/\kappa \cdot \Delta T)\cdot R_i^{1.19} - 1} \tag{2.38}$$

For $\rho = 10^3$ kg/m³ the constant has the value $C = (3.6/4\pi)\cdot(4\pi\rho/3)^{0.73} = 126.3$. The insulation thickness ΔR becomes infinite for a limiting radius $R_i = R_{min}$ where the denominator of Eq. (2.33) becomes zero.

$$R_{min} = (k\Delta T/C)^{0.84} \tag{2.39}$$

At this point no amount of insulation is enough to maintain the body temperature at the value T_i. The position of R_{min} depends sensitive on $\kappa \Delta T$. Smaller values of $\kappa \Delta T$, or an increased metabolic rate (which increases the constant C) would shift the curve toward the left.

Humming birds have such an increased metabolic rate, and thus can be smaller. Figure 2.16 shows ΔR as function of R_i, calculated for air insulation (down feathers), and the temperature difference $\Delta T = 33°$. ΔR approaches ∞ for $R_i = R_{min} \approx 2.5$ cm. An infinite insulation thickness is useless for any animal. Of interest is the insulation aspect ratio $\Delta R/R_i$, which is also shown in Fig. 2.17. A practical minimum aspect ration is $\Delta R/R_i \approx 1$, which is attained at $R_i = R^* = 3$ cm. This practical minimum radius is an order of magnitude larger than the size of a bee, with

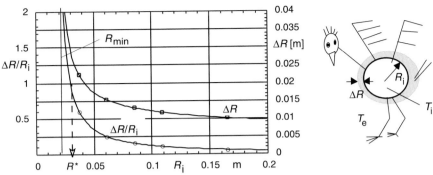

Fig. 2.17. Insulation thickness ΔR as function of body radius R_i

a typical body dimension $R \approx 0.3$ cm, leading to the conclusion that birds cannot be smaller than bees. Warm-blooded animal of sub-centimeter size can never exist in a cold environment, because the physiology is ultimately constrained by the immutable laws of physics.

The limiting radius arises because the radial heat losses cannot exceed the metabolic heat production if the animal is to maintain its internal temperature T_i. Previous studies of size limits on homeotherm animals [Balmer, and Strobusch, 1977, Turner, and Schroter, 1985] applied the concept of the *critical radius, R_c*, known from the engineering literature [for instance see White 1984]. R_c is calculated by equating the radial heat flow in the insulation (characterized by the thermal conductivity, κ), to the heat loss at the outer surface of a hot object (characterized by the surface heat flow coefficient, h_s). The critical radius depends strongly on the surface heat transfer coefficient, which may vary by more than one order of magnitude, see Table 2.7. The critical radius concept is useful for engineering situations, where the net heat flux can be treated as an independent quantity of arbitrary magnitude. However, in applications to living organisms this concept misses the essential point that the total heat production is tied to the animal size by the metabolic rate Γ. Therefore the critical radius does not yield a practical limit for warm-blooded animals.

While thermodynamical principles limit the size of warm-blooded animals, there may be other factors that also demand a size minimum for animals in a particular niche of the biosphere.

2.4.4 Water, the Magic Stuff

Without water there is no life. Therefore all organisms from the plankton at the bottom of the food chain to the killer whales at the top must exist by using the properties of water. Materials can be characterized by their state variables: temperature T, their pressure p, and their density ρ, and their internal energy U (or their enthalpy $h = U + (p/\rho)$). These state variables can be given for all the four phases: solid, liquid, and gas, and plasma.

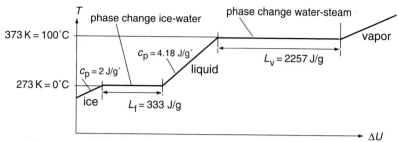

Fig. 2.18. Phases, specific heats, and latent heats of water

Water turns into ice (solid) when cooled bellow 273 K (0°C) releasing the heat of fusion $L_f = 333$ J/g, and water turns into steam (vapor) when heated above 373 K (100°C), consuming the heat of vaporization $L_v = 2257$ J/g. To heat 1 g of water by 1 K (or by 1°C) requires the specific heat 1 cal/g = 4.18 J/g.

Gases change their density when pressure is applied, such substances are called compressible. Water is incompressible (it does not yield to pressure) however, its density ρ depends on the temperature, see Fig. 2.19.

Water of 4°C has the highest density, and thus sinks to the bottom of lakes and oceans. Life is safe down under. Sound caries far, but it is dark deep down, see Sect. 7.3.4. Water is denser than ice $\rho_{ice} = 0.92 \, \rho_{water}$. Hence ice floats on top of the water. For example 92% of an iceberg is submerged. The density of ocean water also depends also on the salinity.

Water contains only about 2–5% of the oxygen content of air. However cold water at 0°C (close to sea ice) has about 60% more oxygen content than tropical water of 25°, see Fig. 2.21.

In summer months the sun shines 24 hours every day in the high Arctic. With plenty of light and oxygen in the surface waters ocean plankton flourishes in the Arctic and Antarctic waters in summer months. The plankton is the basis of the whole food chain. Plankton feed krill, which in turn supports small fish. Small fish attract sea birds, large fish, and marine mammals, all of which migrate to these productive regions in the summer months. Thus the Arctic and the Antarctic waters in summer are the most productive regions in the ocean.

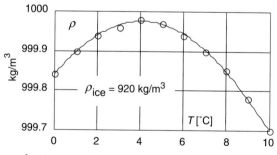

Fig. 2.19. Density ρ of water near 0°C

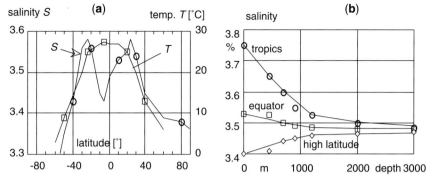

Fig. 2.20. Temperature and salinity of sea water as function latitude (**a**), and depth (**b**)

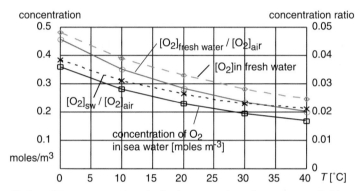

Fig. 2.21. Ratios of O_2 concentrations in freshwater – air, and sea water – air [data adopted from Denny 1993]

2.5 Other Physical Quantities

This chapter demonstrates how animals have learned to use the macroscopic physical effects of energy conversion and temperature control. These physics "tricks" have enabled animals to fill new niches in the biosphere, and to occupy vast tracts of land from high latitudes on the earth, where the temperatures might sink below freezing, to the scorching deserts where the temperature approaches boiling.

Insects, amphibia, and reptilia make use of thermodynamic principles a) in order to acquire the temperature of their environment, b) to get hotter than their surrounding when it is frigid, or c) to get colder when it is too hot. However they have not yet perfected the ability to keep a constant body temperature. Single cell organisms are incapable of using these physical effects.

The homeothermal status of birds and mammals appears to be a rather recent invention in the evolution of life on earth.

Long before animals had mastered these thermodynamic principles organisms had emerged out of the water and established themselves on the land. The evolutionary step onto land necessitated the mastery of different branches of physics. In water large and soft bodies can support themselves by buoyancy forces alone. Terrestrial and airborne animals must deal with gravity and other forces: namely elastic, and electrostatic forces, surface tension, friction, and forces that are associated with motion. Animals on the land and in the air need strong materials to support and contain their bodies, and enable them to move around. Forces and materials are described in the next chapter.

Table 2.9. Frequently used variables of Chap. 2

variable	name	units	name of unit
A	area	m^2	square meter
a	metabolic rate constant		
C	specific heat	$J/(kg\,°)$	Joule / kg & degree (K)
d	distance	m	meter
E_{el}	elastic energy	J	Joule
E_{kin}	kinetic energy	J	Joule
E_{pot}	potential energy	J	Joule
f	specific muscle power	$Pa=N/m^2$	Pascal
h	height	m	meter
h_s	surface heat flow coefficient	$W/(°m^2)$	Watt/degree (K) & square meter
I	intensity of radiation	W/m^2	Watt/square meter
J	mass flow rate	kg/s	kilogram/second
k	Spring constant	N/m	Newton/meter
L	length	m	meter
L_f	latent heat of fusion	J/kg	Joule/kilogram
L_v	latent heat of vaporization	J/kg	Joule/kilogram
$Q, \Delta Q$	heat	J	Joule
Q'	heat flux	W	Watt
R	radius	m	meter
R_{th}	thermal resistance	$°/W$	degree (K)/Watt
s	distance	m	meter
S	solar constant	W/m^2	Watt/square meter
t	time	s	second
$U, \Delta U$	internal energy	J	Joule
u	velocity	m/s	meter/second
W	work	J	Joule
α_r	absorptivity		
ε	emissivity		
η	efficiency		
κ	thermal conductivity	$W/(°m)$	Watt/degree (K) & meter

Problems and Hints for Solutions

P 2.1 Diet

Suppose a 65 kg person lives on a diet consisting of prime beef (h_{PB}= 4 MJ/kg), honey h_H=14·10^6 J/kg) and corn flakes (Δh_{cf}=15 MJ/kg). How much corn flakes must this person eat per day if the diet consists of 20% by weigh of beef, 20% honey and 60% corn flakes? Assume that the average metabolic power consumption is a factor 2 above the resting metabolic rate.

P 2.2 Staircase Walker

A person (M=60 kg) runs up staircase, climbing a height of h=7 m in 5 seconds. a) How much mechanical energy is produced? b) What is the mechanical power generated in this climb? c) Why does the person breathe heavily and why does his heart rate go up as he climbs the stairs? d) If this person would carry a suitcase of m=12 kg, and generate the same mechanical power as he climbs how long would it take him to get up the stairs? e) If this person has a mechanical efficiency of 30%, what is the total metabolic power, and total metabolic energy, and how much energy goes into heat. f) By how much does his body temperature go up? (Assume a specific heat of C=4.18·10^3 J/kg.) g) Do the experiment yourself: Measure the height of one stair in a staircase, and count the number of stairs to find the height Δh of a storey. Then time yourself as you run up the stairs.

P 2.3 The Tour de France

The tour de France lasts for cycling 22 days, at 6 h/day (in 6 weeks). A cyclist (M=70 kg) drinks on average 7.0 l water during the 6 hours of cycling. a) Assume that this water is entirely used for evaporation cooling. How much energy is dissipated? b) If he has a mechanical efficiency of η = 25% what is his total energy production and his muscle power output (W). c) The cyclists has an average speed of u = 47 km/h, how much distance Δs does he cover daily? d) Suppose the rolling resistance is negligible; what is the average drag resistance? e) What is his energy cost per meter $E=\Delta Q/m$? f) Compare his cost of transport $COT = E/M$ with that of a runner and a bird, Fig. 5.9.

P 2.4 Perspiration Cooling

A 30 kg dog runs uphill raising his center of mass by 120 m in 2 minutes. In order to prevent overheating the dog sticks out her tongue and cools by evaporation. How much water will the dog lose? Assume that the dog has a mechanical efficiency of 26%, and maintains her body temperature, i.e. she gets rid of all the excess heat by perspiration cooling.

P 2.5 Cool Drinks

Suppose you hold in one hand a glass with a drink in which ice cubes are floating. a) How much heat is flowing from your hand to the drink by conduction? Assume an average temperature of $T \approx 28°C$ in your hand. Find the contact area, by wrapping a paper towel around the glass and grabbing it with a wet hand, so that you can outline the "hand print" with a pen, and then measure the area approximately. Assume that the capillaries are $\Delta x = 3$ mm away from the surface, taking $\kappa = 0.08$ W/mK for the thermal conductivity of the tissue. b) The heat lost by conduction is replaced by heat convected into the hand by the blood. How many degrees will the temperature of this nearby blood drop, if the blood flow rate through the hand is $m' = 1.5$ g/s?

P 2.6 Sunbirds

A sparrow (mass $M = 0.020$ kg $= \rho \cdot V$) soaks up sunlight on a frosty but sunny morning. a) How much solar power Q'_{abs} does the bird absorb, and how much energy [Joule] will the sparrow collect during 10 minutes? Hint: First determine the Volume $V = (4\pi/3)R^3$ of the sparrow, treating it like a spherical object of density $\rho = \rho_{water} = 1000$ kg/m^3. Since the sun is still low in the sky assume that the sunlight has only 1/5 of its full power (at full power $I = S = 1.37$ kW/m^2). Obviously the bird presents an object like a circle to the sunlight. Assume that the absorptivity of the bird with its feathers ruffled is $\alpha_{rad} \approx 0.8$. b) Compare your calculated value Q'_{abs} with the metabolic heat production of the bird. c) Why do Cormorants spread their wings after diving?

P 2.7 A Hippo-thetical Question

A hippopotamus of $M = 900$ kg has been sleeping in a thicket and it waddles into the open late in the morning. The hippo stands in the bright light of the sun overhead for 20 minutes before it starts its active day. a) How much radiation energy does the hippo absorb in the 20 minutes? Model the hippo body as a cylinder of radius R and length $L = 4R$ supported by 4 chubby legs each having the mass of $M_{leg} = 50$ kg. The solar constant is $S = 1.37$ kW/m^2. Assume an absorption factor $\alpha = 0.65$. Hint the average density of the animal is similar to that of water. b) It is getting very hot and the hippo decides to take a bath in a pool of $T = 22°C$. How much heat does the lose in the pool by thermal conduction through its bum? Assume area $A_b = 0.4$ m^2, thickness of bum fat insulation $\Delta R = 0.08$ cm, thermal conductivity $\kappa = 0.11$ W/mK. c) Thereafter the hippo rolls in the mud, accumulating a healthy layer of wet dirt all over its body. Then standing in a gentle afternoon breeze the hippo cools off by evaporating some water from the mud. How much water must be evaporated to lower its average body temperature by 2.5°?

Sample Solutions

S 2.1 One day has 86 400 seconds. Two times the resting metabolic rate is $\Gamma = 2\Gamma_0 = 2 \cdot 3.6 \cdot 65^{0.75} = 165$ W $= \Delta Q$/day. Sixty percent of this power namely $\Gamma = 98.9$ W $= 98.9$ J/s $= 98.9$ J/s $\cdot$ 86 400 s/day $= 8.54 \cdot 10^6$ J/day is supplied by corn flakes that have an energy content $1.5 \cdot 10^7$ J/kg corn flake $\cdot 8.54 \cdot 10^6$ J/day $= \Delta M$ corn flakes/day $\cdot 15 \cdot 10^6$ J/kg. In this equation the mass of corn flakes, ΔM, is unknown; solve for ΔM to get: $\Delta M = 8.54 \cdot 10^6$ J/day/($15 \cdot 10^6$ J/kg corn flakes) $= 0.57$ kg corn flakes /day.

S 2.3 a) When 7 kg water is dissipated the energy $\Delta E = 7\,\text{kg} \cdot 2.3 \cdot 10^6\,\text{J/kg} = 1.61 \cdot 10^7\,\text{J}$ is released. b) Assume that this energy is entirely used to remove the internal energy, ΔU, which is released as a byproduct of generating the work ΔW in a metabolic process that consumes the energy $\Delta Q = \Delta U + \Delta W$. At the efficiency $\eta = 0.25\%$ one has $\Delta W = 0.25\,\Delta Q$ and $\Delta U = (1 - \eta)\Delta Q = 0.75\,\Delta Q$. Then the total energy conversion is $\Delta Q = \Delta U / 0.75 = 1.61 \cdot 10^7 / 0.75 = 2.15 \cdot 10^7\,\text{J}$, and $\Delta W = 0.25\,\Delta Q = (0.75/0.25)\,\Delta U = 5.37 \cdot 10^6\,\text{J}$. This work is delivered in 6 hours, which implies the mechanical power $P = \Delta W / 6\,\text{h} = 248\,\text{W}$. c) At the average speed 47 km/h the cyclist covers $\Delta s = 6\,\text{h} \cdot 47\,\text{km/h} = 282\,\text{km} = 2.82 \cdot 10^5\,\text{m}$ in 6 hours. d) Interpret the drag force F_D as the mechanical energy expended per meter of travel $F_D = \Delta W / \Delta s = 5.37 \cdot 10^6\,\text{J} / 2.82 \cdot 10^5\,\text{m} = 19\,\text{N}$. The cost of transport is defined as $\text{COT} = \Delta W / M \cdot \Delta s = 0.27\,\text{J/m kg}$. This value is about a factor of 10 lower than the cost of transport for a runner, see Fig. 5.9.

S 2.5 Assume you measure an area of $A = 25\,\text{cm}^2 = 0.0025\,\text{m}^2$. The heat loss is $Q' = A\,\kappa\,\Delta T / \Delta x = 0.0025 \cdot 0.08 \cdot 28 / 0.003 = 1.87\,\text{W}$. b) Assume that the specific heat of blood is about equal to that of blood $C_{blood} \approx C_{water} = 4.18\,\text{J/°g}$. Heat transfer by conduction: $Q' = C\,m'\,\Delta T$. The temperature drop is $\Delta T = Q' / C \cdot m' = 1.87\,\text{Js}^{-1} / (4.18\,\text{J/°g} \cdot 1.5\,\text{g} \cdot \text{s}^{-1}) = 0.3°$.

3. Form and Forces

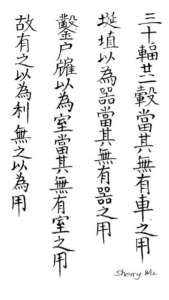

Thirty spokes meet the wheel's hub,

It is the center hole that makes it useful.

Clay is the substance of vessels,

it is the void within that makes them useful.

Houses are built with walls and doors

It is the space within that makes them useful.

Therefore matter is needed

But the essence lies in the form.

Chinese wisdom from the Tao Te King.

Forces Set the Body Form

Life requires matter, energy and information. Energy is absorbed through the food. The metabolic energy consumption is closely tied to the body mass M. Every organism is bound to earth by gravity $F_g = Mg$. The gravity with the gravitational constant $g_{earth} = g = 9.81$ m/s² holds us tighter to the earth surface than the astronauts were held onto the lunar surface, since g_{moon} is only one sixth of g_{earth}. Animals come in all sizes with small or large body mass M. But how should the body of a fish, ox, or pigeon be formed? What shape should it adopt? Suppose each animal is an optimized solution to its niche in the biosphere. Then the body form must have evolved in response to the external and internal forces that act on the organism. To appreciate body forms one must first know how forces act, and what kind of forces animals might encounter. Then one can recognize the elegance of the design of large and small structures.

In this chapter it is shown how animals have skillfully used to their advantage all the forces that nature provides, and how they have adapted their body shapes to make best use of the laws of statics and dynamics. Occasionally one can even see that a particular structure or shape represents an absolute optimum: a minimum of energy expenditure or a maximum of strength.

The chapter begins with a description of some important features of forces: compression, tension, static equilibrium, and how to display them in the free body diagram, Sect. 3.1. Then it is reported how muscles generate forces and what efficiencies muscles can achieve, Sect. 3.2. Static and dynamic forces that animals might encounter are described in Sect. 3.3 and 3.4. Simple body forms are discussed in Sect. 3.5. Some features of large structures are explained in Sect. 3.6. There are certain rules how bodies must be scaled up from small to large sizes, Sect. 3.7. As an example of the origin of the strength of materials, the properties of spider silk are discussed in Sect. 3.8.

3.1 How to Deal with Forces

Statics affects the shape, and size of animals. On first view one might think that the body form is entirely arbitrary. For instance why do fish not have propulsive devices at the front and at the rear of their bodies, so that they might swim equally well forward and backwards?

On closer inspection one learns that nothing in life is purely accidental. Size and form have evolved in response to the particular conditions surrounding each animal giving it that trifle of advantage that makes it fittest in its niche. One part of the surrounding is the inanimate world of gravity, water, wind, seasons, weather, and geographic location. The other part is the rest of the biosphere. Animals that "understand" statics have bodies that are only strong at places where strength is needed, and have shapes which offer the least resistance to motion. They use materials that are matched to the magnitude and direction of the local forces: flexible cables (sinews) or skin, where tension forces occur, and strong but possibly brittle structures (bones) where compression forces are acting.

In the pervious chapter, forces were encountered in the context of energy, and power. A force F pushing an object through the distance Δs at the speed u consumes the energy $E = F \cdot \Delta s$. This action requires the power input $P = F \cdot u$. This is always true. However, if one wants to understand why animals have acquired their distinct body forms, and how they move so elegant and efficiently, one must learn about the nature of static and dynamic forces. When an object is not moving all forces are in balance. This is a problem of statics.

3.1.1 Forces in Static Equilibrium

An animal can sit motionless only if all acting forces and moments are in balance. To see what this statement means one must remember certain properties of forces:

- They can create compression or tension.
- Forces are vectors. Vectors can be added and multiplied.
- A force times displacement in the force-direction is work $W = F \cdot \Delta s \cdot \cos(F, \Delta s)$.
- A force F acting with a moment arm r generates a torque $\tau = F \cdot r \cdot \sin(F, d)$.

  The symbol $(F, \Delta s)$ designates the angle between the force and the displacement, and (F, r) is the angle between the force and the moment arm.

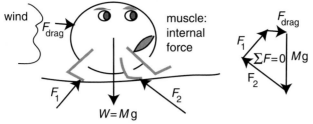

Fig. 3.1. Principles of statics: forces and moments in balance

These principles are outlined in Fig. 3.1. An animal which wants to sit motionless must ascertain that all forces are in balance. Gravity is the most obvious force. Gravity will squash a collection of cells and soft organs into the shape of a pancake unless it is offset by an appropriate arrangement of internal forces, employing structural members that can withstand compression, tension, and bending moments. Many macroscopic forces can be traced back to the microscopic forces between molecules, ions, and electrons. More about the inter molecular forces will be said in Sect. 3.5.5.

in static equilibrium the sum
of all forces vanishes

and the sum of all moments vanishes

$$\sum_{i=1}^{n} F_i = F_1 + F_2 + \ldots + F_n = 0$$

$$\sum_{i=1}^{n} r_i \cdot F_i = r_1 \cdot F_1 + r_2 \cdot F_2 + \ldots + r_n \cdot F_n = 0$$

If the sum of all forces and moments acting on a body is not zero the object cannot be at rest. It must move. More about this is said in Chaps. 4, 5, and 6.

3.1.2 Compression and Tension

Forces always appear in pairs. For example when one pushes with the fist onto a book resting on a table, the table pushes back at the book. The book is not moving, because the force exerted by the table and the force applied by the fist are exactly equal. Pushed from both sides the book is under *compression*.

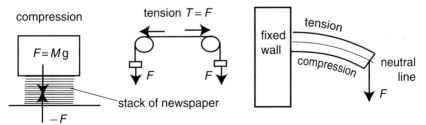

Fig. 3.2. Direction of forces

If a rope tied to a tree and then pulled, the rope is stretched from both sides. This is called *tension*. A cantilevered beam supports tension as well as compression. Tendons, ligaments, and skin, like rope, can support only tension (you cannot push on a rope!) while bones can support compression, tension, and torsion. Bones, skin and tendons under tension can withstand lateral forces.

3.1.3 The Free Body Diagram

In order to understand how animals stand up and move one must locate the forces that act on the whole body or its appendages. Think for instance of the mechanism of holding up the head, mass m which is pulled towards the center of the earth with the force mg, acting at the center of mass of the head, Fig. 3.3a.

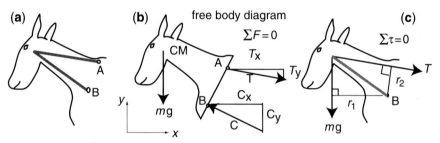

Fig. 3.3. (a) Horse head with supporting members. (b) Free body diagram of a head showing force components. (c) Moments about point B. Center of mass CM

To help visualize the situation it is useful to draw a free body diagram, and then list the force components in all direction to see how they are balanced. To simplify the discussion assume that the head is supported only by the neck bones which are under compression C, and by a tendon that is under tension T. Disregard the additional support which comes from the internal pressure and the tension produced by the skin. This could be done by making a cut through the points A and B, Fig. 3.3b. Then draw the forces T and C and their components in the x- and y-direction. Further locate the center of mass CM of the head where gravity mg pulls down. Show the moments about B, Fig. 3.3c. The following relations apply:

- The sum of the forces in the x-direction (to the right) is $\Delta F_x = 0 = T_x - C_x$
- The sum of the forces in the y-direction (up) is $\Sigma F_y = 0 = -T_y + C_y - mg$
- The sum of the moments about point B is $\Sigma \tau_B = 0 = T \cdot r_2 - mg \cdot r_1$

Note that C_x is counted negative since it points into the negative x-direction.

3.2 Muscles and Tendons

All large animals that move have muscles. A muscle exerts a tensile force when it contracts. Think of the biceps that lifts your hand, or think of the stomach muscles that can lift your legs when you lie flat on your back. When a muscle contracts its length shrinks and its diameter increases, because the muscle volume must remain constant. Therefore the muscle exerts simultaneously an axial tension and a radial compression force. This compression force pushes against your intestines, which in turn push against your spine as you lift your legs. Muscle force is used for chewing, running, flying, pointing ears, or rolling the eyes. Muscles pump blood or restrict the blood flow. Muscle force is transmitted by tendons onto fixed points on bones in order to move the limbs. Muscle force arises from the bending of countless molecular structures in the muscle cells.

3.2.1 Muscle Force

A muscle pulls with a force F (measured in Newton) that depends on its cross section A_m. In slow motion the muscle tissue of all animals generates about the same stress called *specific muscle force*, $f_o = F/A_m \approx 2 \cdot 10^5$ N/m². The force F_m generated by a particular muscle is the product of the specific muscle force and the cross section.

$$F_m = f_o A_m \tag{3.1}$$

This force may lift a weight. The weight $W = Mg$ of the object is the product of its mass M, and the gravitational acceleration g.

It is easy to determine the approximate strength of a biceps. Model the muscle as it is stretched between the shoulder and the forearm, holding up a weight, Fig. 3.4b. Assume the muscle has a cross section of $A_m \approx 50$ cm² $= 5 \cdot 10^{-3}$ m². Then the total force with which the muscle pulls on the tendon is $F = A f_o = 5 \cdot 10^{-3}$ m²$\cdot 2 \cdot 10^5$ N/m² $= 1\,000$ N. Actually the tendon of the biceps is attached at a distance $d \approx 0.06$ m from the elbow "hinge", Fig. 3.4b. The hand, which is a distance $L = 0.5$ m off the elbow joint carries some weight $W = Mg$. The balance of moments (law of levers) requires $d \cdot F = L \cdot W$. Here we have neglected the weight of hand and lower arm. This equation of moments can be used to determine how much weight the hand could hold up:

$W = d \cdot F/L = 0.06$ m $\cdot 1,000$ N/0.5 m $= 120$ N. The mass of this weight is found from $W = Mg$, or $M = W/g$. Approximate g $= 10$ m/s² to get $M = 12$ kg. If one lifts this weight by the distance $\Delta x = 0.3$ m the mechanical work $E_{mech} = \Delta x \cdot W = 0.3$ m $\cdot$ 120 N $= 36$ J is produced. If this was accomplished within $\Delta t = 0.5$ s the power generated by the muscle would be $P = E_{mech}/\Delta t = 36$ J/0.5 s $= 72$ W.

The numerical value of f_o is easy to remember. Its magnitude is double the atmospheric pressure on the surface of the earth.

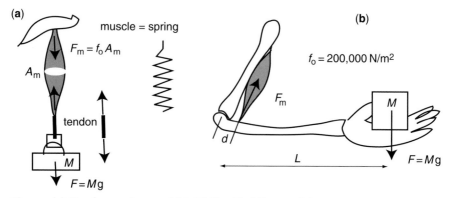

Fig. 3.4. (a) Muscle carrying a weight. (b) Hand holding a weight

Otto von Guericke in his famous experiment 1648 demonstrated the atmospheric pressure by the Magdeburg hemispheres, two empty hemispheres initially held together with a seal and then evacuated. The whole system was about the size of a soccer ball. Its cross section had an area of about $A \approx 0.1$ m^2. With vacuum on the inside the atmospheric pressure from the outside generated a force of $F = pA = 0.1$ m$^2 \cdot 10^5$ N/m$^2 = 10^4$ N. Eight horses could not pull it apart. Had he replaced the hemispheres by a living muscle he would have achieved the same effect by a muscle strand of only half the cross section area A.

A biceps that "holds" a weight without actually moving the weight, Fig. 3.5a, does work. In contrast an elastic material that is stretched, produces a tension without doing work. The comparison between muscle and spring is only partly true, because a spring does not do any work when holding a weight, while the muscle carries out small vibrations around the rest position of the weight in order to hold it at a constant height. These vibrations arise because the position is not statically fixed, but rather constantly adjusted (dynamical stability). To generate any force the cross bridges must be in motion, like a car that can be held motion-

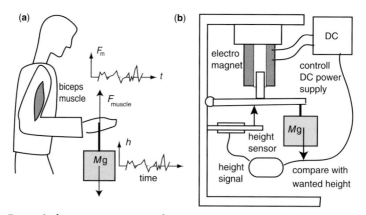

Fig. 3.5. Dynamical support servo control system

less on a slope without using the brakes, by applying enough gas so that the car does not roll backwards but does not drive forward either. The muscle is at work while generating the stress f_o. It consumes power. Some muscles work continuously, like the heart, while others are only called into action once in a while. Muscles under heavy load fatigues rapidly.

3.2.2 A Simple Model of the Muscle

Muscles are made of cells, which have two major functions. First, the cells contain the chemical machinery that transforms food (glucose and oxygen) into the specific fuel that is used in the contractile process (ATP).[1] This metabolic process has an efficiency of typically 50%. Secondly, they contain long filament structures, called *myofibrils*, that actually carry out the energy transformation between the chemical energy stored in ATP and mechanical work output.

Muscle cells are very long. They run the full distance from end to end in a muscle. For instance the biceps, which flexes the elbow, is about 15 cm long. It is made of muscle cells that are all 15 cm long, and have a diameter of 20–100 μm. These very long cells surround the myofibrils, which generate the force. The muscles are connected by tendons to the fixed points on the bone.

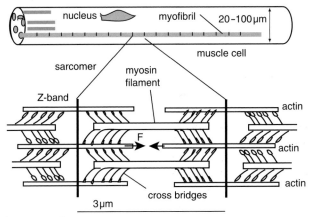

Fig. 3.6. Muscle construction

A muscle cell contains along its full length many myofibrils, like a rope made out of short fiber strands, each about 1 μm in diameter occupying 50% of the muscle cell volume. Each muscle cell contains tens of myofibrils. In a polarized light microscope one can discern a banding pattern spaced 3 μm along the length, indicating a repeating arrangement of protein based filaments called *muscle sarcomere* which provide the basic mechanism of muscle action.

---

[1] In this macroscopic descrition of the muscle one does not have to know anything about the biochemistry of ATP and its conversion to ADP.

Each myofibril contains a large number of sarcomeres arranged in series. For instance the 15 cm long biceps muscle contains along its length $5 \cdot 10^4$ sarcomeres. Each sarcomere contains a highly ordered array of two types of muscle filaments, actin and myosin, arranged in an overlapping pattern. Force is generated by interactions between these overlapping filaments and movement is created by the sliding of these filaments over each other. The boundaries of the sarcomere are established by transverse structures, called the z-bands. The z-band is an anchoring structure that has attached to it two sets of actin filaments which are unidirectionally polarized as indicated by the arrows, and face in opposite directions. The myosin filaments are bipolar filaments that sit in between the actin filaments and overlap with them. The myosin filaments have arms called *cross bridges* (the ends of myosin molecules) that reach out and interact with the actin filaments. The arms are "polarized" so that they can transmit forces towards the center of each myosin filament. Each myosin filament holds two z-bands together, like a link in the myofibril "chain". The force that a single myofibril can generate is therefore equal to the weakest link in this chain.

Cross bridges generate the motion in the muscle motor. At higher magnification it can be seen that the actin filaments are actually made up of linear arrays of globular (i.e roughly spherical) actin molecules and that the myosin filaments contain rod like myosin molecules, Fig. 3.6. The bob or head at the end of the cross-bridge portion of the myosin molecules contains an enzyme that catalyzes a hydrolysis reaction involving the fuel for muscle movement, ATP. The bob is also able to bind to specific sites on the surface of the actin molecules when inorganic phosphate is around. Through these interactions, called the muscle cross bridge cycle, the muscle is able to contract and generate force.

3.2.3 The Muscle Cross Bridge Cycle

The generation of force involves a cycle as illustrated in Fig. 3.7. Starting at the top, a cross bridge with ATP sits unattached but quite close to one of the many binding site of an actin filament, which are typically $\Delta x \approx 10$ nm apart. In this state the cross bridge can hydrolyze ATP to ADP. Inorganic phosphate atoms, labeled Pi,

Table 3.1. Cross bridge rowing cycle

stroke	muscle	rower in rowboat
a	power stroke: release ADP, myosin head rotates moving through $\Delta x = 10$ nm, generating the force $F = 3 \cdot 10^{-12}$ N	rower pulls oars rearward through water and pushes boat forward.
b	attach ATP, detach myosin head	lift oars out of water
c	hydrolyze ATP, and bend back, activating the myosin	bring oars forward
d	Release Pi, myosin attaches to actin	dip oars into water

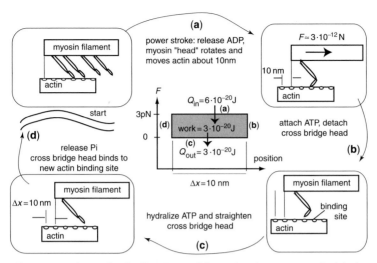

Fig. 3.7. Illustration of cross-bridge "rowing cycle". During the power stroke (**a**), the myosin moves $\Delta x \approx 10$ nm relative to the actin with a force of $F \approx 3 \cdot 10^{-12}$ N per cross bridge. The other stations in the cycle are explained in Table 3.1

which remain bound to the protein in this process, activate the cross bridge so that it can bind to a site in the actin.

When binding takes place the phosphate is released. With the release of the ADP from the myosin the cross bridge structure is altered in some unknown way to cause a large scale change in shape. This shape change rotates the cross bridge head, and generates a force of $F_c \approx 3 \cdot 10^{-12}$ N that moves the actin by $\Delta x \approx 10$ nm relative to the myosin filament.[2] Then the myosin head binds another ATP and this allows the cross bridge head to detach from the actin filament. Thus a new cycle can begin. In this next cycle the cross bridge will attach to the next binding site on the actin filament.

Each cycle therefore moves the actin relative to the myosin by the distance Δx. Because of the bipolar arrangement of the sarcomere, the two z-bands move together in a coordinated fashion, shortening the whole sarcomere by $2 \Delta x \approx 20$ nm. There are about $3 \cdot 10^5$ sarcomers per meter.

Since the sarcomeres are arranged in series along the muscle filament the simultaneous shortening of many sarcomeres yields a significant length reduction of the whole muscle. Consider the shortening of a muscle of $L = 0.1$ m length. Thus 10 steps of $\Delta x \approx 10$ nm at each end of a sarcomere reduces its length by 200 nm = 0.2 μm. When summed up over the $5 \cdot 10^4$ sarcomeres of this muscle, the whole muscle cell changes its length by 10 mm. The cross bridge cycle may be compared to a rowing cycle with the strokes a, b c, and d described in Table 3.1 and Fig. 3.7.

---

[2] Strick et al. [2001] quote $F_c \approx 4 \cdot 10^{-12}$ N, and $\Delta x = 11$ nm.

3.2.4 Muscle Efficiency

The muscle converts ATP and water into ADP according to $ATP + H_2O = ADP + Pi + \Delta Q$, with the heat of reaction $\Delta Q = \Delta h \approx 6 \cdot 10^{-20}$ J/reaction (namely about 38 kJ/mol). The process generates the force $F_c \approx 3 \cdot 10^{-12}$ N per cross bridge, where the cross-bridge head pushes the actin section by a distance of $\Delta x \approx 10$ nm. This represents the work $W = F_c \Delta x = 3 \cdot 10^{-12}$ N $\cdot 10^{-8}$ m $= 3 \cdot 10^{-20}$ J. The difference $\Delta Q - W \approx 6 \cdot 10^{-20}$ J $- 3 \cdot 10^{-20}$ J $= 3 \cdot 10^{-20}$ J must be carried off as heat. The efficiency of generating mechanical energy is then

$$\eta_{ATP} = W/\Delta Q \approx 50\% . \tag{3.2}$$

In order to obtain the overall efficiency of the muscle one must include the metabolic process where glucose and oxygen *burned* in order to convert *ADP* and Pi into the muscle fuel *ATP*. If this process has an energy efficiency of about $\eta_{met} = 50\%$, then the overall efficiency of generating work from glucose is $\eta = \eta_{ATP} \cdot \eta_{met} \approx 0.5 \cdot 0.5 = 25\%$. The force – generating – muscle cycles may be compared to Carnot cycles of thermodynamic engines, which have typical efficiencies of $\eta \approx 20$ to 30%.

If a muscle generates a force of $F_c = 3 \cdot 10^{-12}$ N per cross bridge, and the muscle produces the stress force

$$f_0 = 2 \cdot 10^5 \text{ N/m}^2 \tag{3.3}$$

there must be $n = 2 \cdot 10^5$ N/m^2/$3 \cdot 10^{-12}$N $= 6.7 \cdot 10^{16}$ cross bridges/m^2. The muscle stress force has the same dimension as pressure, however the muscle *pulls* like a rope, whereas pressure *pushes*.

When a muscle of length L, area A and mass $m = \rho AL$ contracts through the distance ΔL in the time interval Δt, it shortens at the velocity $u = \Delta L/\Delta t = a \Delta t$. The muscle force $F = fA$ creates an acceleration $a \approx F/m \approx Af/m = f/\rho L$ that depends on the muscle mass $m = AL\rho$. To compare different muscles one uses the normalized contraction velocity $v = u/L = \Delta L/L \Delta t = \varepsilon/\Delta t$. The muscle stress $f = F/A$ has its maximum value f_0 for very slow contraction speeds. The stress f decreases with in-

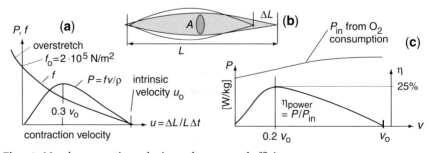

Fig. 3.8. Muscle contraction velocity and power, and efficiency

creasing contraction velocity, see Fig. 3.8. All muscles have a maximum contraction velocity, called the intrinsic muscle velocity $v_0 = \varepsilon_0 / \Delta t$ at which f drops to zero, and a characteristic contraction time $\Delta t_0 = \varepsilon_0 / v_0$. This "no load" contraction time Δt_0 yields a limit for maximum muscle-end-speed u_0:

$$ u_0 \leq \frac{f_0}{\rho} \cdot \frac{\varepsilon_0}{v_0} \cdot \frac{1}{L} = C_C \cdot \frac{1}{v_0 \cdot L} . \qquad (3.4) $$

Equation (3.4) implies that short muscles (small L) contract faster that long ones. Typically muscles operate with strain ratios $\Delta L / L = \varepsilon_0 \approx 10\%$. Taking $f_0 = 2 \cdot 10^5$ N/m² the contraction constant becomes $C_c = 20$ m²/s² Humming birds with muscled lengths $L \approx 0.01$ m achieve intrinsic muscle velocities $v_0 \approx 25$ s^{-1}. This yields top muscle-end speeds of $u_0 \approx 20/(25 \cdot 0.01) = 80$ m/s. In contrast humans, with $v_0 \approx 3$ s^{-1}, and a typical muscle length $L \approx 0.15$ m only reach $u_0 \approx 0.45$ m/s. The muscle power output $P = F \cdot v$ is largest at a typical contraction velocity $v_1 \approx 0.3\, v_0$. The metabolic power input P_{in}, which may be determined from the oxygen consumption, rises with the muscle velocity. Therefore, the power efficiency $\eta = P/P_{in}$ peaks with $\eta \approx 25\%$ at an intrinsic velocity which is smaller than v_1.

3.2.5 Cold and Warm Muscles

Generally the life functions proceed faster at elevated temperatures. This can be quantitatively seen from measurements of the muscle contraction velocity at different temperatures [Wakeling andJohnstone 1998]. At a temperature of $T = 20°C$ a fish muscle produces typically $P \approx 143$ W/kg. A comparison was made of the behavior of Arctic fish, temperate fish, and tropical fish. The peak contraction velocity is shown as function of temperature in Fig. 3.9a.

Cold muscles contract much slower than warm muscles. The contraction velocity is a function of temperature within one species, Fig. 3.9a. However, a comparison of the muscle performance of Antarctic, temperate, and tropical fish shows that all of them essentially reach the same stress $f \approx 2 \cdot 10^5$ N/m², Fig. 3.8b. Since

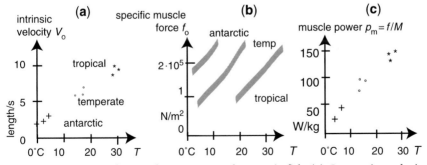

Fig. 3.9. Comparison of tropical, temperate and antarctic fish. (a) Contraction velocity, (b) muscle stress, (c) muscle power

cold muscles contract more slowly the specific muscle power $p_m = P/M = vf/\rho$ of Antarctic fish is much smaller than the power generation of tropical fish, Fig. 3.8c. Since warm muscles have a much higher power output than cold muscles, there could be a strong incentive for animals to operate at an elevated, constant body temperature. Here one sees a physical reason for animals to become warm-blooded (homeotherm), with all its consequences, as described in Chap. 2.

3.2.6 Muscle Connections

In order to generate the motion of limbs or inner organs muscle force is transmitted through tendons (collagen) to other parts of the body. Tendons are generally over designed so that they safely hold excess forces. A typical tendon tensile strength is about $2 \cdot 10^8$ N/m².

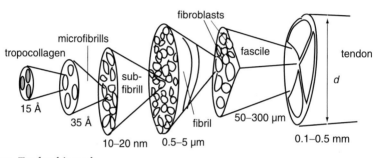

Fig. 3.10. Tendon hierarchy

Tendons are elastic, they stretch under load, like springs or very stiff rubber bands. Tendons have a complicated structure, Fig. 3.10. They are composed of bundles, called *fasciles*, that contain many fibrils. These are composed of *tropocollagen* molecules, which are bundled into *microfibrils*, which in turn are collected into *subfibrils*.

Elastic forces play an important role for the body design and locomotion of animals. Elastic members can store elastic potential energy. Animals use tendons to store some energy during certain phases of periodic motion processes (running, swimming, flying) and recover this energy in other phases in order to reduce the total energy cost of locomotion. Some of the biological materials rival the strength of technical materials (see Table 3.3).

3.3 Static Forces That Animals May Encounter

Forces determine the posture, shape, and motion of animals. There are many different forces, some of them are always there, like gravity, electrostatic forces, pressure, buoyancy, and surface tension. Other forces, like elastic tension or compression, appear only when objects are distorted. A third important group of forces

Table 3.2. Static forces

name of force	equation	parameters	
muscle force	$f_o \approx 2 \cdot 10^5\,\mathrm{N/m^2}$, $F_m = f_o A$	A	muscle cross section area in m²
gravity	$F_{gr} = Mg$	M	mass in kg, $g = 9.81$ m/s²
pressure	$F_{pr} = pA$	p	pressure in N/m², A area in m²;
	in the ocean $p = \rho g y$	y	depth in m, ρ density in kg/m³
buoyancy	$F_{bu} = (\rho_w - \rho) V g$	ρ_w	water density, V volume in m³
elastic	$F_{el} = k\,\Delta L$, tension T	k	spring constant N/m
	stress $\sigma = T/A = Y\,\Delta L/L$	ΔL	elongation m, Y Young's modulus
capillary	$F_{cap} = \gamma \int ds$	γ	surface tension of fluid N/m
		$\int ds$	total length of surface line m
electrostatic	$F_{el.ste} = q_e E$	q_e	electric charge, in coulomb C
		E	electric field in Volt/m
static friction	$F_{fr} = \mu_s N$	μ_s	coefficient of static friction,
		N	normal force

only come into existence when there is motion. These include the Bernoulli force, sliding friction, lift, various forms of drag, and centrifugal forces. These dynamic forces will be discussed in Sect. 3.4.

3.3.1 Pressure

Animals are always exposed to the force of gravity, which pulls every part of the body down, and they are subject to the pressure force exerted by the atmosphere onto every segment of surface area of the skin. The pressure at sea level is about $p_a = 1$ atmosphere(atm) $= 1.013 \cdot 10^5$ N/m². The total pressure force acting onto an area A is then

$$F_p = p_a A .$$
(3.5)

The palm of your hand has a surface area of typical $A_h = 8$ cm $\cdot$ 10 cm $= 0.008$ m².

Therefore the air presses onto the palm of your hand with the force $F_p = p_a A = 1.013 \cdot 10^5$ N/m² $\cdot$ 0.008 m² $= 810$ N. However, one does not feel this force because the air pressure acts with equal force on the back of the hand, and the internal pressure in the tissue is also 1 atm.

The pressure depends on the altitude. At sea level the pressure is equal to one atmosphere. At an elevation y above sea level the pressure is lower, under water it is higher. The pressure at a depth d, under water, arises from the weight of the water column and the air above the object $p = \rho g d + p_o$. The weight of a water column with cross section $A = 1$ m² is $W = \rho g d$. The resulting water pressure

$$p_w = W/A = \rho g d$$
(3.6)

is called the hydrostatic pressure.

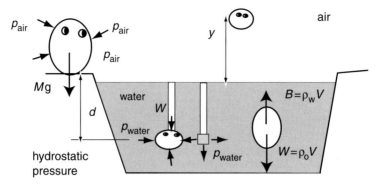

Fig. 3.11. Pressure at various positions

For instance the pressure at the depth $d=120$ m due to the water column is $p_w = 1000$ kg/m$^3 \cdot 9.81$ m/s$^2 \cdot 120$ m $\approx 1.2 \cdot 10^6 \approx 12$ atm. The total pressure p at this depth is p_w plus the atmospheric pressure (1atm) above the water: $p_w + 1$ atm ≈ 13 atm. If the pressure varies on the surface the total force is found by adding up the forces section by section (integrating over the surface area)

$$F = \int_{\text{surface-area}} p(A)\, dA . \tag{3.7}$$

Figure 3.12 shows an example. A membrane curves under the load of water. For each surface area element ΔA on the membrane, one can calculate the weight force ΔW of the water column resting on it. The loads $\Delta W_1 = \Delta A\, \rho\, g\, d_1$, and $\Delta W_2 = \Delta A \rho\, g\, d_2$ differ since the depths d_1 and d_2 are not the same: The pressure varies locally on the membrane.

Hydrostatic pressure affects animals that live at great depths in the ocean, because the pressure in the ocean increases by 1atm for every 10m of depth. Objects at 1000 m depth must hence be able to withstand a pressure of 100atm. If the internal pressure in an organism rises exactly as the external pressure grows there is no problem, and the skin of the animal does not have to be very strong. The actual load on the skin depends on the difference of pressures on the outside and the inside $F = (p_0 - p_i)A$. When a fish with a swim bladder is brought up from a great depth to the surface the animal is suddenly decompressed. The swim bladder expands rapidly, it "explodes".

On top of a mountain the pressure is smaller than at sea level, because the column of air above a certain point is the smaller the higher the elevation h. The variation of pressure with height is

$$p = p_0\, e^{-Ch} . \tag{3.8}$$

The parameter C has the value $C = \rho_0 g/p_0 = \mu g/(R_g T_0)$ where μ is the molecular weight in kg, ρ_0 is the density at ground level, and $R_g = 8.31$ J/mole° is the gas constant. For $T_0 = 273$ K one has $C = 1.27 \cdot 10^{-4}$/m.

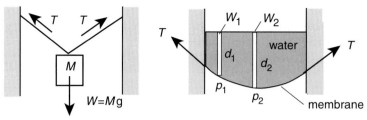

Fig. 3.12. Pressures and tensions in a membrane

The pressure in a gas can be given as function of the number of molecules n per unit volume, and the temperature T measured in Kelvin, or as function of density, temperature, and molecular weight, or by the volume V and number of moles N_m located in the volume, and the temperature

$$p = nk_BT, \quad p = \rho R_g T/\mu, \quad \text{or} \quad p = N_m R_g T/V \tag{3.9}$$

where $k_B = 1.38 \cdot 10^{-23} J/K$ is Boltzmann's constant. As the pressure decreases with altitude so does the number n of molecules per m³. The air gets thinner, and the lungs of birds must work harder to extract the oxygen for the metabolism.

3.3.2 Buoyancy

Objects that are partially or fully submerged experience a buoyancy force B related to the pressure distribution on their outside see Fig. 3.11. This force points up. It is the difference in weight of the object, $W = \rho_o Vg$, and the weight $\rho_w Vg$ of the displaced water.

$$B = \rho_w Vg - \rho_o Vg = (\rho_w - \rho_o)Vg \tag{3.10}$$

The density of water depends on its temperature, see Fig. 2.19. Since water of 4°C has the largest density, such cold water sinks to the bottom of the ocean. The deep sea water has the constant temperature of 4°C. An object with $\rho_w = \rho_o$ is neutrally buoyant. It floats as if gravity did not exist. Therefore astronauts have training sessions submerged in water tanks in order to learn to cope with zero gravity.

What counts for the buoyancy force is the average density of an object. An air bubble adds volume without adding much mass. Fish tissue has an average density that is slightly larger than the density of water. Therefore fish would sink. Sharks compensate lack of buoyancy by their swimming activity. Many fish regulate their average body density with the help of a swim bladder: an empty swim bladder reduces the volume thereby giving a fish its highest density, and the fish sinks. A full swim bladder gives the fish its lowest average density, the animal ascends. To stay at a certain level the fish regulate the size of their swim bladder to the point where they are neutrally buoyant. A ship settles into the water to a depth where the weight of the displaced water is equal to the weight of the ship.

Buoyancy force acts on objects in the atmosphere as well. However, since the air density is so low compared to that of body tissue, no animal uses buoyancy to stay aloft. Remember $\rho_w = 1\,000$ kg/m³, whereas $\rho_{air} \approx 1.2$ kg/m³. The density ρ of a gas of molecular weight μ may be derived from the gas law

$$\rho = \mu p / R_g T. \tag{3.11}$$

Only zeppelins and balloons have managed to use this effect. They are filled with gases lighter than air: helium, hydrogen or hot air, and they need huge volumes to carry small payloads. The density scales as the pressure: The higher balloons get, the more volume they must have to support their weight with the buoyancy force.

3.3.3 Elastic Forces

All materials are elastic. Biological examples of elastic materials are tendons, spider webs, bladders, skin, bones, wood, and technical devices like springs, tygon tubing, rubber gloves, and tennis balls.

The more the length L of an elastic material is changed either by stretching or compressing, the stronger it resists with an opposing force. The force, in tension T, or in compression C is shown as function of the length change ΔL in Fig. 3.13. Within the linear regime the tension force T can be given as function of the spring constant k with the units N/m and the elongation ΔL, units m:

$$T = k \Delta L. \tag{3.12}$$

For larger elongations the tension grows as $T \propto \Delta L^2$ or with a higher power in ΔL, until the object breaks apart when the breaking strength is reached.

A tension T acting on cylindrical object of length L, and cross section area A generates the stress $\sigma = T/A$, and it causes the object to increase its length by a certain amount ΔL. The relative elongation $\varepsilon = \Delta L/L$ is called the *strain*. The elasticity of a material can be characterized by the modulus of elongation Y. Table 3.3 shows Y for various biological and technical materials. Within a linear regime stress, strain, and modulus of elongation are related by

$$\sigma = T/A = Y \Delta L/L. \tag{3.13}$$

By comparison of Eqs. (3.12) and (3.13) the spring constant k can be related to the modulus of elongation and the dimensions A, and L, of the cylindrical elastic material:

$$k = YA/L. \tag{3.14}$$

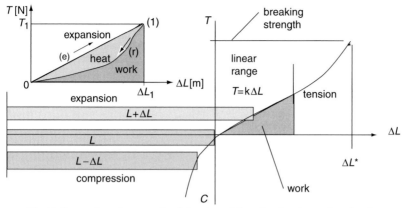

Fig. 3.13. Elastic forces. Insert: expansion(e)–release(r) cycle of an imperfect elastic material showing recoverable work, and energy lost as heat. See also Fig. 3.26

Elastic forces can always be related to mechanical work W. The product $T \cdot dL$ represents the work dW needed to stretch an elastic object by the infinitesimal distance dL. The total work needed to stretch the member by the distance ΔL_1 is

$$W = \int_0^{\Delta L_1} T \, dL = \frac{1}{2} k \cdot \Delta L_1^2 = \frac{1}{2} \cdot \frac{L}{A \cdot Y} \cdot T_1^2 \, . \tag{3.15}$$

This work is shown as the area under the $T(\Delta L)$ curve (e) in the insert of Fig. 3.13. A *perfect* elastic medium releases *all* of this energy when the (strain) displacement is gradually removed. When the tension T is replaced by a compression force C Eq. (3.15) also holds for an elastic medium in compression. The stored energy W grows with the square of the applied force. Note that soft elastic media with small Y can store more energy that hard ones (large Y) if both have the same breaking strength, and long springs of small area A can hold more energy than short, wide springs.

If the material is not perfectly elastic only a smaller force (r) is recovered when the tension is released. Thus stretch (e) and release (r) define a loop enclosing an area labeled *heat*, see insert of Fig. 3.13. This area represents the loss of mechanical energy of an imperfect elastic material. Shock absorbers and spider webs are such imperfect elastic materials. The release of elastic potential energy is a dynamical process, which occurs in a certain time interval, see Sects. 5.1.7, and 5.3.5.

The work of fracture is defined as the energy per unit area needed to break the material. It represents the area under the tension curve up to the breaking elongation L^*. The atomic origin of these forces is briefly discussed in Sect. 3.5.5. The maximum tensile strength of steel and spider web is about equal at $2 \cdot 10^9$ N/m², but the modulus of elongation of spider silk is lower by about a factor 100, and the energy to break silk is larger by a factor 10 than that of steel.

Table 3.3. Mechanical properties of biological and technical materials

material	max tensile strength 10^9 N/m²	max compression 10^9 N/m²	Y tension 10^{10} N/m²	Young's compression 10^{10} N/m²	energy to break 10^6 J/m³	work of fracture J/m²	max $\Delta L/L$ %
tendon(collagen)	0.15		0.15		7.5		10
bone	0.16	2.2	2	2	4	1700	2
arterial walls	0.005		10^{-3} to 10^{-4}				100
disc material		0.01	$\approx 5 \cdot 10^{-4}$				25
soft cuticle	0.02	0	10^{-3}				100
hard cuticle	0.1	0.2	1	1	4	1500	10
teeth (enamel)	0.05	0.2	2	3 to 7	0.5		0.2
spider web	2	0 (rope)	0.3		200		40
kevlar	3.6		13		50		3
glass fiber	2.5		7	7	50	10	5
bulk glass	0.1		7		0.07		0.01
elastic (rubber)	0.05		10^{-4}	10^{-4}	≈ 100		800
steel (high tens.)	1.5	≈ 1	22	22	≈ 20	$\approx 10^5$	1
concrete	0.002	0.17		2.3		3–40	
fused silica	0.0048	1.1	7.2				

Example: The tendon in Fig. 3.4a is stretched slightly under the force $T = 1000$ N. How much will the tendon stretch under this load? The stretch ΔL can be found from (3.14). If the tendon has the diameter $d = 4$ mm, the length $L = 30$ mm, and the modulus of elongation $Y = 1.5 \cdot 10^9$ N/m². The amounts of stretch is $\Delta L = (T/A)(L/Y) = (1000/\pi \, 0.002^2) \cdot (0.03/1.5 \cdot 10^9) = 0.16$ mm.

The Achilles tendons of kangaroos can store significant amounts of elastic energy when the foot hits the ground, and release the energy subsequently to help propel the animal onto its next bounce.

Both solids and gases have elastic properties. A rubber balls pumped up with *air* bounces. Likewise a swim bladder filled with air has quite good elastic properties. The spring constant is easily found. Consider a cylinder of volume $V = AL$ with a piston of area A, containing air. We want to calculate the force that holds the piston in place. The gas law (3.9) relates the pressure p and the volume V to the temperature T, namely $pV = pAL = N_m R_g T$. The force acting onto the piston $F = pA = N_m R_g T/L$. The pressure, and hence the force F change when the length is changed. By differentiation with respect to L on finds $\Delta F = pA = -k_{gas} \cdot \Delta L$, where $k_{gas} = \mu R_g T/L^2$ is the spring constant of the gas.

3.3.4 Electrostatic Force

Electrostatic forces mainly act on the molecular level. They are very important for animals, because they provide the holding mechanism at the atomic level of muscles. Electrostatic forces bind the muscle cross bridge heads to the actin sites thereby generating the fundamental muscle force per site $F_c = 3 \cdot 10^{-12}$ N. Electrostatic forces arise when electric charges q_e, measured in Coulomb C, are located in electric fields E, measured in Volt per meter or Newton per Coulomb: 1 V/m = 1 N/C.

$$F = q_e E \tag{3.16}$$

One electron has the charge $q_{el} = e = 1.6 \cdot 10^{-19}$ C. Electrical fields that we encounter in everyday life are small. For instance near a household wire one might find a field $E = 10^{-2}$ V/m. A lightning bolt has about 10^4 N/C. However, the electrostatic force is very important inside atoms. An electron in the hydrogen atom sees the field $E = 6 \cdot 10^{11}$ V/m. If the cross bridge head carries a single electron of $e = 1.6 \cdot 10^{-19}$ C and generates the force $F_c = 3 \cdot 10^{-12}$ N, it must see the field $E = 3 \cdot 10^{-12}$ N/$1.6 \cdot 10^{-19}$ C $= 1.8 \cdot 10^7$ V/m. This is much smaller than the field in the hydrogen atom, but significantly larger than the field in a lightning bolt. The electric field decreases with the square of the distance r (similar as the gravitational field):

$$E = \text{const } q_e / r^2. \tag{3.17}$$

Atomic distances are very small, hence the electric fields in atoms are very large. Cell walls of tissue have a typical thickness of $d = 10$ nm. Living cells separate ions, so that a cell is typically charged to -70 mV. Therefore there is an electric field across the cell wall of $E = 0.07$ V/10^{-8}m $= 7 \cdot 10^6$ V/m $= 7 \cdot 10^6$ N/C. An electron, which is sitting in a channel through the cell wall will experience the force $F = q_e E = 1.6 \cdot 10^{-19}$ C $\cdot 7 \cdot 10^6$ N/C $= 1.12 \cdot 10^{-12}$ N. This is of the same order of magnitude as the cross bridge force in the muscle cycle.

3.3.5 Capillary Forces, a Form of Surface Tension

Surface tension is the force which allows water bugs to walk on top of a water surface, or which permits a needle to "float" on the water. The force arises from the difference of attraction between the fluid, the immersed solid and the air above it, which come together along a line. The force required to lift a needle (length L and mass m) off the water surface is

$$F = \gamma_s 2L + mg. \tag{3.18}$$

Note that $2L$ is the total length of the water line. The parameter γ_s, measured in N/m or in Nm/m² (Joule/area), is the surface tension or surface energy per unit surface area, which depends on the temperature see Table 3.4. Water at room temperature has the surface tension $\gamma_s = 0.073$ N/m.

A water column will rise inside a capillary tube up to a height h where the weight is equal to the vertical component of the surface tension $F \cos \varnothing = W$. Now enter the surface tension on the left side and the weight on the right side of this

Table 3.4. Viscosity and surface tension of various materials

medium	density ρ kg/m³	viscosity η kg/s m	kinematic viscosity $v = \eta/\rho$ m²/s	surface tension γ_s N/m (contact with air)
air	1.29	$18 \cdot 10^{-6}$	$16 \cdot 10^{-6}$	
water (at 0°C)	1000			0.0756
water (at 20°C)	1000	$1.0 \cdot 10^{-3}$	$1.0 \cdot 10^{-6}$	0.073
water (at 60°C)				0.0662
water (100°C)	958.4		$2.9 \cdot 10^{-7}$	0.0589
blood		$4.0 \cdot 10^{-3}$	$\approx 4.0 \cdot 10^{-6}$	
engine oil		$200 \cdot 10^{-3}$	$\approx 2.0 \cdot 10^{-4}$	0.032
olive oil				0.032
soap solution				0.025

equation to get: $\gamma_s \cdot 2\pi r \cos \phi = \rho(\pi r^2 h)g$, which can be solved for the height h of the fluid in the capillary

$$h = \frac{2\gamma_s \cdot \cos \phi}{\rho g r}.$$ (3.19)

Capillary action is the force that lets a paper towel soak up a puddle of water. Capillary action is another transport process that requires no external energy input. The strongest capillary action is produced in "capillaries" with convex walls, that are formed between solid rods, see Fig. 3.14. Here the effective radius r becomes very small. Capillary action lets water rise in the plants. Capillary forces lift fluids without any biological cost on the part of the organism.

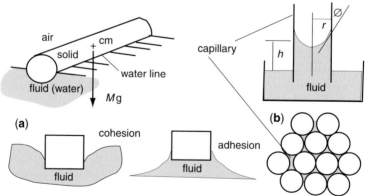

Fig. 3.14. (a) Surface tension, (b) capillary action in a tube and between solid rods

For instance water in a capillary of $r = 0.1$ mm at maximum angle $\phi = 0°$ or $\cos \phi = 1$, will rise to the level $h = (20 \cdot 0.073 \text{ N/m} \cdot 1)/(1000 \text{ kgm}^{-3} \cdot 9.81 \text{ ms}^{-2} \cdot 0.001 \text{ m}) = 0.149$ m.

A whole range of animals, namely all the bugs and water spiders that "walk on water" make extensive use of surface tension. There is a whole new range of measures and countermeasures to utilize or spoil this effect. Insects that want to walk on the water surface must stay below a certain size and it seems that this size limit has been reach by some species.

3.3.6 The Maximum Size of Water Striders

There is a maximum diameter r_m for a simple object like a rod of length L and radius r and density ρ that can be supported by surface tension. Assume that the rod is nearly fully submerged. Then the weight W is supported by surface tension S and by buoyancy B. The equilibrium condition is

$$Mg = S + B = \pi \rho \, r^2 L g \leq 2 L \gamma + \pi \, \rho_w r^2 L g \,. \tag{3.20}$$

The length L can be canceled in every term, and the relation may be solved for r

$$r \leq \sqrt{\frac{2\gamma}{\pi g (\rho - \rho_w)}} \,. \tag{3.21}$$

For an organic object with $\rho = 1500$ kg/m^3 this maximum radius is $r_m = 2.9$ mm. For a steel needle the minimum radius is about $r_{st} = 0.75$ mm. If a water strider rests on 4 forelegs of length L and radius $r \ll l$ the surface tension force becomes $S = 8 L\gamma$. The buoyancy force of 4 forelimbs that are nearly fully immersed, is $B = 4 \rho g \pi r^2 L$ and its equilibrium condition is $W = Mg \leq 4 \rho \pi r^2 L g + 8 L\gamma$. We introduce the aspect ratio $b = L/r$ to get $W \leq 4 \pi \rho_w L^3/b^2 g + 8 L\gamma$. This cubic equation could be used to derive the length of limbs needed to support an arbitrary body weight W. In the approximation that the buoyancy forces on the 4 limbs do not significantly contribute to the lift, one has the body design condition

$$W_{max} = M_{max} g \approx 8 L \gamma \,. \tag{3.22}$$

With $L = 0.05$ m, and $\gamma = 0.072$ N/m the maximum body mass becomes $M_{max} = 8 \cdot 0.05 \cdot 0.072/10 = 0.0028$ kg.

3.3.7 Friction

Friction arises when surfaces slide across each other. Smooth surfaces have little friction, rough surfaces have more friction. Lubrication hides some of the roughness. Animals could not walk or run without friction, and wheels could not accelerate or stop cars and bicycles.

The static friction force depends on the static friction coefficient μ_s and the normal force $F_n = Mg \cos \o$. Friction prevents objects from sliding on inclined planes if the friction force

$$F_{st} = \mu_s F_n \qquad (3.23)$$

is equal to or larger than the tangential force $F_t = Mg \sin \o$. Feet and wheels slide on wet and icy roads, because wet roads and icy patches have a much smaller friction coefficient than dry roads. Friction gradually wears down the contact surfaces. Tendons and muscles that are made to slide, must be housed within tissue that has very low friction coefficients, and that can be easily repaired to replace the wear and tear.

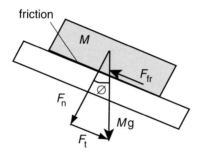

Fig. 3.15. Static friction $F_{fr} = -Mg \sin \o$

Once an object starts to slide the friction force gets smaller, because then the friction coefficient changes from μ_s to the smaller value μ_d called the dynamic friction. The dynamic friction force is

$$F_{dy} = \mu_d \cdot F_n . \qquad (3.24)$$

3.4 Dynamic Forces

Static forces like gravity, pressure, and surface tension, are always there. Dynamic forces, Table 3.5, arise only when the momentum Mu of an object of mass M is changed. Such momentum change can happen in various ways.

The velocity u could be altered in magnitude or direction. This acceleration could happen in a part of a body, like an arm that is swung in a circle above the head, or acceleration could occur in the fluid or air surrounding an object, thereby generating lift or drag forces. The mass M is changed for instance when a burst of fluid that is ejected by a squid.

Table 3.5. Dynamic forces

name of force	equation	parameters
sliding friction	$F_{sl} = \mu_{sl} N$	μ_{sl} coefficient of sliding friction, N normal force
Bernoulli force	$\Delta p = \frac{1}{2}\rho\{u_1^2 - u_2^2\}$	pressure difference at points 1 and 2
centrifugal force	$F_{cent} = Mu^2/R$	M mass, u velocity of object, R path radius
Stokes friction for $Re = \rho R_s u/\eta \ll 100$	$F_{Stokes} = 6\pi R_s \eta u$ (sphere)	R_s radius of sphere, η viscosity ρ density, Re Reynolds number,
hyro drag, for $Re = \rho R_s u/\eta \gg 100$	$F_h = \frac{1}{2}C_D A \rho u^2$	C_D hydro drag coefficient, A frontal cross section area
skin friction, fast motion	$F_{sf} = \frac{1}{2}C_{sf} S_w \rho u^2$	C_{sf} skin friction coefficient, S_w wetted surface area
ventilation drag	$F_{vent} = \text{const } u^4$	u velocity relative to fluid
lift	$F_L = \frac{1}{2}C_L S \rho u^2$	C_L lift coefficient, S wing surface area
Magnus effect	$F_{mag} = \text{const } u \times \omega$	ω angular velocity vector of spinning object, × vector product sign
thrust	$F_{th} = u_e m'$	u_e ejection velocity of thrust, generating fluid, m' ejected mass flow rate kg/s
electro-magnetic	$F_{el. magn} = j \times B$	j current vector, Amp B magnetic field vector, tesla

3.4.1 Bernoulli Force

A fluid of density ρ flowing at some velocity u through a pipe has a local static pressure p and a dynamic pressure $0.5\,\rho u^2$. The sum of both is a constant. Therefore one can relate the local pressures p_1 and p_2 at different locations, 1, 2 to the local velocities u_1 and u_2 at those positions.

$$p_1 + \frac{1}{2}\rho \cdot u_1^2 = p_2 + \frac{1}{2}\rho \cdot u_2^2 \tag{3.25}$$

If the velocity goes up, the pressure must go down. This phenomenon is sometimes called the Bernoulli effect. The fluid velocity changes when a flow is locally restricted.

For instance if a pipe has a narrow section, Fig. 3.16a or if a flow passes over an obstacle Fig. 3.16b, the fluid will speed up. If the speed is known at two different locations, (1) and (2), one can derive the pressure difference $\Delta p = p_2 - p_1$.

$$\Delta p = p_2 - p_1 = \frac{1}{2}\rho \cdot u_1^2 - \frac{1}{2}\rho \cdot u_2^2 \tag{3.26}$$

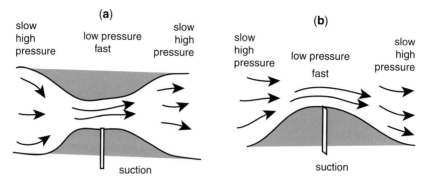

Fig. 3.16. (a) Bernoulli effect: pressure reduction at the throat of a nozzle. (b) Pressure reduction at the top of a hill

Δp is negative if u_2 is larger than u_1. This "under-pressure" will suck external fluid or objects into the flow field. Jet pumps work with this principle. Figure 3.16b could represent a barnacle in a water current or a molehill in a surface wind. The Bernoulli effect then provides some "free" ventilation for these animal structures. The Bernoulli effect also explains some of the lift force on animal wings.

3.4.2 Centrifugal Force

An object of mass M moving on an arc, of radius R at the (constant) speed u must change its the direction of travel continuously. This change of direction represents an acceleration directed towards the axis of rotation. Therefore, the force

$$F_c = M u^2 / R \tag{3.27}$$

must be applied in order to keep the object on its circular path. The reaction to this force is experienced as a centrifugal force.

The centrifugal force is important for a monkey that swings by its arms from branch to branch. Large monkeys actually limit the maximum swing angle because their finger muscles are not strong enough to hold on to a branch at the bottom of the swing. No mechanical energy is stored or converted into other forms by the centrifugal force, because the force F_c is directed at right angle to the direction of motion.

3.4.3 Drag

An object moving at some velocity u relative to a fluid (liquid or gas) causes the fluid nearby to change its velocity due to the drag force F_{drag}. Drag impedes the motion. Drag disappears when the motion stops. All drag forces can be understood within the framework of energy conversion.

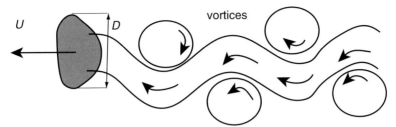

Fig. 3.17. Vortex street

When the object is moving relative to a fluid (or gas) it "forces" segments of the fluid to "get going". The object drags along fluid elements as they pass through a region close to the object. The closer the fluid passes to the object the more it is affected. If the object is stationary the fluid is slowed down. If the fluid is at rest the object drags the fluid along. This steady acceleration requires a constant energy input $E/\Delta s$ per meter of travel. $E/\Delta s$ is the drag force F_{drag}.

The consumed mechanical power P_{mech} can be written as $P_{mech} = F_{drag} \cdot u$. Depending on which segment of the fluid is accelerated one distinguishes skin friction F_{sf} hydrodynamic drag F_h, wave drag F_w, and ventilation drag F_{vent}. All these forces have their own characteristic functional dependence on the geometry of the body and its velocity u relative to the fluid.

If the flow velocity is slow enough so that the flow is *laminar*, the drag force is a linear function of the velocity. $F_{drag, lam} = const \cdot u$. Laminar flow is found when the Reynolds number $Re = Du/v = \rho D u/\eta$ is smaller than about 100. D is a typical lateral dimension of the object. The kinematic viscosity v, and the dynamic viscosity η are related as $v = \eta/\rho$. Values are given in Table 3.4 The laminar drag force F_{Stokes} acting onto a sphere of radius $R_s = D/2$ is known as the Stokes friction

$$F_{Stokes} = 6\pi R \eta u. \tag{3.28}$$

This friction force can be neglected in flows where the Reynolds number is larger than about 100. Then the object experiences hydrodynamic drag F_h, which is related to the appearance of a vortex street trailing the object, Fig. 3.17. The hydrodynamic drag grows with the frontal surface area A and square of the velocity

$$F_h = \tfrac{1}{2} C_D A \rho u^2. \tag{3.29}$$

The drag coefficient C_D is shown as a function of the Reynolds number Re in Fig. 3.18.

The hydro drag arises because the fast moving object generates numerous and sizable eddies, or vortices in its wake. They are shed at the Strouhal frequency $f_{st} = St \cdot u/D$. The Strouhal number St is a dimensionless parameter, which depends on C_D, and the Reynolds number of the flow, and has numerical values in the range $0.12 \leq St \leq 0.23$. For $Re > 10^4$ the relation $St \approx 0.077(C_D + 1)/C_D$ can be derived

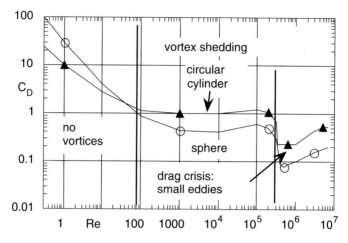

Fig. 3.18. Drag coefficients for cylinder and spheres as function of Reynolds number

[Ahlborn et al. 2002c]. Hydro drag is the principal resistance that fast running animals must overcome.

Hydro drag becomes negligible at small velocities, but a different type of drag is present at all speeds: Skin friction F_{sf}. Fluid molecules attach themselves to the whole surface of a moving body and travel at the same speed u. Far away from the body the fluid does not move at all. In a thin "boundary" layer bl, Fig. 3.19, the relative velocity between the free flow and the body changes steadily from the far field velocity to the body velocity.

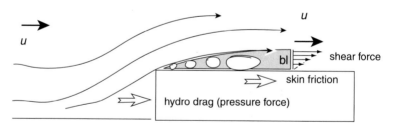

Fig. 3.19. Boundary layer bl, at a moving object

The boundary layer region travels along with the body somewhat like an open cage, and continuously accelerates the stream of fluid which passes though it. This process requires a force, the skin friction F_{sf}. The boundary layer encloses the whole body, thus F_{sf} depends on the wetted surface area S. Since energy is consumed in accelerating the boundary layer flow F_{sf} depends on the square of the velocity u of the object. Formally one writes

$$F_{sf} \approx (1/2)\, S\, C_{sf}\, \rho\, u^2. \tag{3.30}$$

The skin friction coefficient Cs depends on the small-scale structure of flow in the boundary layer. Slimy fish like *Chinook* salmon, and sharks with their micro roof-tile structure appear to have smaller skin friction. One may lump the skin friction and hydro drag together into a compound drag coefficient, C_{sh}. This parameter has a minimum value if the width to length aspect ratio D/L falls in the range $0.1 \leq D/L \leq 0.4$.

3.4.4 The Minimum Compound Drag

Animals swimming well below the surface encounter hydrodynamic drag F_h, which depends on the frontal cross section area $A = \pi R^2 = \pi D^2/4$, and the skin friction drag F_s, which scales with the total surface area $S \approx \Delta L$. These two quantities must be added to obtain the compound drag $F_{sf} = F_s + F_h$. An animal swimming at the velocity u must expend the mechanical power $P = F_{sf}\, u$. When the drag force is small, the power required to swim at a certain speed is also small.

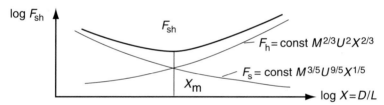

Fig. 3.20. Compound drag force

Animals rarely miss an opportunity to minimize their operating expenses. Fish of the diameter to length aspect ratio $X = D/L \approx 0.25$ experience minimum total drag [see for instance McMahon and Bonner 1983].

Consider fast swimmers of identical mass with different aspect ratios. A fish with spherical form with $D/L = 1$ has the frontal cross section area $A = \pi(D/2)^2$ and the smallest possible surface area $S = 4\pi(D/2)^2$ so that it has the smallest possible skin friction. A long and slender fish has a larger surface area S and a much smaller frontal cross section A. Therefore the slender fish will have a small hydro drag but a larger skin friction. The skin friction drops and the hydro drag rises with aspect ratio X. Therefore the sum of both components, $F_{sf} = F_s + F_h$, must have a minimum.

This minimum can be derived analytically [Ahlborn et al. 2001]. The hydrodynamic drag is $F_h = (1/8)C_D \pi D^2 \rho u^2$ for an object with the frontal area $A = \pi D^2/4$. A smooth cylindrical object moving through water has a typical value $C_D \approx 0.03$. For a cylindrical object the drag can be given as function of the total volume $V = \pi(D/4)^2 L$, and the aspect ratio $X = D/L$. Between these two equations one can

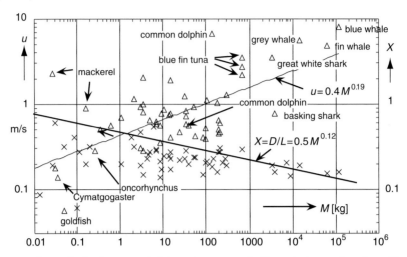

Fig. 3.21. Data of velocity u, symbol $\triangle$, and aspect ratio X, symbol $\times$, as function of body mass, and model prediction (3.34)

express the diameter $D=(4/\pi)^{1/3}V^{1/3}X^{1/3}$, and the length $L=(4/\pi)^{1/3}V^{1/3}X^{-2/3}$ as function of X and L. Therefore Eq. (3.30) can be rewritten as

$$F_h=C_1 V^{2/3} X^{2/3} U^2 . \qquad (3.31)$$

The constant C_1 includes the drag coefficient and the density of water $\rho = 1\,000$ kg/m³, leading to the numerical value $C_1 = 0.5\,C_D(\pi/4)^{1/3}\rho \approx 15$. Equation (3.31) shows that the hydrodynamic drag of a fully submerged vehicle increases with the aspect ratio to the power $X^{2/3}$. Large aspect ratios lead to large hydro drag forces.

The skin friction force can also be expressed as function of X and V, namely[3]

$$F_s=0.37\,(\pi/6)\rho\,v^{1/5}\,D\,L^{4/5}\,U^{9/5}=12.22\,V^{3/5}\,X^{-1/5}\,U^{9/5} . \qquad (3.32)$$

The skin friction force scales with $X^{-1/5}$, large aspect ratios imply small skin friction forces. The total friction force is the sum of skin friction and hydro drag. $F_{sf}=F_s+F_h$. This quantity has a minimum at a certain aspect ratio X_m, which can be found as follows: First the volume $V=M/\rho$ is replaced by the mass M, and the density ρ. Then the minimum aspect ratio X_m, is found by differentiation of $F_{sf}(X)=F_s+F_h$. The result is $X_m=0.14\,M^{-1/13}u^{-3/13}$. The velocity of aquatic animals

---

[3] The skin friction can be thought of as the work needed to accelerate all the new fluid elements that continuously get entrained in the boundary layer, thickness $\delta(L)$. If this process requires the energy per second P, then the swimmer must overcome the skin friction force F_s so that $F_s \cdot u = P$, or $F_s = P/u$.

increases continuously with mass. On average big fish swim faster than small ones, see Sect. 6.2.1. Empirical data suggest a relation

$$u \approx 0.5 \, M^{0.18}. \tag{3.33}$$

It can be used to eliminate u from X_m so that the aspect ratio becomes $X_m = 0.14 \, M^{-1/13} (0.5 \, M^{0.18})^{-3/13}$. When the mass functions are contracted one has:

$$X_m = 0.5 \, M^{-0.12}. \tag{3.34}$$

This model predicts that the aspect ratio of minimum drag should decrease with body mass. Figure 3.21 shows some aspect ratios of aquatic animals as function of their body mass extracted from the literature. Indeed large fast swimmers are more slender than the small ones.

3.4.5 Ventilation Drag

Animals which swim on the surface experience *ventilation drag* F_v in addition to hydro drag and skin friction. Ventilation drag arises because the water piles up on the front side and "ventilates" on the rear side, depressing the water surface by the distance h_v and exposing the rear of the swimming object to air pressure, Fig. 3.22a. Then the hydrostatic pressure on front and back is not balanced.

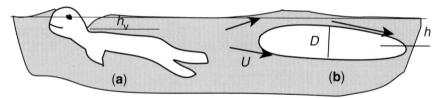

Fig. 3.22. (a) Ventilation drag, (b) Wave drag

The ventilation drag force arises from the difference of pressures on the back and the front surface. It can be shown [Blevin 1984] that the ventilation depth h_v scales with the velocity u as $h_v = \text{const } u^2$. The ventilation drag force F_v scales as the exposed area A_{ex} multiplied with ρu^2. The exposed area $A_{ex} \approx W h_v$ is a function of the width W of the protruding body part times the ventilation depth h_v. Thus the ventilation drag force F_v grows with the 4th power of u

$$F_v = \text{const } u^4. \tag{3.35}$$

Animals try to avoid this drag by swimming fully submerged. However if they swim close to the surface they experience *wave drag*. This force is connected to the acceleration of some volume elements of water located between the body and the surface. These water segments must pass above the body at increased speed, in order to squeeze through the narrow channel, between water surface and body, Fig. 3.22 b. The acceleration requires extra work that appears as an additional drag component. The wave resistance therefore increases as the swimmer gets closer to the surface. The effect becomes important if the immersion depth h is smaller than about 5 body diameters, see problem 3.7 for experimental data.

3.4.6 Lift Force

Flat or profiled surfaces in a steady flow generate lift F_L. This force can be expressed as function of the "supporting" surface area of the wings S and the lift coefficient C_L.

$$F_L = (1/2) C_L S \rho u^2 \tag{3.36}$$

The lift coefficient grows with the angle of attack β, Fig. 3.23a. However, if β gets too large the flow separates from the upper edge of the wing, and the lift disappears. This phenomenon is called stalling.

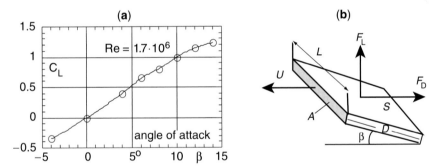

Fig. 3.23. (a) Lift coefficient of air foil as function of the angle of attack β, (b) Frontal surface area A, lifting surface S, of a flat plate mounted at an angle of attack β

Even a cylindrical appendage of diameter D generates lift. However, this lift force is non-steady. It changes its direction approximately as

$$F_{L,ns} \approx F_{Lo} \sin 2\pi f t , \tag{3.37}$$

where the amplitude F_{Lo} is of the same order as the drag force F_D acting on the cylinder. The force points to the side where the latest eddy is produced. The frequency f_{st} of this motion, the Strouhal frequency, is connected to the Strouhal number St.

$$f_{st} = St \cdot u/D \tag{3.38}$$

D/St can be interpreted as the wavelength of the vortex street. Whenever a vortex is shed on one side the lift force pulls in that direction. Insects and very small birds exploit this non-steady lift. They move their wings until one vortex is shed on the upper edge, then reverse the direction of wing motion and shed another eddy on the other edge.

Both lift and drag depend on the density of the fluid, which decreases with altitude. Therefore, runners have an easier time to push their bodies though the air at high altitudes, but birds have less lift high up in the air. More details about lift will be discussed in the Sect. 6.5 on flight.

3.4.7 Magnus Effect

Spinning and moving invokes the Magnus effect. A ball that spins at an angular velocity ω while flying through the air at the velocity u experiences the Magnus effect, which acts like a lift force F_L at right angle to both, the velocity u, and the axis of rotation (angular velocity vector ω). The Magnus force can be symbolically written as the vector product $F_L = \text{const } U \times \omega$.

3.4.8 Jet Thrust Force

Many aquatic animals, like squid, nautilus, clams, jelly fish, propel themselves by expelling a jet of water, density ρ_w, at the mass flow rate J. The average outflow velocity u m/s can be found if J kg/s, and the nozzle area A m² of the thruster are known.

$$u = J/(\rho_w A) \tag{3.39}$$

The jet thrust force F_{th} is the product of velocity and mass flow rate:

$$F_{th} = J u = J^2/\rho_w A . \tag{3.40}$$

3.5 Simple Body Forms from Skin Bags to Bones

The first larger animals probably floated in the oceans, neutrally buoyant. Very simple animals in the ocean do not need much supporting structure. They can be designed like jellyfish, which can weigh up to a ton, and exhibit graceful forms as they float around. A jellyfish that is lifted out of the water loses all its form, and when left on the beach dries up into a thin membrane. Animals on the land need first a tough skin to avoid loss of body fluid and second a structure to support their body.

We now like to find out how the body forms of animals on the land and in the water might have evolved, making use of the physical opportunities that exist in the fields of statics, and material strengths. Evolution produced terrestrial organisms with large bodies that can move swiftly. Large bodies require strong construction elements. Then it is of advantage to use materials with high strengths, both in tension and in compression. Right from the beginning one should be aware that materials are generally stronger in compression than in tension. This leads to the invention of hard shells, and bones.

3.5.1 Animals Without Bones – Giant Caterpillars?

Big size is an advantage for an animal provided there is enough food. Large animals have fewer enemies, they can cover a larger area to find food, and they can move swifter. But do animals have to have a complicated body design? Why not use the body plan of a caterpillar? How high could a caterpillar reach up on its rear end?

Consider a model animal that consists of a fluid interior held together by a soft cuticle with the tensile yield strength Y. The compression strength is provided by the pressure in the fluid, the tensile strength comes from the skin alone. We like to find out how large such an animal could possibly be.

The weight of the "water" column of height L generates a hydrostatic pressure $p = \rho g L$ at the base of the neck that must be contained by the tensile forces $T = \Delta r \, \Delta L \, \sigma$ in the skin. The quantities Δr and ΔL are defined in Fig. 3.24, σ is he stress in the skin tissue. From the balance of force in the section shown on the right of Fig. 3.24a one finds

$$2\pi r \, \Delta L \, \sigma = 2r \, \Delta L \, p = 2r \, \Delta L \, \rho g L, \quad \text{or} \quad \sigma = \rho g L \, (r/\Delta r). \tag{3.41}$$

The stress σ must be smaller than the maximum yield strength $Y \approx 10^7 \, \text{N/m}^2$ of the skin tissues. Assume $\Delta r/r = 0.1$, $\rho = 1000 \, \text{kg/m}^3$, $g = 10 \, \text{m/s}^2$. This leads to

$$L \leq Y(\Delta r/\rho g r) \approx 100 \, \text{m}. \tag{3.42}$$

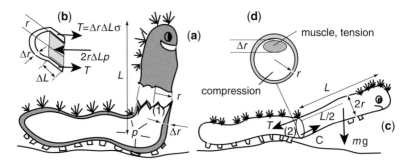

Fig. 3.24. (a) Caterpillar rearing up, (b) Animal leaning forward

Obviously caterpillars never reach this length, and therefore they are in no danger of "bursting" out of their seams at the bottom of their long neck when they rear up. However a similar height limit might affect tall plants. There must be other effects that limit the size of caterpillars. Much larger pressure forces are encountered if the caterpillar leans forward.

Treat the caterpillar as a cylindrical pipe of radius r and length L, filled with water, Fig. 3.24b. Consider moments about the point C at the base of the neck. The moment of the weight force mg of the neck about C with the moment arm $L/2$ is balanced by the moment of the muscle force T acting on the moment arm r.

$$Tr \approx f_0 2\pi r(\Delta r/6)r = mg(L/2) = \pi r^2 L \rho g(L/2) \qquad (3.43)$$

The tension $T = A f_0$ is found by multiplying the muscle section area A by the specific force per unit area $f_0 \approx 200{,}000$ N/m². If there are 6 muscles around the circumference $2\pi r$ in the caterpillar the area is approximately $A \approx 2\pi r \Delta r/6$, Fig. 3.24d. Then one has $T \approx 2\pi r(\Delta r/6) \cdot 2 \cdot 10^5$ N/m². We assume a maximum muscle thickness $\Delta r \approx 0.3\, r$ and solve (3.43) for the maximum supportable length L

$$L \leq 4(r/L) . \qquad (3.44)$$

For an aspect ratio $r/L \approx 1/10$ Eq. (3.44) yields a length of $L = 0.4$ m. If the muscle had a thickness Δr of less than $0.3\, r$ the maximum length would be even smaller. Obviously skin which supports tension and internal pressure that supports compression are not the best construction elements to build big bodies.

The radius r should be increased in order to hold up a heavy head or other appendages. The body mass goes up with r^3, but the muscle strength increases only with r^2. The animal would have to be rather bulky. Furthermore the caterpillar relies mainly on tensile forces. These are not nature's strongest forces. Material in compression can withstand higher loads. There is hence an incentive to invent structural elements that are used in compression: *Bones*.

Apart from the static requirements of large animals, they must also be concerned with the dynamics of motion. To move swiftly they have to speed up and slow down appendages, like legs, flippers, and wings. If an animal wants to impart an acceleration a to a limb of mass m, then by Newton's law a force $F = ma$ is required to move the limb. This force needs only to be small if the mass of the limb is small mass. One can see that small animals with light limbs have a certain advantage. This reinforces the earlier conclusion when it was shown that the specific muscle power of an animal P_{mech}/M, decreases with the body mass as $P_{mech}/M = \text{const}\, M^{-1/4}$. Small animals have more power, pound for pound, and they also need less force. In many respects small is beautiful.

3.5.2 Elephant Trunks and Octopus Arms

Before leaving the subject of simple skin-bag body designs one should acknowledge that some animals have developed structures bases on the design of a caterpillar: octopus arms and elephant trunks are examples. These are flexible and strong construction elements.

Such structures may be more easily replaced than appendages that have internal bones. Octopus often re-grows one of their arms, after a predator attack. Of course octopus lives in the water. Its body is about neutrally buoyant; it needs no muscle forces to hold up its arms.

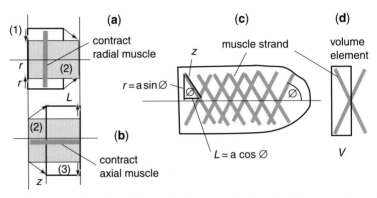

Fig. 3.25. (a) Contraction of radial muscle increases the length, (b) contraction of axial muscle increases the diameter, (c) Muscles oriented in diagonal pattern at angle ø, (d) muscle element

A cylindrical section can be moved with the help of radial and axial muscles, see Fig. 3.25 a, b. The volume of a cylindrical structure is $V = \pi r^2 L$. In order to see how radius and length are coupled, one can take the logarithm of the volume relation and differentiate $dV/V = 2dr/r + dL/L$. If the volume must remain constant one has

$$dL = -\frac{2L}{r} \cdot dr. \tag{3.45}$$

The length increases if the radius is contracted by a radial muscle. If there are several muscles in axial direction the structure bends towards the side where an axial muscle is contracted.

3.5.3 The Spiral Structure of Filaments in Nematodes

Small worms use the tensile forces of their muscle to move around. Lacking bones, they employ the incompressibility of their body fluids as a building element that can transmit forces in compression.

The 1 mm long soil nematode, Caenorhabditis elegans, has an elegant way of using only two sets of spiral muscles to move forwards and twist and turn [Harris, and Crofton 1957]. The contraction of a single muscle in the axial direction would only allow it to increase the diameter, but would not help to push the front end forward. A single muscle in radial direction would push the ends out to increase the length, but would not allow it to change the lengths of the structure.

A spirally wound muscle might achieve both in combination with the elasticity of the skin. The contraction of both axial, and radial muscles changes the volume except for one particular angle, which is found as follows: a volume element ΔV of the animal, Fig. 3.25d, can be expressed as function of the length a of each diagonal muscle segment $\Delta V = \pi r^2 \Delta l = \pi (a \sin \phi)^2 a \cos \phi$. One can find out how the volume changes when the angle is varied by taking the derivative $d \approx V/d\phi$. No change of volume occurs if this derivative is zero. This condition leads to

$$(\sin \phi)^2 = 2(\cos \phi)^2. \tag{3.46}$$

This equation is solved by $(\tan \phi)^2 = 2$. The numerical value is $\tan \phi = \sqrt{2}$ or $\phi = 55°$. For larger angles the volume would decrease when the muscles contract. The observed angles of the muscle filaments of *Caenorhabditis elegans* are $65° \leq \phi \leq 75°$. This angle implies that the internal pressure is increased upon muscle contraction, *and* body fibers are stretched. These two effects amount to a strong restoring reaction, that helps the animal to get back into the relaxed body state.

3.5.4 Hard Shell Critters

Animals that consist only of skin and soft tissue, have limitations in size and mobility. One possible way to overcome these limitations is to evolve a rigid outer shell. Nature has produced an immense variety of insects with hard shells, but only very few large animals like clams, snails, nautilus, and crabs. These big animals are generally sluggish. One of the construction problems appears to be the design of flexible hollow joints through which body fluids can be transported.

The insects, on the other hand have stayed small. But these animals are remarkably successful in many respects. At home on land in water and air they have acquired sophisticated senses, and have even developed social systems.

3.5.5 The Invention of Bones

Bones were invented because materials are stronger in compression than in tension. Think of the force needed to pull a cotton ball apart, and compare that to the weight that you can safely place onto a cotton ball. The strength of the materials comes from the intermolecular forces in the tissue and the bones.

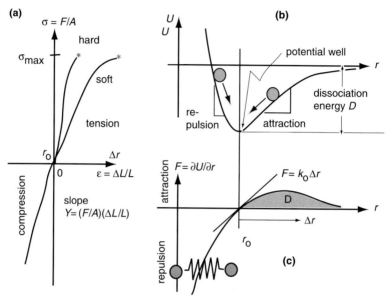

Fig. 3.26. Potential energy (a), and force (b) between two atoms. (c) Stress strain relation of a "macroscopic" object. Spring constant $k_0 = \partial F / \partial L \propto Y \propto k_0$

These forces may be understood on the molecular level. Think of a molecule consisting of two atoms. They are separated by the distance r_0, and connected by an ionic (chemical) bond. When they are compressed they will repel each other. The repulsive force, shown in Fig. 3.26a, grows rapidly as the distance gets smaller. When the atoms are pulled apart, an attractive force comes into play. This force first grows, but then it declines. The net force F namely the sum of the attractive and repulsive forces, is $F=0$ at the distance r_0. A similar representation of the binding force can be obtained from the potential energy curve $U(r)$ shown in Fig. 3.26b, $F=-(\partial U/\partial r)$ is the negative slope of the curve $U(r)$, Fig. 3.26c. Over a distance range where this slope is constant, the force $F=k_0\Delta r$ is equivalent to the elastic force of a macroscopic object with a (linear) spring constant k_0, or a modulus of compression or elongation Y.

When no external force is applied one particle sits at the "bottom of the potential well", Fig. 3.26b. In order to increase the distance r between the particles one must invest energy $E=\int Fdr$. To completely separate both particles one must provide the dissociation energy D. For $r<r_0$ the slope is negative, the force is repulsive (supporting compression), and for $r>r_0$ the slope is positive, the force is attractive (supporting tension). It is easy to see that the slope is steeper for the compression side $r<r_0$ then for the tension side $r>r_0$. Thus compression forces are larger than tension forces. This is the atomic origin for the effect that structural elements are stronger in compression than in tension. Table 3.3 shows elastic constants of materials. Note that the energy to break steel and spider web are of the same order of magnitude. Obviously early animals recognized this physical fact and invented body elements that could support compression: bones and hard shells.

3.6 Large Structures with Bones

Animals with bones and tendons can build quite complicated bodies that are strong, mobile, but not too heavy. The laws of mechanics help them in their various tasks. It starts with the ingestion of food, which may require strong teeth to tear up the prey. Forces can be amplified by levers, and by lateral pull on flexible components. Bodies parts may be build like trusses or suspension bridges, where forces are kept as small as possible, or force amplification principles are applied. Often elastic elements are incorporated to counteract gravity, or to store energy temporarily.

3.6.1 Chewing: Pressure Amplification and Lethal Bananas

The most fundamental motion of animals is the motion associated with ingesting food. Plant eaters only need chewing actions that break off tender leaves or fruit, and then squash it so that the stomach juices can get at all the edible material. Meat eaters first must kill their prey, then open up the carcass, rip off the meat, and possibly even crack bones open to access the nutritious marrow or brain. Meat eaters need strong jaws and powerful chewing muscles.

Jaws are hinged structures that can act like levers. Suppose a chewing muscle generates the force F, and this force is applied at a distance d_0 from the hinge, see Fig. 3.27. The muscle generates the torque

$$\tau_0 = F\, d_0.\tag{3.47}$$

If an object is placed between the back teeth, an equal and opposite torque $\tau_1 = F\, d_1$ is generated. The force acting at position 1 is therefore

$$F_1 = F(d_0/d_1).\tag{3.48}$$

The force at a point closer to the jaw hinge is magnified by the factor d_0/d_1. The teeth at the tip of the mouth, which are farther away from the hinge than the chewing muscle exert a smaller force, $F_2 = F(d_0/d_2)$.

Table 3.6. Biting forces of animals in N. Values are adopted from Padian [1999]

T-rex	alligator	lion	orangutans	dusky shark	wolf	human	Labrador dog
>13 000	13 000	4 200	1 700	1 040	1 300	600	580

Chewing muscle are laid out to produce large forces. Table 3.6 shows biting forces in Newtons. The biting force of Tyrannosaurus Rex was estimated from the biting marks left by T-rex on the bones of its prey, and generating similarly deep indentations on cow pelvis bones. Incisor teeth are generally pointed, like T-rex's awe inspiring "lethal bananas" as Padian terms it. This shape produces another mechanical advantage. The predator must puncture or cut open the skin of his prey. This requires some minimum pressure. The predator exerts a certain force F at the location of a tooth. The pressure at the root of the tooth, $p_1 = F/A_1$, depends on the impact surface area A_1.

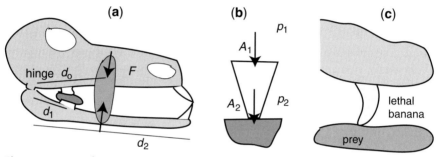

Fig. 3.27. Force and pressure amplifications in chewing machines

The force F acting onto the food, is the same as the force at the root of the tooth, However, the tip of the tooth has a much smaller surface area, A_2 than the base area A_1. Hence the pressure is amplified by the ratio of the surface areas

$$p_2 = (A_1/A_2) p_1. \tag{3.49}$$

A similar force amplification can be generated with tendons, Fig. 3.28a. A string is stretched between two fixed points with some tension T. If a lateral force F is applied for instance at the midpoint, the string is laterally deflected as shown. The tension T in the string can be calculated from the free body diagram Fig. 3.28b. It is easy to see that the relation $F = T \sin\phi$ must hold. Therefore

$$T = F/\sin\phi. \tag{3.50}$$

Since $\sin\phi$ is always smaller than 1, T is always larger than F. One can define an amplification factor $b = 1/\sin\phi$. The smaller the angle ϕ, the larger the amplification factor. For instance the angle $\phi = 5°$ yields $b = 11.5$.

3.6.2 Triangular Elements in Large Structures

The strength of materials like skin, tendon bones and cartilage limits the maximum size of animals. It was already shown that thermodynamics gives a lower limit for the size of warm-blooded animals. A head is held by the spine, a bony structure which mainly supports compression, by tendons and neck muscles which support tension, and by the rest of the tissue, which transmits pressure p. This pressure is born by internal forces within the tissue and by the skin, Fig. 3.28c. Skin is always under tension. In first approximation one can reduce all these forces to one (tendon–muscle) element which is under tension, a second (neck bone) element which is under compression, and a shoulder structure that is under compression and bending. The free body diagram Fig. 3.28d reveals the basic construction element needed for rigidity: a vector triangle consisting of compressive (C) and tensile (T) elements, and the weight mg, Fig 3.28e. If the corners are hinged and the tensile element contains a muscle, the endpoint (1) can be moved. All stable body configurations contain 3-member building blocks which either support compression, or tension. Tension can be provided by muscles tied to bones by tendons, or by elastic bands (cartilage) alone.

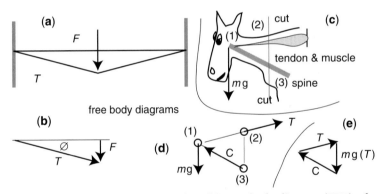

Fig. 3.28. (a) Lateral force acting on tendon. (b) Free body diagram (FBD) of tendon. (c) Head section. (d) FBD of head section. (e) Closed vector diagram

3.6.3 Vertebrae Construction, Bridges with Cable Support

Long before humans constructed bridges with sturdy decks and cable support, vertebrates have used this construction principle. The horizontal and vertical element in the model animal Fig. 3.29 mainly support compression forces. These are the bones. The hanging arches support tension forces. They can be made of tendons and skin. Joints make the skeleton flexible, and the muscles set it into motion. Dinosurs built by this general construction plan may have reached masses up to 30 tons.

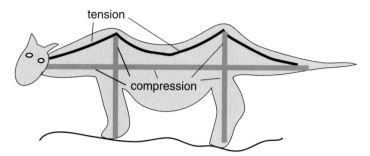

Fig. 3.29. Vertebrate construction plan

3.6.4 Elastic Elements as Support Structures

The important role of tendons for supporting body parts is well documented in petrographs of extinct animals. Figure 3.30a shows a sketch of the fossilized remains of a bird, with its distinctly long neck, arched backwards. When the bird was alive the neck was probably stretched forward like a heron or a goose in flight, Fig. 3.30b. The tension T in the stretched tendon above the neck held up the head. After death, when the bird fell flat on its side, gravity was no longer weighing the head down.

However, the stretched tendon still exerted the force T that curled the neck into the radius R, bending it through the angle $\emptyset$. Let L be the length of the spinal section of the neck. It is the same in the stretched and the curled(unstretched) position, namely $L = R\emptyset$. The un-stretched tendon length $L_0 = L - \Delta L = (R - \Delta R)\emptyset$ is seen in Fig. 3.30c. The stretch ΔL is

$$\Delta L = L - L_0 = R\emptyset - (R - \Delta R)\emptyset = \Delta R \cdot \emptyset . \tag{3.51}$$

In the straight position the tendon of cross section area A holds up the weight mg of the head. The tendon force T generates the stress $\sigma = T/A = Y(\Delta L/L_0)$, where

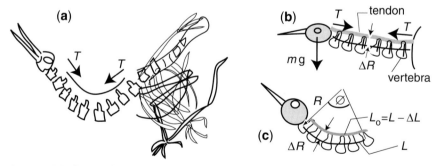

Fig. 3.30. (a) Sketch of fossilized remains of a bird. T indicates the direction of pull of the tendons after the death of the animal. (b) Tendon stretched by gravity, (c) Relaxed tendon

$Y \approx 1.5 \cdot 10^9 \, \text{N/m}^2$ is Young's modulus of elasticity of the tendon. The moment $T \cdot \Delta R$ of the tendon force counteracts the moment mgL of the head and the neck, so that $T = mg(L/\Delta R) = AY(\Delta L/L_o)$. Combine this relation with (3.52) to get

$$mg \approx AY\phi \, (\Delta R/L_o)^2 = A \, X_n^2 \, Y\phi \, . \tag{3.52}$$

Since ϕ, and the neck aspect ratio $X_n = \Delta R/L$ can be measured (3.52) allows drawing conclusions about the mass of the head and the tendon cross section A.

An elastic member like a tendon can also provide the restoring force for simple harmonic motion, where the head and neck move up and down periodically at a resonance frequency f_o. It is then likely that this frequency would be tuned to the frequency of the wings so that the head would oscillate up and down relative to the body with the least possible energy input. The center of mass of large birds is known to move up and down at the wings frequency f_w due to the unsteady lift force. Head and wing frequency could easily be tuned to keep the head steady on level flight. More detail of such motions will be discussed in Sects. 5.1.3, and 6.1.1.

3.6.5 The Secrete of Posture

Animals have the ability to change their body shapes. Muscles and skeletons must be strong enough to guarantee static equilibrium in each position. The load of spine and limbs depends very much on the geometry.

A case in point is the load on the spine for a person that lifts a weight either in the forward prone position or in the upright position. An extra load is put on the spine due to the weight Mg. In the upright position this load is essentially Mg. In the forward prone position the weight is lifted by a pull of the back muscle T, which generates the moment $\tau = Td_1$ that balances the moment of the weight Mgd_2. The required force is

$$T = Mg \, d_2/d_1 \, . \tag{3.53}$$

Typically the moment arm of the back muscle is $d \approx 0.05$ m, while the moment arm of the weight in the prone position is about 0.5 m. Hence the back muscle must pull with a force 10 times larger than Mg. The force T puts a load L onto the spine, namely $L = F \cos \phi$. Since the angle ϕ is quite small $L \approx T = Mg \, d_2/d_1 \approx 10 \, Mg$. This is bad news for the back and is the cause of many back problems.

Actually this calculation yields an upper limit of the force on the spine, because part of the weight is supported by the internal pressure p in the belly. The stomach muscles and the tensions T in the skin of the belly must also support this pressure. Weight lifters always wear a heavy belt that helps to support this internal pressure.

This upper limit of the force on the spine is found as follows. Suppose a person of $M = 70$ kg lifts a weight $m_w = 20$ kg. Assume that the legs of the person have a mass of 10 kg each, so that the hip joints must carry the load of the mass of the upper part of the body $m = 50$ kg.

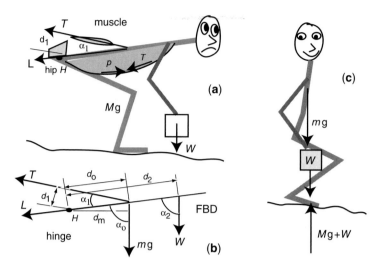

Fig. 3.31. (a) Loading of spine in prone position, (b) free body diagram, (c) carrying a weight in the upright position

In the up right position, the force vector mg describing the weight of the upper body, and the vector describing the additional weight $W=m_w g$ go right through the hip joint. Neither of these vectors generates a torque. The back muscle does not have to pull at all. The load L on the low back vertebrae is $L=mg+m_w g=(50 \text{ kg}+20 \text{ kg}) g \approx 700$ N. In the prone position the mass m of the upper body generates the torque $\tau_m=mg d_o \sin \alpha_o$ about the hip joint, because it acts with the moment arm $d_m=d_o \sin \alpha_o$. Similarly the weight W supported at the hip H generates the torque $\tau_w=W d_2 \sin \alpha_2$, because it acts with the moment arm $d_2 \sin \alpha_2$. The sum of these torques must be balanced by a torque $\tau_1=T d_o \sin \alpha_1$ generated by the tension T in the lower back muscle, which has only the small moment arm $d_1=d_o \sin \alpha_1$. Note that the angle α_1 can be found from $\tan \alpha_1 \approx d_1/d_o$. This leads to the equation $T \cdot d_1 = mg d_o \sin \alpha_o + m_w g d_2 \sin \alpha_2$. Assuming some typical values $d_o=0.5$ m, $d_1=0.05$ m, one finds $\alpha_1 \approx \text{arc} \tan(0.05/0.5) = 5.7°$, $d_2=0.6$ m $\alpha_o=\alpha_2 \approx 90°$. The tension in the muscle becomes $T \approx (1/0.05)$ $(50 \cdot 0.5 + 20 \cdot 0.6) \cdot 9.81 \approx 7400$ N. The component $T_b=5000$ N of this tension is due to the body weight itself $T_w=2400$ N is due to lifting the weight. The tension T can be split into two components, the vector $T_\perp=T \sin \alpha_1$ which generates the torque, and the vector $L=T \cos \varnothing_1$ which puts a load onto the lower back vertebrae. The extra load placed onto the vertebrae due to the weigh $m_w g$ is $L_w=2400 \cos 5.7° = 2388$ N. This is much bigger than the extra load $m_w g \approx 200$ N placed onto the hip joint in the upright position. Don't do it. Never lift heavy loads while bending forward.

3.6.6 Impediment by Gravitation on Other Planets

If there was life on other planets, the animals would likely acquire different sizes because the gravitational acceleration g is not the same as on earth.

Table 3.7. Gravitation on the planets

planet	Mercury	Venus	Earth	Mars	Jupiter	Saturn	Uranus	Neptune
g/gearth	0.37	0.85	1	0.37	2.51	1.07	0.83	1.14
g[m/s²]	3.63	8.34	9.81	3.6	24.6	10.5	8.14	11.2

Animals living on Mars would appear to have 3 times less weight $W = Mg$ as on earth. Therefore, their bones could be much lighter. Also they could run faster, but they would walk more slowly, as we will see in the next chapter.

3.7 Scaling Up

Large size is desirable for many reasons. However, the burden of a big body falls onto bones, joints, muscles, and tendons. To study the consequences of growing big, one can extrapolate from the shape and form of a small animal and scale up to larger sizes.

The different approaches of geometrical scaling, maximum tension scaling, and equal bending scaling lead to different allometric relations for limb dimensions. Geometrical scaling, where every body part is enlarged in the same ratio, turns out to be unrealistic, because it results in too high stress levels. The other two scaling methods assume that stress levels and bending moments must remain constant, because the maximum size ultimately depends on the materials from which a body is built. The results are summarized in Table 3.8.

Table 3.8. Scaling relations of limb dimensions, Body mass M, limb mass m, limb diameter d, cross section area of limb A, limb length L

limb dimension	geometric	maximum tension	equal bending
L	$\propto M^{1/3}$	$\propto m^{1/5}$	$\propto m^{1/4}$
d	$\propto M^{1/3}$	$\propto m^{2/5}$	$\propto m^{3/8}$
d	$\propto L$	$\propto L^2$	$\propto L^{3/2}$
A	$\propto M^{2/3}$	$\propto m^{4/5}$	$\propto m^{3/4}$

3.7.1 Geometric Scaling

Could a 50-ton dinosaur look like an enlarged horse? Length dimensions scale as $L = \text{const } M^{1/3}$, and areas scale as $A = \text{const } M^{2/3}$. With these relations one could compare the body dimensions of a dinosaurs ($M_d = 50,000$ kg) to that of a horse ($M_h = 500$ kg).

Compare the compression stresses $c = M/A$ in the knee joints of the animals. Assume each knee of the horse has a typical cartilage surface area of $A_k = 0.0015$ m². The four knees of the animal with a total surface area $A_h = 4 A_k$ must carry the load $g M_h$ of the body. This generates the static compression stress $c_h = g M_h/A_h = 5\,000/6 \cdot 10^{-3} \approx 8 \cdot 10^5$ N/m², which is well below the safe limit of 10^7 N/m². The stress will of course be much higher when the animal is running.

Geometrical scaling of the knee cartilage surface area implies $A_h/A_d = (M_h/M_d)^{2/3}$. Then the compression stress scales as $c = M/A = \text{const } M/M^{2/3} = \text{const } M^{1/3}$. With this relation one can compare the compression stresses in the knees of horse and dinosaur knee $c_d/c_h = (M_d/M_h)^{1/3}$, or

$$c_d = c_h (M_d/M_h)^{1/3} \approx 4.6\, c_h . \qquad (3.54)$$

Geometrical scaling yields an allometric relation where the stress increases as $M^{1/3}$. The stress in the dinosaur's knee joint would be a factor 5 larger than in the horse's knee. This stress could be tolerated if the animal would never run fast. If the animal had similar habits as today's horses the dimensions of the skeleton would have been distorted in order to maintain a similar stress in the skeletal structure. The knee area would increase more rapidly than body length L. In fact, to keep the static compressive stress identical in both animals the knee areas must scale as the body masses: $\sigma_d = M_h/A_h = M_d/A_d$, so that the area for four legs becomes

$$A_d = A_h (M_d/M_h) = 0.6 \text{ m}^2 . \qquad (3.55)$$

Hence each knee would have a cartilage surface area of $A = 0.6 \text{ m}^2/4 = 0.15 \text{ m}^2$, corresponding to a length dimension (diameter) of $2r_{knee} \approx 0.38$ m. We conclude that animals in general should not show geometrical scaling. A better approach may be weakest link scaling.

3.7.2 Weakest Link Scaling

The "safe" tensile strength found in today's animals could be used to take a guess at the actual size of an extinct animal who's bones have been recovered. Suppose a knee joint a diameter of $2r = 0.2$ m has been found, and one would like to estimate the body mass of the animal.

First, consider the stresses in the cartilage of the knee joint. Assume that the cartilage tissue of each of the leg bone joints (knee, or ankle) of an animal of mass M has a top surface area of $A = 4\pi r^2$. The weight Mg generates the pressure

$$p = \frac{F}{A} = \frac{Mg}{4\pi r^2} \ . \tag{3.56}$$

This pressure acts onto the soft disk shaped tissue in the knee joints. The internal forces in this tissue, Fig. 3.32, consist of the pressure p, which wants to split the disc apart and the internal tension τ that holds the disk material together. The material will not break apart as long as $p \le \tau$.

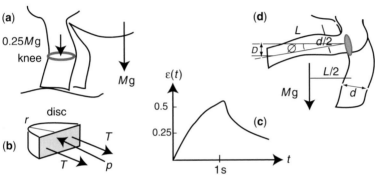

Fig. 3.32. (a) Tension in knee joint. (b) Section of soft tissue in the knee lining. (c) Time response of cartilage. (d) Bending stress

Suppose this tissue has a maximum yield strength $Y \approx 10^7 \text{ N/m}^2$, then one has $p = Mg/\pi r^2 \le \tau = Y$, or $M \le Y 4\pi r^2/g$. For a knee joint with $r \approx 0.1$ m the maximum mass M which can be supported is

$$M \le \frac{4\pi Y r^2}{g} \approx \frac{4\pi \cdot 10^7 \cdot 0.1^2}{10} = 1.2 \cdot 10^5 \text{ kg} \ . \tag{3.57}$$

The supportable mass M scales with the square of the radius r of the cartilage disc. Of course there must be a safety margin, because the animal could not risk to split its cartilage in the knee joint whenever it stamps its foot. With a safety margin of a factor 10 the animal could have a mass of $M \approx 12,000$ kg. This safety margin would assure that the knee could withstand impact forces associated with acceleration and deceleration, which appear for short times and can be many times the static weight. Additional limits on the size would come from the ability to collect enough food, the compressibility of the ground, which might not be able to support such giants.

Such estimates apply to land animals only. Aquatic animals can become much bigger, because buoyancy forces support them. The blue wale with its mass of ≈ 100 tons is an example.

For the dynamic loading one must also consider the time dependent elastic behavior of cartilage. This tissue responds well to short time loads, where it can withstand stresses of 10^7 N/m^2, but it gets weaker under long time loads.

The time dependent stress E becomes a time function $E(t) = \sigma / \varepsilon(t)$, which can be characterized by a short time behavior by $E_{1s} \approx 10^7$ N/m^2, and the long time behavior $E_\infty \approx 10^6$ N/m^2, Fig. 3.32c. Some other material also have different short time and long time elastic behaviors. You can make a simple experiment at home with cornstarch. Mix some cornstarch with water in a metal bowl. When you gently push your hand into the slurry, the hand will easily reach the bottom with little force applied. Try next to pound your fist quickly into the slurry, some water may slosh into your face but the mixture will resist your fist like a rock. Another example is the "softness" of water. It flows gently around the body when one steps into a swimming pool, but an object, dropped from a height of 100 m, hits as if the water was a brick wall.

3.7.3 Maximum Tension Scaling

Skeletal structures are exposed to compression, tension, and bending forces. Bending is a combination of stretch and compression. It yields a different allometric relation.

Consider the model, Fig. 3.32d of a limb of mass m, diameter d, and length L that is held sideways. It is depressed by its own weight mg acting with the moment arm $L/2$. The mass of the limb scales as $m = \text{const } d^2 L$. Therefore the torque exerted by the weight of the limb is $\tau_w = \text{const } d^2 L^2$.

The leg is held up by a tensile force F, which is produced by tendons and muscles acting with a moment arm $r = d/2$. The force F can be expressed as some maximum tension σ_m multiplied by a cross section area which is proportional to d^2. Hence the muscles exert the torque $\tau_m = \text{const } d^3 \sigma_m$. In equilibrium this torque must be equal to $\tau_w = \text{const } d^2 L^2$, so that $d^3 \sigma_m = \text{const } d^2 L^2$. If the maximum stress σ_m is the same for all animals, it can be treated as a constant, resulting in $d = \text{const } L^2$. This dependency can be used to eliminate d from a relation for the mass of the limb $m = \text{const } d^2 L$. This leads to $m = \text{const } L^5$, or $L = \text{const } m^{1/5}$, and

$$d = \text{const } L^2 = \text{const } m^{2/5}. \tag{3.58}$$

so that the surface area d^2 of cartilage in the knee joint must scale as $A = \text{const } d^2 = \text{const } m^{4/5}$. If one further assumes that the limb mass m is a constant fraction of the body mass M, one can compare horse and dinosaur to see how limb cross section areas scale with body mass. If the cartilage area of the knee scales as horse and dinosaur dimensions must be related as

$$A_d = A_h (M_d/M_h)^{4/5} = 6 \cdot 10^{-3} \cdot 40 = 0.24 \text{ m}^2. \tag{3.59}$$

This number is nearly twice as large as the surface area calculated under the assumption of geometrical scaling.

3.7.4 Elastic Similarity Scaling

For another estimate or the knee joint surface area we consider the deflection D in Fig. 3.32d. The deflection angle ø is approximately found from $\tan g \phi = D/L \approx \Delta L/r = 2\Delta L/d$. This bending comes about because the upper edge of the limb is stretched and elongated by some distance ΔL, while the lower edge is compressed. Let the stretching be an elastic process where the elongation is related to the applied force $F = k\Delta L = Y(A/L)\Delta L$. The spring constant k contains Young's modulus Y, the cross section area $A = \text{const } d^2$ of the stretched member, and the length of the stretched member. Again we balance the torque of the weight of the member $\tau_w = m g (L/2) = \text{const } d^2 L^2$ with the torque of the applied force $\tau_m = Fd/2 = \text{const } Y (d^2/L)\Delta L(d/2)$ to yield $Yd^3\Delta L/L = \text{const } d^2 L^3$ or

$$Yd\Delta L = \text{const } L^3 . \tag{3.60}$$

For similarity bending one can assume that the bending angle ø remains constant. This implies $\Delta L/d = \text{const}$. Then Eq. (3.60) reduces to the similarity of constant stretch.

$$L = \text{const } d^{2/3} \tag{3.61}$$

As before, the limb mass scales as $m = \text{const } d^2 L$, so that $m = \text{const } d^{8/3}$, or $d = \text{const } m^{0.375}$. Then the surface area scales with limb mass as $A = \text{const } d^2 = \text{const } m^{0.75}$, and provided $m = \text{const } M$, the knee area of the dinosaur must have the size

$$A_d = A_h (M_d/M_h)^{3/4} = 6 \cdot 10^{-3} \cdot 32 = 0.19 \text{ m}^2. \tag{3.62}$$

All these estimates point to a surface area in the range of $\approx 1/4$ m² which yields a surface stress of $\sigma = 5 \cdot 10^5$ N/0.25 m² $= 2 \cdot 10^6$ N/m², which is probably safe for a slowly moving animal. These scaling relations are shown in Table 3.8.

3.8 Strong Materials in Biology

The strength of biological and other materials arises primarily from the bonds that hold atoms together. Most biological materials however do not achieve the theoretical maximum strength. Spider dragline is one of the few exceptions, as it achieves a strength that approaches the limit set by the chemical bonds. To understand these limits and why they are mostly not reached one has to look at the principle of bonding. Figure 3.26b shows the bond energy potential U versus atomic separation r that characterizes a chemical bond. Figure 3.26c shows the derivative dU/dr, which gives the bond force. The minimum of the function $U(r)$ corresponds to zero net force. Here the repulsive force (which dominates at close range) equals the attractive force of the chemical bond, which arises (for ionic solids) from the electrostatic interaction between $(+)$ and $(-)$ charged particles.

Table 3.9. Work of fracture per unit mass W/ρ

material	yew wood	spring steel	keratin (horn)	collagen (tendon)	bone	rubber	spider silk
W/ρ [MJ/kg]	900	130	30,000	2,500	1.500	10^5	$\approx 2 \cdot 10^9$
max $\Delta L/L$	0.009	0.003	0.3	0.08	0.02	8	0.4

　　The bonding interaction is responsible for the strength of the material, because materials break when atoms are pulled apart in a failure process. Note in Fig. 3.26b that the net force first rises linearly from zero as the separation r increases, then it levels off and quickly falls back to zero, the bond is broken. One strength quantity that can easily be measured is the work of fracture per unit mass W/ρ. Table 3.9 shows some data.

3.8.1 Surface Energy γ and Breaking Strength

If the distance r was increased by a mechanical force that pulled the atoms apart, the bond breaking would be a mechanical failure. Figure 3.33 illustrates the failure in a perfect elastic solid. Shown is the model of a perfect crystal with the layer spacing a between atoms in the vertical direction. (In real life this spacing could be determined with X-ray diffraction.) If we knew the chemical nature of the solid, we also could determine the theoretical stiffness, and the strength of the material from first principles.

　　A piece of material of cross section A, which is elongate by $\varepsilon = \Delta L/L$ experiences the stress $\sigma = F/A = Y\varepsilon$, where Y is the modulus of elongation. The area un-

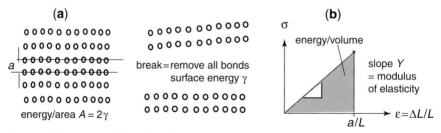

Fig. 3.33. (a) Perfect solid and surface energy γ, (b) energy/volume

der the $\sigma(\varepsilon)$ curve, Fig. 3.33b, represents the energy W per unit volume $V = A\,a$ needed to break the material; force/area has the same dimension as energy/volume.

$$\frac{W}{A \cdot a} = 0.5\sigma \cdot \frac{\Delta L}{L}, \text{ or } \frac{W}{A} = \frac{a \cdot \sigma \cdot \Delta L}{2L} \tag{3.63}$$

where $\varepsilon = \Delta L/L$ can be replaced by σ/Y. Let γ be the separation energy per area. It is an empirical quantity that can be measured. Typical values are $\gamma = 1$ J/m². When the material breaks, two surfaces are created each one having the surface energy γ, hence $W/A = 2\gamma$.

Substituting this relation and $\Delta L/L = \sigma/Y$ into (3.63) yields the maximum breaking strength σ_{max}

$$\sigma = \sigma_{max} = 2 \cdot \sqrt{\gamma \cdot Y/a} . \tag{3.64}$$

3.8.2 The Strength of Real Materials

Underlying the breaking strength derivation is the assumption that the failure in the material has occurred within a single layer of the crystal lattice, and that with this failure only the bonds that previously spanned this layer have been broken. This theoretical breaking strength is however rarely approached in real materials. Table 3.10 shows data for the strength of some technical and biological materials. Obviously the biological materials fall well short of the theoretical materials. Two questions come to mind. 1) Why are the biological materials so weak, and 2) why do spider silk and kevlar come much closer to the theoretical levels?

Table 3.10. Surface energy γ and maximum breaking strength σ_{max}

material	surface energy γ [J/m²]	σ_{max} theoretical [N/m²]	σ_{max} measured [N/m²]
iron	2	$4 \cdot 10^{10}$	$\approx 10^9$
glass	0.5	1.6	$5 \cdot 10^7$ to $4 \cdot 10^9$ for fiber
MgO	1.2	3.7	
Al$_2$O$_3$	1.0	4.6	$2 \cdot 10^9$ to $1.5 \cdot 10^{10}$ for whisker
muscle			$\approx 2 \cdot 10^5$

The first question has an easy answer. No real material achieves the perfection of molecular structure assumed in the analysis. All real matter, especially large blocks of it, has *flaws* and *defects* when viewed at the level of individual atoms. Flaws exist on the scale of atomic lattice and even more frequently on a much larger scale. These flaws cause redistribution of stress and stress amplification, which makes the fracture process quite different from the models used to calculate the theoretical maximum stress. Instead of all bonds breaking simultaneously to fail in a single catastrophic effect the flaws cause stress concentration, and the materials fail by the sequential breaking of bonds.

Stress may be illustrated by stress lines, or stress trajectories, as indicated in Fig. 3.34. The closer the lines are together, the higher is the local stress. Near the notch in Fig. 3.34a the stress must be much larger than the average stress σ_a, because all the stress lines from the top have to squeeze around the notch. Near the tip of the notch the stress may be increased 100 fold. The atoms near the tip of the flaw must see this increased stretch. They break apart once the tension exceeds the theoretical strength. Then the flaw propagates inwards, and as it gets deeper, more stress lines crowd into the still coherent cross section and the local stress increases further. The material will break at increasing speed.

This concentration of stress lines is the reason that biological materials are so much weaker than their theoretical strength. The actual process is however more complicated than this. For many materials the strain energy, $\Delta W = 0.5\,\sigma_{av}\,\varepsilon$, stored throughout the sample at the average stress σ_{av}, at which the flaw growth starts, is sufficient to account for all the bonds that must be broken to form the fracture surface. These materials fail catastrophically (shatter), because the crack propagates across the sample at speeds that approach the speed of sound in these materials. Materials that fail this way are extremely brittle, and they are usually of limited value in the construction of living or man-made materials. Materials with large cross sections can generate large stress concentrations. It is not possible for a glass fiber of 1 µm diameter to have a flaw of a few hundred nm. Therefore the effect of flaws on the material strength can be reduced by limiting the size of the material thus limiting the size of the flaws.

Figure 3.34b shows measured maximum stress as a function of the diameter of different fibers. Clearly the thin fibers are much stronger. The very fine glass fibers approach the theoretical maximum strength. For material strength small is beautiful.

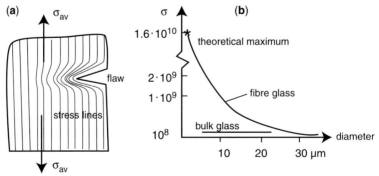

Fig. 3.34. (a) Stress concentration near notch in crystalline lattice. (b) Maximum stress σ as function of fiber diameter

3.8.3 Why Are Spider Silk and Kevlar so Strong?

The notion that small fiber diameter will improve strength is probably one part of the explanation of the quite remarkable strength of spider's dragline. These fibers are typically 1 to 5 μm in diameter, and their strength approaches that of the very strongest man-made super fibers.

The extreme toughness of silk arises from an optimal combination of rigid polymeric crystals and amorphous polymeric linking domains. The stress strain curves for spider silk and kevlar, Fig. 3.35a, are quite different, although both have similar maximum strength σ_{max}. Kevlar stretches very little, $(\varepsilon_{max}=\Delta L/L_{max} \approx 4\%)$, whereas spider silk stretches to $\varepsilon_{max} \approx 35\%$. As a consequence the energy required to break spider silk $W/\rho = 200$ MJ/m³, the hatched area under the stretch curve $\sigma(\varepsilon)$, is much larger for spider silk than for kevlar, and is nearly the highest for any known material. This makes spider dragline the toughest known material even though it is significantly weaker than Alumina whiskers, see Table 3.10.

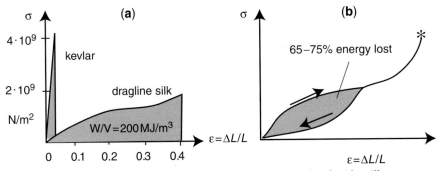

Fig. 3.35. (a) Stress-strain of kevlar and spider silk. (b) Load cycle of spider silk

Toughness is a measure of the ability to resist breaking, which is equivalent to the aptitude to absorb energy prior to breaking. Toughness and strength are not the same, and toughness is probably the most important property for spiders in their use of dragline silk. Kevlar is entirely crystalline, and like other crystalline materials, it cannot be stretched significantly before is fails. Dragline materials have been optimized by nature to their functions in the spider's web and safety line.

An orb web is a planar structure that has been optimized to capture flying insects, which the spider eats. The structure has to absorb the kinetic energy of the flying prey without breaking. It is of course essential to do it with the least amount of structural material. Spiders need to build large nets in order to catch enough food, and spider silk is pure protein, a valuable commodity to all animals. Also spiders make a new net every day and would have to consume huge amounts of proteins for their webs were they not able to make very fine fibers of high performance materials. Some spiders actually eat their olds web.

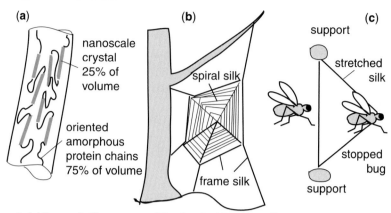

Fig. 3.36. (a) Internal silk structure, (b) orb web, (c) spider silk stopping a fly

The orb web is actually made from two kinds of silk. One is dragline whose properties and structure were discussed above. This silk forms the frame and radial strands of the web. They serve as the support for a rubber-like sticky silk that is laid down as a spiral, Fig. 3.36b. We will only consider the dragline here. Its main function is to absorb the energy of impact of the flying prey, which arrives essentially at right angle to the strands, Fig. 3.36c. This geometry imposes some interesting physics associated with high deformation and energy absorption.

First, consider the situation where a flying bug has been stopped by the deflection of the silk strand. If the silk was highly elastic like all true crystalline substances, then the spider would have created a perfect sling shot: when the bug is stopped, its kinetic energy would be converted into elastic energy of the stretched silk. Then the strand would bounce back and fire the bug back into the direction where it came from at the initial impact velocity. This is obviously not what the spider had in mind, and it is not what actually happens.

The mixture of amorphous and crystalline structure in silk makes this material highly viscoelastic: part spring, part fluid. This viscoelastic behavior is illustrated in Fig. 3.35b. The "recovered" stress is much smaller than the strain generated during the first stretching. The σ-curve is a loop encircling an area that corresponds to the lost work. 65 to 70% of the kinetic energy of the bug is absorbed through molecular friction (loss), and only 30−35% is available for elastic recoil. Thus, the flying bug will likely be held in the glue strands.

3.8.4 The Optimum Stretch of Spider Silk

The extensibility of dragline (frame silk) appears to have been optimized through evolution to work best in side-on loading, where the force F is applied at right angle to a string. The tension T, which appears inside the string, is much larger than the force F itself. From the geometry, Fig. 3.37b one finds the tension as function of angle ø and force ; $F/2 = T \sin ø$, or

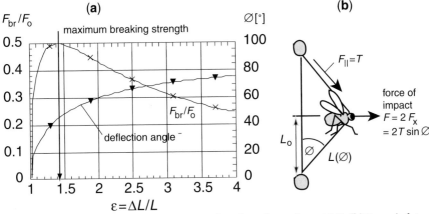

Fig. 3.37. (a) Breaking strength in a string as function of stretch $\varepsilon = \Delta L/L$. (b) Force in laterally loaded string

$$T = \frac{F}{2 \sin \phi}.$$ (3.65)

If the string stretches little, the angle ø is kept small and $1/(2 \sin \text{ø})$ becomes a large number. Therefore, the tension in an inelastic string can be much larger than the impact force F.

If the string is quite elastic, it just stretches under the impact. For angles ø > 30° the denominator of Eq. (3.65) is larger than 1, so that the tension is smaller than the impact force F. If however the string is stretched too much the cross section decreases, (since the total mass of the string must be constant) so that less force can be supported. There is an optimum deflection angle where the lateral force $F_{\parallel}$ is a maximum.

This maximum can be found as follows. As the angle ø increases the length of the silk increases $L = L_0/\cos$ ø. As the silk is stretched, the area decreases since the volume must stay constant, $A = V/L = (V/L_0) \cos$ ø. Assume tension in silk is maximum T_{max}, then the silk can support the "longitudinal" force.

$$F_{\parallel} = A T_{max} = (T_{max} \cdot V/L_0) \cos \text{ø}$$ (3.66)

The impact force that can be absorbed is $F = 2 F_{\parallel} \sin \text{ø} = 2(T_{max} V/L_0) \cos \text{ø} \cdot \sin \text{ø}$. If one further substitutes $\cos \text{ø} = L_0/L$, $\sin \text{ø} = (1/L)\sqrt{(L^2 - L_0^2)}$, and further introduces the elongation $\varepsilon = L/L_0$, the maximum force can be given as

$$F = 2 \frac{T_{max} \cdot V}{L_0} \cdot \frac{\sqrt{\varepsilon^2 - 1}}{\varepsilon^2}.$$ (3.67)

The maximum braking force F_{max} derived by this model can either be given as optimum stretch $\varepsilon_{max} = L_{max}/L_0$, or as optimum angle ϕ_{max}. The maximum force normalized by $2(T_{max} \cdot V/L_0) = \sin \phi \cdot \cos \phi$ is plotted in Fig. 3.37 a. The optimal extension where the fiber can support the maximum lateral force without breaking occurs for $L/L_0 = \sqrt{2}$, or $\varepsilon = 42\%$, or $\phi_{max} = 45°$. At this deflection the strands support exactly 50% of the breaking strength $2F_0$ of the two strands which are stretched in the longitudinal direction.

3.8.5 The Dragline as Safety Line

Spiders also match the size and properties of the dragline for the use as a safety line. They manufacture their drag lines with a cross section area A that will make it an effective safety line to catch the spider itself, should it fall. Since spiders grow about 1000 fold during their lives, they must ensure that the dragline changes in size in an appropriate manner. Figure 3.38 shows experimental values of breaking force F_{br} as function of the spider body mass M. Experimental values of the breaking force F_{br} scale with body mass M as

$$F_{br} = 10 \, M^{3/4}. \tag{3.68}$$

This implies that the statistic safety factor SF (breaking force F_{br}/body weight M) scales with the body mass as, see Fig. 3.38

$$SF = F_{br}/Mg = 1.02 \, M^{-1/4}. \tag{3.69}$$

As a consequence of the negative exponent in this allometric relation the statistic safety factor falls as the animal grows. For the smallest animals the safety factor is very large (about 20 to 30), and it is not clear why small spiders make such large overgrown fibers.

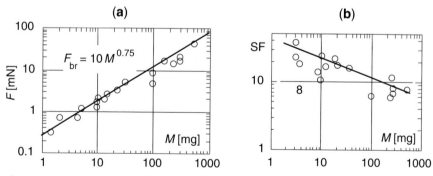

Fig. 3.38. (a) Breaking force of dragline silk. (b) Statistic safety factor SF

Table 3.11. Frequently used variables of Chap. 3

variable	name	units
A	area	m^2
B	buoancy force	N
C	compression	N
c	compression stress	kg/m^2
E	electric field	V/m
F_D	drag force	N
F_L	lift force	N
f_o	specfic muscle force	N/m^2
F_c	cross bridge force	N
h	height	m
j	current density	Amp
m	Mass of part of the body	kg
p	pressure	N/m^2
S	surface area	m^2
SF	safety factor	
St	Strouhal number	
T	tension	N
T	temperature	o
u	velocity	m/s
V	volume	m^3
v_o	specific muscle velocity	1/s
W	weight	N
X	aspect ratio	
Y	modulus of elongation, or yield strength	N/m^2
$\varepsilon = \Delta L/L$	elongation	
γ	surface energy	J/m^2
γ_s	coefficient of surface tension	N/m
η	viscosity	$kg\,m^{-1}\,s^{-1}$
μ	molecular weight	kg/mol
μ_d	dynamic friction coefficient	
μ_s	static friction coefficient	
v	kinematic viscosity	m^2/s
ρ	density	kg/m^3
σ	stress	N/m^2
τ	torque	Nm

Problems and Hints for Solutions

P 3.1 Gastrolyths
Crocodilians have a heavy tail bringing their center of mass close to the hind legs. When they are submerged, gas in their lungs or stomach could give them extra buoyancy and make their front rise. It is hypothesized that they swallow stones (gastrolyths) in order to offset the buoyancy force. An alligator, mass $M = 30$ kg resting under water has developed an air pocket of volume $V = 700$ ml at a distance $d_1 = 0.15$ m to the front of the center of mass.

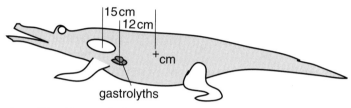

Fig. 3.39. Crocodile with gastrolyths

a) What is the weight W of the stones the crocodile has to swallow to offset the buoyancy of the gas? (density $\rho_{\text{stone}} = 2.4 \cdot 10^3$ kg/m^3). b) Suppose the crocodile would dive to a depth of $d = 6.0$ m. What would happen to the gas in the lung? c) Would the crocodile have to change its' ballast and if so, how much ballast would it need at this depth?

P 3.2 Old Birds
Examine Fig. 3.40 and assume that the Pterodactyl was about the size and mass of a duck.

a) Estimate by how much the tendon of the neck is shortened in the position shown, compared to its length L when the neck is stretched. Assume $Y \approx 10^8$. b) What cross section area A for the tendon is required to hold the head? c) How much elastic energy $\Delta E = (1/2)k\,\Delta L^2$ is stored in the tendon? The spring constant k is defined by $F = mg = k\,\Delta L$.

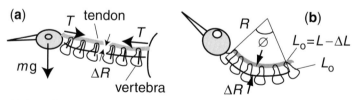

Fig. 3.40 Birds neck. (a) Stretched forward with tendon balancing the weight of the head. (b) Neck of birds carcass lying on its side: head is pulled backwards by tendon

P 3.3 Force Amplification in a Dog's Bite

a) Calculate the forces F_1 and F_2 that a Doberman dog can exert with its front and back teeth. Assume that the chewing muscles on either side of the mouth have a cross section of $A = 20$ cm^2, and that the dog chews on a piece of meet which protrudes on both sides of his mouth. Assume $d_0 = 1.0$ cm, $d_1 = 2.5$ cm, $d_2 = 6.0$ cm.

b) Assume that the teeth are pointed. Each tooth makes an indentation of 1.5 mm across when the dog gently touches an object. What pressure can the front and the back teeth apply on the surface of a bone?

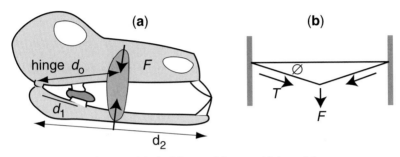

Fig. 3.41. Force amplification. (a) Skull bones, (b) rope with lateral force

c) Forces can be amplified by pulling in the lateral direction. Calculate the tension force T which is generated when a muscle pulls with a force $F = 6$ N in the direction shown in Fig. 3.41 b, so that the angle becomes $\varnothing = 15°$. Give the amplification factor $\alpha = T/N$. d) Some muscles (pinnate muscles) are arranged to pull at an angle so as to amplify their force. Find three examples of animals with such pinnate muscles and estimate their amplification factor α.

P 3.4 Toeholds

Determine the tension T in your Achilles tendon when you are standing with the toe of one foot on a book. For this calculation you must know your weight and measure the distance a from the toe to the ankle joint and b from the ankle joint to the heel where the tendon is attached.

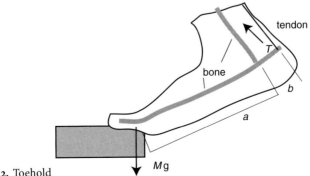

Fig. 3.42. Toehold

a) Estimate the area cross section area of the muscle ($A_m = w \cdot d$) that holds the tendon. Measure its width w, and depth d to an accuracy of ± 1 cm. Calculate the specific muscle force $f = T/A$ and compare this value to the standard value $f_0 = 2 \cdot 10^5$ N/m². Discuss the uncertainty of the result. (An uncertainty of $\pm 20\%$ is quite acceptable for these rough calculations.) b) Determine the area A_t of the tendon by measuring its diameter, assuming a circular cross section. Calculate the stress $\sigma = T/A_t$ in the tendon, and determine the safety margin, namely how much smaller σ is than the maximum yield strength of tendon material quoted in Table 3.3.

P 3.5 Weight Lifting
An person lifts a 6 pound weight by moving the forearm (mass $m = 2.5$ kg) $n = 90$ times in 2 minutes from the vertical down position to horizontal position (a). Assume that the center of mass cm of the forearm is located a distance $s_{cm} = 0.23$ m from the elbow joint, the center of mass of the weight (held by the hand) at $s_w = 0.42$ m, and the tendon connection at $s_t = 0.03$ m (tendon diameter $d_t = 0.003$ m). Model the biceps as a cylinder of max length $l_0 = 0.15$ m, and diameter $d_0 = 0.06$ m, in the stretched position. a) What is the tension force F_t and the stress σ in the tendon in the horizontal position? b) What is the length contraction ΔL of the muscle? c) How much mechanical work W is done? d) What is the maximum weight this person could hold in position (b)? e) How much metabolic energy ΔQ is expended. f) Calculate the mass m_b of the biceps and determine by how many degrees ΔT, the temperature in the muscle would go up if no heat was carried away. g) If the heat was removed by sweating alone how much water would have to be evaporated? h) How else could the heat be removed? How much water would have to be evaporated?

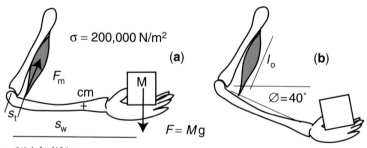

Fig. 3.43. Weight lifting

P 3.6 Red Fingers
Swirl one arm around as fast as you safely can. Measure the length of your arm L, and the time τ_{10} it takes for 10 revolutions, so that you can determine the period τ.

a) Calculate the average speed of your fingers. b) Determine the radial centrifugal acceleration, and the additional pressure in the blood vessels in your fingers due to the motion. Compare this pressure to the systolic pressure generated by your heart ($\Delta_{ph} \approx 120$ mm Hg), and comment why your fingers are red.

P 3.7 Near Surface Drag

Natasha, a good swimmer, wants to quantify the effect of near surface drag.[4] For that purpose she swims with the dolphin kick at various depths y under the water surface. She maintains a constant depth by watching a horizontal line on the pool wall.

Table 3.12. Laps times Δt, and pulse rate F_h of Natasha while swimming at depth y under water

y [m]	Δx [m]	Δt [s]	F_h [beat/min]
0	25	26.24	140
0	25	26.37	136
0.5	25	25.45	140
0.5	25	25.15	140
1.0	25	23.85	140
1.0	25	27.73	140
1.5	25	22.95	144
1.5	25	22.95	140
2.0	22.86	21.90	140
2.0	22.86	21.56	144
2.5	22.86	21.60	140
2.5	22.86	21.44	140
3.0	22.86	21.15	136
3.0	22.86	21.23	140

For every length she swims she measures her travel time with a stopwatch on her wrist, see Table 3.12. The pool is slightly shorter below a depth of 1.50 m. She also measures her pulse rate F_h, and only uses runs where it stays close to the same value, 140 beats per minute.

a) Calculate the average speed per run. b) Assume a body shape like a flattened torpedo of cross section area $A = 0.07$ m², and a drag coefficient $C_{D3} = 0.05$ to calculate the average drag force F_3 at a depth of $y = 3.0$ m. This is also the average propulsion force generated by the swimmer. c) Since the heart rate is about the same at all depths one can assume that the propulsion force too is the same at all depths. Calculate the drag coefficient ration C_{Dy}/C_{D3} as function of depth y.

P 3.8 Body Exercise

Measure the force of your biceps. Take a pail and fill it with water until pail and water together are so heavy that you can barely hold it with your hand. Measure the weight on a bathroom scale. Say the scale reads 7 kg. The force exerted by the pail in your hand is then $F\,mg = 7\,\text{kg} \cdot 9.81$ m/s² $= 68.7$ N. Your arm is a lever, and the pail acts with the moment arm $d_1 =$ distance pail handle – elbow joint, while the biceps tendon, which pulls up with the force F_t, has the much shorter moment arm d_2.

---

[4] Student research project by Natasha Szucs, University of Bristish Columbia Zoological Physics April 2002.

Both distances can be easily measured with a ruler. The balance of moments yields $F_t = mgd_1/d_2$. The tendon only transmits the muscle force, therefore the muscle generates the force $F = F_t$. Measure the diameter D of your biceps, again using a ruler. The muscle cross section area is $A_m = \pi (D/2)^2$. Then find $\sigma = F/Am$. Instead of filling the pail until you can barely hold the pail, you can make the pail heavier and slide it on the forearm from the elbow joint towards the hand till it becomes too heavy. Measure the distance d_1 where you can just hold it. This experiment can be easily performed in the classroom, using a student volunteer.

Sample Solutions

S 3.1 Gastrolhyths

a) Take moments about the center of mass: $mg \cdot 0.12 \text{ m} = V(\rho_w - \rho_{air}) g \cdot 0.15 \text{ m}$, solve for $m = (0.15 \text{ m}/0.12 \text{ m}) \cdot 0.7 \text{ l} \cdot (1 \text{ kg/l} - 0.00129 \text{ kg/l}) = 0.874 \text{ kg}$. b) Pressure increases by 10^5 N/m^2 for every 10 meters of depth h. At the surface $p_0 = 10^5 \text{ N/m}^2$. At $h = 6.0$ m the total pressure is $p_6 = p_0 + (h/10 \text{ m}) \cdot 10^5 \text{ N/m}^2 = 1.6 \cdot 10^5 \text{ N/m}^2$, from the gas law $pV = \text{const}$ one has $V_6 = V p_0/p_6 = (1/1.6)V = 0.7 \text{ l}/1.6 = 0.44 \text{ l}$. This smaller volume of air generates a smaller lift. c) At $h = 6$ m depth the animal would only need $m_6 = (0.15/0.12) \cdot 0.44 \text{ l} \cdot (1 \text{ kg/l} - 0.00129 \text{ kg/l}) = 0.54 \text{ kg}$ of stones.

S 3.5 Weight Lifting

a) Take moments in position B. Solve $F_m \cdot 0.03 \text{ m} = g \cdot (2.5 \text{ kg} \cdot 0.23 \text{ m} + 6 \text{ lb} \cdot 0.45 \text{ kg/lb} \cdot 0.42 \text{ m})$ for the muscle force $F_m = 536$ N. The stress in the tendon of radius $R = 0.003$ m/2 is $\sigma = F_m/\pi(0.0015)^2 = 7.65 \cdot 10^7 \text{N/m}^2$. b) The length contraction of the muscle between positions A and B is $\Delta s = d \cdot \sin 40° = 0.003 \cdot \sin 40° = 0.00192$ mm. c) Mechanical work for lifting the weight one time: the weight is lifted by $\Delta h = L \cdot \sin 40° = 0.27$ m. The center of mass is lifted by $\Delta h_{cm} = s_m \sin 40 = 0.23 \sin 40° = 0.148$ m. Work for a single lift $\Delta W = 0.148 \text{ m} \cdot g \cdot 2.5 \text{ kg} + 0.27 \cdot g \cdot 6 \text{ lb} \cdot 0.45 \text{ kg/lb} = 10.8$ Nm, for 90 lifts of the weight $\Delta W_{90} = 90 \Delta W = 970$ J. d) Maximum weight held in position B: Muscle radius $R_0 = 0.03$ m. Muscle area $A = \pi 0.035^2 \text{m}^2 = 3.85 \cdot 10^{-3} \text{m}^2$. Maximum muscle force $F = A \cdot f = 3.85 \cdot 10^{-3} \text{m}^2 \cdot 2 \cdot 10^5 \text{N/m}^2 = 769$ N. e) Take the moments about the elbow joint $769 \text{ N} \cdot 0.03 \text{ m} = g \cdot (2.5 \text{ kg} \cdot 0.23 \text{ m} + m_b \cdot 0.042 \text{ m})$, and solve for the biceps mass $m_b = 5.5$ kg. e) Of the metabolic energy ΔG only the fraction $\eta \approx 0.25$ is turned into mechanical work. $\Delta W = 970 \text{ J} = \eta \Delta Q$. Hence $\Delta G = W_{90}/0.25 = 3880$ J. Then $\Delta Q = \Delta G - \Delta W = 2910$ J appears as heat. f) The muscle mass is $m_b \approx \rho \pi R^2 l_0 = 1000 \text{ kg/m}^3 \cdot \pi \cdot 0.035^2 \cdot 0.15 = 0.58$ kg. The heat ΔQ leads to an increase of temperature ΔT. Caloric equation $\Delta Q = C m_b \Delta T$. Specific heat $C = 4.18 \text{ kJ/kg}°$. Therefore $\Delta T = \Delta Q/Cm_b = 2910 \text{ J}/(4.18 \cdot 10^3 \text{ J/kg} \cdot 0.58 \text{ kg}) = 1.2°$C. g) Evaporation of mass m_s of sweat at $L_v = 2300$ J/g: $\Delta Q = m_s L_v$, or $m_s = \Delta Q/L_v = 2840 \text{ J}/(2300 \text{ J/g}) = 1.23$ g of sweat. h) The heat could be moved away in the blood by convection.

4. Fluids in the Body

Panta rei   *Everything flows*
Heraclit   540–480 BC

Supplying the Body with Nutrients and Oxygen

Big animals consist of a myriad of cells. All living cells require nutrients and oxygen to perform the functions of life. Cells extract these substances from their surrounding by diffusion. This holds for single celled organisms as well as for the specialized cells that make up the body of any animal. Therefore, big organisms had to find ways of transporting oxygen and nutrients directly to the doorstep of every cell and removing the waste products; this had been achieved through convection. Blood is the conveyor of substances of all higher animals. The convection system, the cardiovascular apparatus, consists of a pipe system with veins, arteries, a pump (the heart), a gas exchanger (the lung), blood storage reservoirs like the spleen or other tissue, and various valves to control the flow volume.

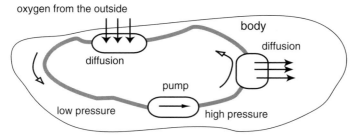

Fig. 4.1. Circulatory system with convection and diffusion components

This convection system must inundate the whole body with oxygen and food so that each cell is surrounded by a nutritious sea of body fluids. If the body would consist of just a few big cells this convection system could be quite simple. It would only need a few branches. However, if the cells are small, the arteries and veins must divide and subdivide many times to reach every single cell. Living cells tend to be small. Therefore our cardiovascular system is complicated. Cells need to be small because they feed by diffusion.

Chapter 4 begins with determining the maximum size of cells that can live by diffusion alone, Sect. 4.1. Larger organisms need pipe flow to feed the body. The cardiovascular system and the breathing apparatus are modeled in Sect. 4.3. These fluid flow systems are perfectly matched to the metabolic demands of the body. The physics principles needed to make these quantitative predictions, namely flow in pipes with friction (Hagen Poiseuille) and without friction (Bernoulli) are reviewed, and the differences between laminar and turbulent flows are described in Sect. 4.2.

4.1 Motion in Concentration Gradients

All animals, big or small, consist of cells. Cells must eat and excrete, that is they put the *good* molecules inside and push the *bad* ones out. This requires motion of materials. Cells are enclosed by walls, thickness d, which contain gates and, pumps that maintain the proper concentrations n_i of ions and neutral particles inside. Concentration gradients drive the process of diffusion and osmosis. The flow of material is large when d is small and the concentration n_i of a particular molecule inside is very much different from the concentration n_o on the outside. The quantity $(n_o - n_i)/d$ is called the concentration gradient.

Unfortunately diffusion is a relatively slow process. The father away from a surface some tissue is located, the slower the rate at which it can obtain molecules by diffusion through the surface. Only very small animals can operate with diffusion alone. Larger animals need a convection system for gases (respiration) or liquids (primary and secondary system) to transport materials through their body, but they always rely on diffusion to move molecules into individual cells.

4.1.1 Diffusion

Diffusion is a very convenient transport mechanism for atoms and molecules in gases and liquids, because it proceeds *without external energy input* as long as there are many particles in one region and only few in the neighboring region.

Diffusion is the process in which particles try to *get away from the crowd*. It is basically an atomic collision process, in which particles travel freely through the distance λ (mean free path) at some average thermal speed a until they make the

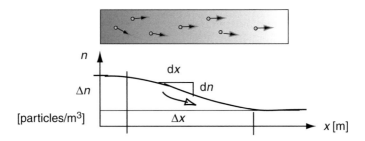

Fig. 4.2. Diffusion in concentration gradients

next collision. A quantitative measure of the particle concentration is the number density n particles/m^3, which is a function of the position x, see Fig. 4.2. Important for diffusion is the distance Δx between adjacent regions. Let region 1 have the number density n_1, and region 2 the number density is n_2. The quantity $(n_1-n_2)/\Delta x = \Delta n/\Delta x$ is called the concentration gradient. A measure for the diffusion flux is N', which is the number of particles per second that pass through a surface area A separating the regions of high and low concentration. The diffusion flux is

$$N' = AD\Delta n/\Delta x, \tag{4.1}$$

where D is the diffusion coefficient, or diffusion constant which depends on the medium, its temperature, and the size of the diffusing particles. Diffusion can also be related to the partial pressure gradient, since in a gas the number density n_i is related to the partial pressure p_i and temperature T,

$$p_i = n_i k_B T \tag{4.2}$$

where k_B is the Boltzmann constant. Here one must remember that each constituent i of a gas has a partial pressure p_i, which contributes to the total pressure, $p = \Sigma p_i$. For instance $p_{air} = p_{oxygen} + p_{nitrogen} + p_{p\,trace\,gases}$.

Pressure may be given in several different units: atmospheres, pounds per square inch, or N/m^2. These units are easily converted. $p = 1$ atm $= 1.013 \cdot 10^5$ N/$m^2 = 760$ mmHg $= 760$ torr. The partial pressure of oxygen in atmospheric air is $p_{O_2} \approx (760/5) = 152$ torr. One mole of gas at standard pressure and temperature (STP: $T = 273$ K, $p = 1$ atm) occupies a volume of 22.4 liter $= 2.24 \cdot 10^4$ cm^3. It contains $6 \cdot 10^{23}$ molecules, 20% of which are oxygen. Hence every cm^3 of air holds $6 \cdot 10^{23}/2.24 \cdot 10^4 = 2.67 \cdot 10^{19}$ molecules, which contain $n_{O_2} = (1/5) \cdot 6 \cdot 10^{23}/2.24 \cdot 10^4$ $cm^3 = 5.4 \cdot 10^{18}$ oxygen molecules per cm^3, or $5.4 \cdot 10^{24}$ oxygen molecules/m^3.

Table 4.1. Diffusion coefficients D adapted from Denny [1993]. Note that D has the same dimension as the kinematic viscosity v

substance	medium	molecular weight	diffusion coefficient D m^2/s
tobacco mosaic virus	water	40,000,000	$0.08 \cdot 10^{-10}$
human serum albumin	water	69,000	$0.61 \cdot 10^{-10}$
sucrose	water	342	$5.2 \cdot 10^{-10}$
glycine	water	75	$10 \cdot 10^{-10}$
oxygen	water	32	$18 \cdot 10^{-10}$
oxygen	air	32	$0.2 \cdot 10^{-10}$

4.1.2 Osmosis

Osmosis acts somewhat like diffusion. It is based on the fact that some molecules are attracted to molecules of a different species, as if there were nooks and crannies within the structure of on kind of molecules where other can fit into.

When different substances are separated by a semi-permeable membrane, one substance may attract the other as if there was a pump pushing one into the other. The action is the same as water being sucked up by blotting paper. The water will actually rise in the blotting paper to a certain height. The height of a water column can always be characterized by the pressure at its bottom, $p = \rho g h$. Similarly, one can define an osmotic pressure as the affinity of one substance to another. Osmotic pressures can range up to several atmospheres.

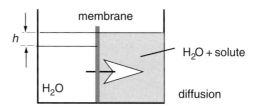

Fig. 4.3. Osmotic pressure

4.1.3 The Size of Body Cells

Diffusion moves molecules whenever there is a concentration gradient, $\Delta n/\Delta x$. Diffusion *goes by itself.* However, the metabolic demands of an organism is so high that a diffusion flux can supply living tissue only over short distances R_d. This distance is also a measure for the radius of an animal that can supply its whole body by diffusion alone.

In order to find the size limit R_d above which animals will need convection to feed the cells, we approximate the organism as a sphere of radius R, with the mass

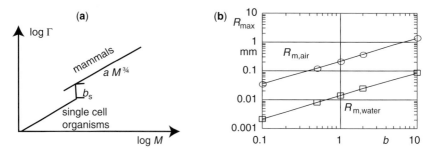

Fig. 4.4. (a) Metabolic rate Γ of single cells, which lies by a factor b_s below the mouse to elephant curve, M body mass. (b) Maximum cell size R_{max} as function of the activity factor b_s

$M = \rho(4\pi/3)R^3$, where $\rho \approx 1000$ kg/m^3 is the density of water. The number of oxygen molecules N' needed for the metabolism depends on the *consumption* of the animal, characterized by the typical metabolic rate Γ. It will be shown in Sect. 4.3.3 that mammals consume oxygen molecules at the typical rate $N'_{mam} = 2.21 \cdot 10^{18} b \Gamma_0$ O$_2$ molecules/sec, where Γ_0 is the metabolic rate described by Kleiber's famous mouse to elephant curve, and b is the activity factor.

Of interest is the specific oxygen demand of the animal $d_{mam} = N'/M$. Single cell organisms have an oxygen consumption rate $d = d_{mam}/b_s = N'/b_s M$ that is smaller than the mammal rate by some factor b_s, as indicated in Fig. 4.4a. Therefore, the specific demand of the single cell organism is $d = 2.2 \cdot 10^{18} \cdot 4\,M^{3/4}/b_s M = 8.8 \cdot 10^{18} M^{-1/4}/b_s$, where $M = 1000$ kg/m$^3 \cdot (4\pi/3)R^3$, namely

$$d = \frac{1.09 \cdot 10^{18}}{b_s \cdot R^{3/4}} \left[\frac{O_2 \text{ molecules}}{\text{kg} \cdot \text{s}}\right]. \tag{4.3}$$

This demand must be met by diffusion. Diffusion through the skin of the spherical animal (surface area $A = 4\pi R^2$) supplies the oxygen flow $N'_{diff} = AD\, dn/dr = 4\pi R^2 D(dn/dr)$. We divide this diffusion flux by the mass of the spherical organism $M = \rho(4\pi/3)R^3$ and thereby obtain the specific supply $s = N'_{diff}/M = (3/\rho R)D(dn/dx)$. The diffusion constant for oxygen in water is $D = 18 \cdot 10^{-10}$ m^2/s, and $\rho = 10^3$ kg/m^3, hence $s = (5.4 \cdot 10^{-12}/R) \cdot (dn/dx)$.

The gradient dn/dx can be approximated as $dn/dx \approx \Delta n/\Delta x$. Oxygen will reach the center of the sphere only if $\Delta x \approx R$, Fig. 4.5a. When $R > \Delta x$ the center of the sphere cannot be supplied with oxygen, Fig. 4.5b. The oxygen concentration difference Δn can be obtained by realizing that at a pressure of $p = 1$ atm there are $n_{air} = 5.4 \cdot 10^{24}$ O$_2$ particles/m^3 in the ambient air, see Sect. 4.1.1. Assuming that the oxygen concentration is very small at the center of the sphere one can set $\Delta n \approx n$. Most single cell organisms live in water and not in air.

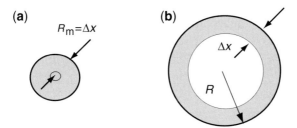

(a) $R_m = \Delta x$

(b) Δx R

Fig. 4.5. Single cell in a sea of Oxygen. (a) Organism can be supplied by diffusion. (b) The organism is too big for diffusion to supply the interior with O$_2$

The oxygen concentration n_w in water is much smaller than in air, and n_w depends on the water temperature (see Fig. 2.21). A typical value is $n \approx 0.03\, n_{air}$. Then one can give the supply of oxygen by diffusion in air and water as

$$S_{air} = \frac{3.0 \cdot 10^{13}}{R^2} \left[\frac{O_2 \text{ molecules}}{kg \cdot s} \right] \tag{4.4}$$

and

$$S_w = 0.03 \, s_{air} = \frac{9.1 \cdot 10^{11}}{R^2} \left[\frac{O_2 \text{ molecules}}{kg \cdot s} \right]. \tag{4.5}$$

The oxygen demand d of the tissue and the diffusive supply s are plotted in Fig. 4.6. One can see that for very small animals the diffusion supply is much greater than the tissue demand. However, for large radii the diffusion rate is too small. The crossover point is at the radius R_m where $s=d$, or $R_{m,air}^{5/4} = b_s \cdot 3.0 \cdot 10^{13}/1.09 \cdot 10^{18}$, and $R_{m,w}^{5/4} = b_s \, 9.1 \cdot 10^{11}/1.09 \cdot 10^{18}$. This yields the maximum cell radius for which the whole cell receives oxygen at the average metabolic rate Γ.

$$R_{m,w} = 1.38 \cdot 10^{-5} \, b_s^{4/5}, \text{ and } R_{m,air} = 2.2 \cdot 10^{-4} \, b_s^{4/5} \tag{4.6}$$

The first relation can be used to calculate the maximum radius for a mammalian cell, where the metabolic factor $b_s = 1$. The mammalian body is a watery environment, hence $R_{m,w} = 13.8$ μm. Cells that have a smaller metabolic consumption (where b_s is larger than 1) can be larger. Bacteria generally stay below this maximum size. Typically a bacterium has $R \approx 1$ μm. If a particular cell needs more oxygen, it must be smaller than R_m.

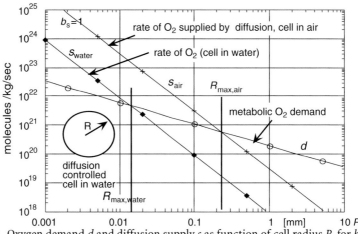

Fig. 4.6. Oxygen demand d and diffusion supply s as function of cell radius R, for $b_s = 1$

Of course animals need not be spherical. Animals of cylindrical shape or flat body design have more surface area per unit mass. The surface to volume ratio is still larger, if the organism has vented holes, such as the tracheal system of the insects.

4.2 Convection and Pipe Flow

The simplest convection system is found in animals that only have one opening which alternately functions as the mouth or the anus. The first improvement is a flow-through system with separated mouth and anus organs. Still, major parts of a larger body can not be reached by nutrients from a single duct. The next improvement is a convective transport system, such as the blood system with many branching pipe connections and one or more pumps and valves. All vertebrates employ the hemoglobin molecule as the shipping crate to transport oxygen in the primary (blood) system as well as various different molecules to move other components.

The motion of the transfer fluids involves diffusion, and convection through pipes. Important physical parameters are the mass flow (or flow rate) J, the flow velocity u, the pressure drop Δp along a pipe section, the density, and the radius R of the flow duct. The densities of air and water differ by about a factor 1000: $\rho_{water}=1000$ kg/m^3, and $\rho_{air}=1.29$ kg/m^3. Reference quantities of fluid flow are the Reynolds number Re$= uR/v$, where v is the viscosity, and the speed of sound $c_s=\sqrt{(\gamma R_g T/\mu)}$, where $R_g=8.31$ J/mol$\cdot$° is the universal gas constant, γ is the adiabatic exponent, and μ the molecular weight. For instance, for air one has $\gamma_{air}=1.4$ and $\mu_{air}=0.029$ kg/mol, hence the sound speed in air $c_{s,air}=\sqrt{(1.4\cdot 8.31\cdot T/0.029)}$, or

$$c_{s,\,air} \approx 20 \sqrt{T} \text{ m/s} . \tag{4.7}$$

In addition one may need to find the hydrostatic pressure $\Delta p = \rho\, g\, \Delta h$, and use the gas law, which connects the temperature T, the volume V, the number of moles N_m enclosed in the volume, and the pressure p:

$$p V = N_m R_g T . \tag{4.8}$$

Slow flow is laminar and easy to push through a conduit. Fast flow turns turbulent, requiring a larger pressure difference to maintain it. Therefore organisms try to keep the flow laminar. Laminar flow in pipes has Reynolds numbers below 2300. Very fast flow, where the speed u exceeds the velocity of sound c_s, is not of interest in this chapter.

Hier sitzt der Mann auf seinem Sitze	*Here rests a man with happy zeal,*
und isst zum Beispiel Hafergrütze.	*starting to eat his porridge meal.*
Der Löffel führt sie an den Mund	*The spoon convects it to his lips*
sie rinnt und rieselt durch den Schlund,	*whence oatmeal down his gullet slips.*
und wird, indem sie weiter läuft,	*As it drops down by gravity*
sichtbar im Bäuchlein aufgehäuft.	*it piles up as nutritious sea.*
So blickt man klar wie selten nur	*Thus you obtain some rare displays*
ins innere Walten der Natur.	*of nature's complicated ways.*

Wilhem Busch   (translated by the author)

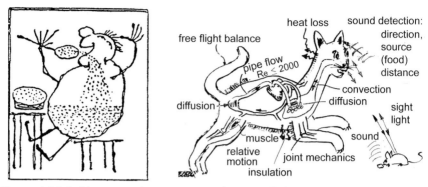

Fig. 4.7. (a) Primitive convection system according to Wilhelm Busch [1884]. (b) Advanced convection system

4.2.1 Pipe Flow and Bernoulli Equation

The motion of fluids and gases can be described with three physical principles: (i) the conservation of mass, (ii) the conservation of momentum, and (iii) the conservation of energy, summarized in Table 4.2.

A fluid flowing through a pipe of radius R always clings to the wall of the pipe and acquires the largest velocity in the middle. This *velocity profile* is shaped by viscous forces, which involve the viscosity η of the fluid. However, it is often practical to define an average velocity u_{av}. Of primary interest in a pipe system of cross section area $A = \pi R^2$ is the mass flow J kg/s. The volume flow φ m³/s traveling through the pipe per second[1] is found by dividing J by the density ρ

$$J = \rho \, u_{av} \, A \text{ kg/s, and } \varphi = u_{av} \, A_1 = J/\rho \text{ m}^3\text{/s .} \tag{4.9}$$

Arteries may become constricted by disease, and change their cross section area rapidly from some value A_1 to the constricted value $A_2 < A_1$. Thereby the flow velocity must change locally. The same mass flow J must pass through every section of a pipe. Then the flow velocities are related as $J = \rho \, u_{av,1} \, A_1 = \rho \, u_{av,2} A_2$, or

$$u_{av,2} = u_{av,1} \, A_1/A_2 . \tag{4.10}$$

This relation is the conservation of mass. It shows that the blood must flow faster through the constriction, and thus become turbulent.

The pressure in a pipe is closely related to the flow velocity, and the density. Note that the hydrostatic pressure p measured in N/m² = (kg m/s²)/m² has the same dimension as dynamic pressure $0.5 \, \rho u^2$ (kg/m³)(m/s)². These two quantities are connected by the conservation of momentum, known as the Bernoulli equation[2]:

---

[1] This equation was used already in an example dealing with heat convection.

[2] The Bernoulli equation is an approximation of Newton's equation for fluids, called the Navier Stokes equation. It can be used when viscous forces in the flow can be neglected.

$$p_1 + 0.5\,\rho\,u_1^2 = p_2 + 0.5\,\rho\,u_2^2 . \tag{4.11}$$

The Bernoulli equation connects the parameters p_1, u_1, p_2, and u_2 at two different locations in the flow. The simple message of the Bernoulli equation is that pressure must be low when the flow velocity is high. When the velocity is zero the fluid attains the stagnation pressure.

The Bernoulli equation can be used to determine the speed at which fluid escapes from a pressurized vessel. It could be the speed of water ejected by a jelly fish or the speed of blood squirting out of a blood vessel that is punctured.

Example: Suppose it is observed that the cherry size jellyfish *polyorcas* ejects water at a speed of $u_{ej} = 8.0$ cm/s. Since the water is essentially at rest inside the body cavity, $u_{in} = 0$ Eq. (4.11) reduces to $p_{in} + 0 = p_{out} + 0.5\,\rho\,u_{ej}^2$. The animal has to generate the excess pressure $\Delta p = p_{in} - p_{out} = 0.5 \cdot 10^3 \cdot 0.08^2 = 3.2$ N/m^2 in order to produce the outflow.

Animals use flow to move materials rapidly though the body. The flow is generally conducted through arteries and veins. High flow velocities require large pressure differences, which are maintained by special organs: hearts.

When fluid moves slowly through a vessel the flow is laminar. However, when fluid propagates too quickly the flow turns turbulent. Animals try to avoid turbulent flow in their arteries because the pressure drop Δp is larger and it takes more power to move turbulent flow through a pipe section.

Table 4.2. Equations of mass, momentum, and energy, pressure p, Reynolds number Re, viscosity ν

type	flow without viscous forces	flow with friction
mass flow	$J = \rho\,u_{av,1} A_1 = \rho\,u_{av,2} A_2$ [kg/s] u_{av} = average velocity	$J = \rho\,u_{av,1} A_1 = \rho\,u_{av,2} A_2$
volume flow	$\varphi = u_{av} A$ [m^3/s]	$\varphi = u_{av} A$
momentum equation	Navier Stokes (N-S) equation $\rho\,(du/dt) = \rho\,(\delta u/\delta t) + \rho\,u\,grad\,u = \Sigma F = -\,grad\,p + $ viscous force	
Approximations of the N-S equation	Bernoulli equation, incompressible flow like water, u_1, u_2 are average velocities $p_1 + 0.5\,\rho\,u_1^2 = p_2 + 0.5\,\rho\,u_2^2$	drag force acting onto sphere, radius R; Re < 100, Stokes friction $F_D = 6\,\pi\,R\,\eta\,u$ drag force on object of cross section area A, Re > 100, hydro drag $F_D = \frac{1}{2}\,A\,C_D\,\rho\,u^2$
energy	kin energy per unit volume $0.5\,\rho\,u^2$ potential energy/unit volume $\rho\,g\,y$	
laminar & turbulent in pipe	pipe: laminar flow Re $= 2Ru/\nu < 2300, \rightarrow$ radius R, length L, pressure drop Δp turbulent flow Re > 2300, $\rightarrow$	Hagen Poiseulle: $J = (\pi R^4/8\nu)(\Delta p/L)$ flow resistance $\lambda_{HP} = 8\nu L/(\pi R^4)$ Darcy-Weissbach $J = 2\,\pi\,R^{5/2}\,\sqrt{[\rho\,\Delta p/(\lambda\,L)]}$

4.2.2 Laminar and Turbulent Flow

Flow through a pipe can be slow and uniform, called laminar, or fast and erratic called turbulent. In both cases the fluid sticks to the walls; the flow velocity at the wall is zero. Laminar flow in a pipe has a velocity profile $u(r)$ that looks like a parabola, Fig. 4.8 a. The farther away from the wall the faster the fluid can move. This velocity profile is *steady* – namely it is always the same, and comes about because adjacent layers of the fluid cling to each other due to the action of the viscosity η.

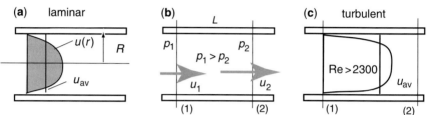

Fig. 4.8. Velocity and pressure profiles in pipe flow. (a) Laminar flow, (b) Average values, (c) turbulent flow

The adjacent radial layers of fluid exert onto each other forces in axial direction $F_z = A\eta\, du(r)/dr$, which must be overcome by the pressure differences between the inflow and the outflow end of the pipe. The forces depend on the kinematic viscosity $v = \eta/\rho$. Water has $v_w = 10^{-6}\,\mathrm{m^2/s}$, blood is more viscous it has $v_{bl} = 4 \cdot 10^{-6}\,\mathrm{m^2/s}$. Other values for viscosity are given in Table 3.4

Laminar flow in a pipe is smooth, see Fig. 4.9. However, some regions in flow close to stationary objects may have turbulent boundary layers. Turbulent flow, Fig. 4.10, is made up of countless *eddies* embedded between segments of meandering *rivulets* and stagnant fluid [Loewn et al. 1986]. Therefore the velocity is not always the same at a given place. Turbulent flow in a pipe, Fig. 4.8 c may be characterized by an average flow velocity profile that is broader than in laminar flow, Fig. 4.8a. The eddies contain a great deal of internal energy, stored in rotational motion, and pressure - volume work.

For all flow situations one can characterize the fluid state by a velocity and a length dimension combined into a convenient dimensionless number: the Reynolds number Re. One can think of this number as the ratio of inertia forces F_{inert} and viscous forces F_{vis}.

$$Re = Du/v = F_{inert}/F_{vis}\,. \tag{4.12}$$

u is a typical velocity, and D is a typical distance, measured at right angle to the flow. The typical distance could be the diameter of bacterium that is swimming in a fluid, or the diameter of an artery through which blood is streaming. Flow surrounding an object like a bird wing of span width D becomes turbulent for $Re \le 50$. Flow in a pipe of diameter D becomes turbulent if the Reynolds number exceeds the critical value $Re_{crit} = u D/v \approx 2300$.

Fig. 4.9. Obstacle at the wall of a flow field. Exposure time 1/40 s, $u = 37$ cm/s. Photo Friedrich Ahlborn [1918]

Objects, which are placed into a laminar pipe flow, may render the flow turbulent. This could for instance happen when an artery has a constriction, or a blood clot clings to the wall of a blood vessel. The turbulence might just affect a small region of the flow or it could stretch across the whole flow field. Remember that friction forces can be neglected for $\text{Re} \gg 1$, and inertia effects can be neglected for $\text{Re} \ll 1$. Some typical Reynolds numbers encountered by animals are shown in Table 4.3.

Table 4.3. Reynolds numbers of some moving objects

animal	Boeing 707	whale	dolphin	Canada goose	swallow	crane fly	sperm
Re	$3 \cdot 10^8$	$\approx 5 \cdot 10^7$	$\approx 10^6$	$\approx 5 \cdot 10^5$	$\approx 5 \cdot 10^4$	$\approx 10^2$	$\approx 10^{-4}$

Animals avoid turbulent flow in their circulatory systems, because much larger pressure differences are needed to push turbulent fluids through a pipe. Should

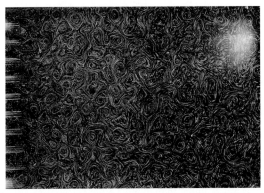

Fig. 4.10. Turbulence behind a grid. Exposure time 60 s. Diameter of the grid bars $D = 12.6$ mm, mesh size $L = 51$ mm $u \approx 15$ cm/s. $\text{Re} \approx 7{,}500$. Loewen, et al. [1986]

the blood flow become turbulent, the heart would have to use more power in order to pump the blood through the primary system. Similarly an object that "swims" through a fluid experiences a much larger drag force once the flow surrounding it becomes turbulent.

4.2.3 Pressure Drop in Blood Vessels

The Bernoulli equation shows that one needs a pressure difference to generate fluid flow. In narrow pipes, like blood vessels, the flow is affected by friction. In order to determine the pressure drop and the power that is needed to push a steady, friction dominated flow through a pipe one can use another approximation of the Navier Stokes equation, namely the Hagen Poiseuille relation, where only the viscous forces and pressure forces are retained.

Consider a fluid of viscosity v flowing through a pipe of radius R and length L driven by the pressure difference Δp. Due to the viscose forces the velocity $u(r)$ varies laterally as shown in Fig. 4.8a. However, one can always define an average velocity u_{av} to characterize the total mass flow J. The Hagen Poiseuille equation gives this mass flow for laminar $(2Ru/v < 2300)$ and steady flow as

$$J = \rho u_{av} \pi R^2 = \frac{\pi R^4}{8v} \cdot \frac{\Delta p}{L} . \tag{4.13}$$

The Hagen Poiseulle equation can be used to determine the pressure drop Δp needed to push a certain blood flow volume through an artery.

$$\Delta p = \frac{J \cdot L \cdot 8v}{\pi R^4} \tag{4.14}$$

Example: the neck of a giraffe may have the length $L = 4.0$ m. Assume that the blood flows at a velocity of $u = 0.3$ m/s, through an artery of $r = 3$ mm radius, which is an area $A = \pi \cdot (0.003 \text{ m})^2 = 2.8 \cdot 10^{-5} \text{ m}^2$. This yields a blood flow volume of $\varphi = uA = 8.4 \cdot 10^{-6} \text{ m}^3/\text{s}$, or 8.4 cm³/s, and the mass flow $J = \rho\varphi = 8.4 \cdot 10^{-3}$ kg/s. When numbers are entered for J, L and R in (4.14) one finds $\Delta p = \{8.4 \cdot 10^{-3} \text{ kg/s} \cdot 4.0 \text{ m} \cdot 8 \cdot 4 \cdot 10^{-6} \text{ m}^2/\text{s}\}/(\pi \cdot 0.003 \text{ m}^4) = 4.210^3 \text{ N/m}^2$. Such a pressure drop corresponds to about 4/100 of an atmosphere. This calculation is correct only if the giraffe stretches its neck horizontally forward. If the head is held up one must also consider the hydrostatic pressure difference. The blood in the artery of the upright held neck is like a column of fluid of the height $\Delta y = L$. For a neck length of $L = 4$ m the hydrostatic pressure difference $\Delta p = \rho g \Delta y$ between bottom and top is $\Delta p_{hs} = 10^3 \text{ kg/m}^3 \cdot 9.81 \text{ m/s}^2 \cdot 4.0 \text{ m} = 3.92 \cdot 10^{-4} \text{ N/m}^2 \approx 0.4$ atm. The blood should arrive at the head with a pressure $p_{head} = 1.0$ atm, so that the tissue does not implode. Then the heart must provide the pressure $p_{heart} = \Delta p + \Delta p_{hs} + p_{head} = 0.04 \text{ atm} + 0.4 \text{ atm} + 1.0 \text{ atm} = 1.44$ atm. This internal pressure is much higher than in most other vertebrates. For instance, the human heart pumps blood into the aorta at $p_{heart} \approx 1.1$ atm.

The Hagen Poiseulle equation is also useful to determine the metabolic costs of pumping a mass flow J of blood through blood vessels. The volume flow through the pipe is $\varphi = J/\rho$. The power P (measured in Watt) that is needed to drive the mass flow through the pipe is the product of pressure drop and flow volume

$$P = \Delta p \cdot \varphi = \frac{J^2 \cdot L \cdot 8v}{\rho \cdot \pi \cdot R^4} .$$

(4.15)

To pump blood into the head, when held horizontally, the giraffe must expend the power $P = \Delta p \varphi = 4.2 \cdot 10^3 \text{ N/m}^2 \cdot 8.4 \cdot 10^{-6} \text{ m}^3/\text{s} = 3.6 \cdot 10^{-2} \text{ W}$.

The quantity $\lambda_{HP} = 8v L/(\pi R^4)$ represents the mass flow resistance. The flow resistance is akin to the electrical resistance in an electrical circuit, and the power P is the equivalent to the Ohmic loss in an electric circuit. Part of the metabolic rate Γ is the cost of pumping the blood through all the veins and arteries of the body.

4.2.4 Flow Control in Blood Vessels

The veins and arteries in animal bodies are laid out to supply blood to all organs under all conditions. When a muscle is suddenly strained, or the brain has to solve a difficult problem, the blood flow must be increased. When the extremities loose too much body heat the blood supply must be reduced to prevent deadly hypothermia. This throtteling is accomplished by muscles that change the vessel diameter.

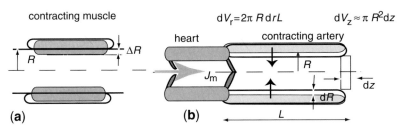

Fig. 4.11. (a) Section of blood vessel with muscle contracted to reduce the blood flow. (b) Simple heart with attached artery

A small change of radius of a pipe produces a large change in flow rate provided the pressure drop Δp along the pipe is kept constant, Fig. 4.11a. To show this effect we write The Hagen Poiseulle law for the mass flow rate in the form $J = C \cdot R^4$, where the pressure drop Δp, the length of the pipe L, and the viscosity v have all been lumped into the constant C. Take the natural logarithm, namely $\ln J = \ln C + \ln R^4 = \ln C + 4 \ln R$, and differentiate to get $\Delta J/J = 4 \Delta R/R$. For instance if the pipe radius is reduced by only $\Delta R/R = 10\%$, the mass flow will be reduced by

$\Delta J/J = 4 \cdot 10\% = 40\%$. This effect gives rise to an easy way to throttle flow in the primary system: reduce the radius of a blood vessel by constricting some muscles.

An asymmetric muscle or two muscles that contract subsequently will move fluid in a preferred direction. This is the peristaltic motion of the esophagus, or the action of muscles surrounding the blood vessels to assist the heart. In fact a section of pipe with an asymmetric muscle is a primitive heart. One case in point is the action of the aorta. Heart and aorta are show schematically in Fig. 4.11b. The aorta is elastically expanded by the discharge from the heart. Subsequently it contracts due to its own elasticity. When the radius contracts by the small amount dR the small volume $dV_r = 2\pi R\,dRL$ is lost, and blood is expelled. It occupies the volume $dV_z = \pi R^2 dz$. By conservation of mass one has $2\pi R\,dRL = \pi R^2 dz$, or $dz = (2L/R) \cdot dR$. If this contraction occurs in the time interval dt one can relate the contraction velocity $u_r = dR/dt$ to the flow velocity $u_z = dz/dt$ of the escaping fluid.

$$u_z = d_z/dt = (2L/R)u_r \tag{4.16}$$

4.2.5 Strokes

Suppose the cross section area of veins or arteries is reduced through deposits such as cholesterol. Then the flow resistance increases and the mass flow rate will go down. An increase in pump pressure by the heart will maintain the same flow volume. However, due to the conservation of mass the flow velocity must go up, see Eq. (4.9). If the velocity gets too high the flow will turn turbulent, and the pressure must again be increased to get the flow through. This likely changes the flow from laminar to turbulent, where dead water and eddies occur behind the obstacles, so that further deposits may accumulate. This is a vicious circle that often ends in a stroke or in the a rupture of the vessel. Some migraine headaches have a similar cause.

4.2.6 Why Turbulent Flow Is Bad

The Hagen-Poiseuille (4.13) relates the mass flow rate J to the pressure difference Δp. Similar as in Ohms law for electrical currents one can define the Hagen-Poiseuille pipe resistance is

$$\lambda_{HP} = \frac{8v \cdot L}{\pi \cdot R^4} . \tag{4.17}$$

Turbulent flow through a pipe is described by the Darcy Weissbach equation:

$$\frac{\Delta p}{L} = \frac{\lambda}{2R} \cdot \frac{1}{2} \cdot \rho \cdot u^2, \text{ or } u = \sqrt{\frac{4R \cdot \Delta p}{\lambda \cdot \rho \cdot L}} . \tag{4.18}$$

The pipe resistance coefficient λ depends on viscosity v, the surface roughness, and the Reynolds number, see Fig. 4.12. From the Darcy-Weisbach relation one finds the mass flow rate $J = \underline{u}\,\rho\,\pi\,R^2$ for turbulent flow

$$J = 2\pi \cdot R^{5/2} \cdot \sqrt{\frac{\rho \cdot \Delta p}{\lambda \cdot L}} \;. \tag{4.19}$$

The two different scaling relations are summarized in Table 4.4.

Table 4.4. Mass flow as function of pressure difference

laminar	turbulent
$J \propto r^4\,\Delta p$	$J \propto r^{2.5}\,\Delta p^{0.5}$

This scaling shows that turbulent flow is harder to control than laminar flow: To double the mass flow rate in laminar flow one can either double the pressure drop or increase the radius by 25%. In turbulent flow one would have to quadruple the pressure drop or increase the radius by 40%. For this reason animals maintain conditions of laminar flow in their circulation systems. Then the flow resistance of the pipe system is small, and the mechanical cost of pumping the blood is kept at a minimum.

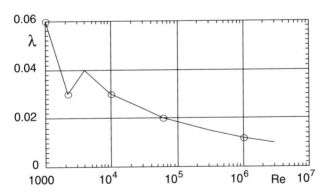

Fig. 4.12. Pipe resistance λ given in $m^{-1}s^{-1}$ for smooth pipes as function of the Reynolds number Re

4.3 The 'Highway System' of the Body

Nutrients and oxygen are carried through the body by the cardiovascular system. Higher vertebrates have two main circuits driven by a double pump, and numerous parallel- and series branches, Fig. 4.13a. One can compare this fluid system to an electrical circuit, Fig. 4.13b. Fluid flow and electrical currents have some things in common. The electrical current I corresponds to the fluid mass flow J, the voltage drop V compares to the pressure drop Δp, and the flow resistance λ is analog to the electrical resistance R_{el}.

The flow processes of heat Q', mass, J, particles n', and electrical current I fit into the general scheme of fluxes and forces first encountered in Sect. 1.2, when discussing flow processes that "go by themselves". Each flow is driven by a force or potential: ΔT for heat flux, Δp for mass flux, Δn for particle flux, and voltage drop ΔV for electrical currents I. Each flow encounters a resistance and incurs a loss of potential. The potential loss is the price for maintaining the flow. Each process increases the entropy of the universe, as a consequence of the action induced by the potential.

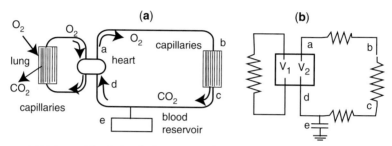

Fig. 4.13. (a) Primary blood circulation system, (b) electrical circuit equivalent

4.3.1 Pressure and Velocity in the Arteries

Blood is the transport medium of the cardiovascular system. Approximately 7% of the body mass of an animal is blood and is pumped through the body by the heart. The mass of the heart varies from about 0.8% of body mass for birds to $\approx 0.6\%$ for mammals, to $\approx 0.2\%$ for cold blooded vertebrates. The pressure in the arteries varies with the heart-beat between the systolic (high) value and the diastolic (low) value. Consequently the blood flow velocity fluctuates as well. Figure 4.14 shows the velocity as function of time. In this vessel the peak (systolic) velocity reaches about $u_{max} \approx 60$ cm/s, and the minimum (diastolic) velocity is $u_{min} \approx 30$ cm/s. The average blood velocity is approximately $u_{av} \approx 40$ cm/s.

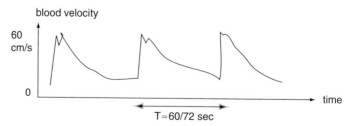

Fig. 4.14. Blood velocity in one of the arteries of the author's head measured with laser Doppler anemometry

The average pressure declines steadily from point (a), see Fig. 4.13 at the outlet from the heart to (b) before it enters the capillaries system to point (d) at the inlet to the heart. This pressure difference pushes the blood through various parts of the system. Figure 4.15 shows the pressure schematically.

The cardiovascular system is only one part of the metabolic circuit. The other part is the breathing apparatus. Both are tuned to accommodate the shipping crates for oxygen: the hemoglobin molecules in the red blood cells.

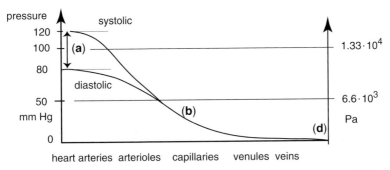

Fig. 4.15. Pressure in the blood system, 1 atm = 760 mm Hg = 760 torr = $1.013 \cdot 10^5$ N/m

4.3.2 The Hemoglobin Connection

Animals move around, reproduce and grow, copy their own building plan, perceive their environment to extract benefits or to cope with its dangers, and they communicate. Such activities require energy, which is produced at the metabolic

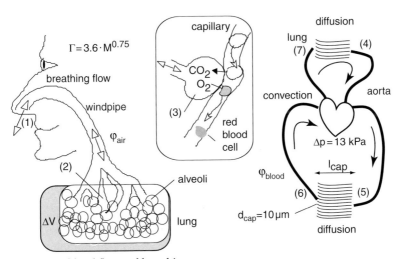

Fig. 4.16. Tissue, blood flow and breathing

rate that depends uniquely on the body mass M. Metabolic energy production arises from the reaction of fuel and oxygen.

Oxygen is convected in the body by the blood, where it is bound by reversible adsorption to hemoglobin molecules residing in the red blood cells. The blood is charged with oxygen in the alveoli of the lung at a rate determined by diffusion. The lungs are purged and refilled at a steady breathing rate f. The need for supplying every cell of the body with a steady flow of oxygen therefore connects breathing and heartbeat to the metabolic rate $\Gamma = b \cdot 3.6\, M^{3/4}$. The metabolic rate sets engineering design criteria for the number of capillaries, the volume flux of the blood, the size of the largest blood vessels, the power of the heart, the surface area of the lung, the size and number of the alveoli, and the breathing frequency. These body parameters will be derived in the following sections [Ahlborn et. al 1998]. The starting point is the specific metabolic rate Γ/M, discussed in the first chapter, $\gamma = \Gamma/M = 3.6\, b M^{-1/4}$. Remember that in all allometric relations the factor $a = 3.6$ in front of the mass M carries a dimension, and it is derived assuming that the mass is given in kg.

The metabolic power is supplied by the release of the enthalpy of reaction $h \approx 30$ kJ/g of hydrocarbon molecules. Therefore, the fuel consumption J_{fuel} of an animal of body mass M is

$$J_{fuel} = \frac{\Gamma}{\Delta h} . \tag{4.20}$$

Equation (4.20) is the basis for all allometric relations derived here.

4.3.3 How Much Oxygen Does the Body Need?

In order to maintain the metabolic rate, all animals must transport oxygen to each and every cell in their bodies, and they must remove CO_2.

In single cell organisms and in very primitive multi-cellular animals this transport occurs by diffusion alone. Diffusion is a convenient process since it does not cost energy; however, diffusion requires large concentration gradients, and it works only over small distances. No larger animal could have come into existence if diffusion was the sole mechanism of material transport. Animals could grow in size only after animals had "discovered" a highly efficient transport vehicles for oxygen and carbon dioxide: the hemoglobin molecule. These massive molecules are imbedded in red blood cells, which in turn are suspended in the blood fluid and are distributed by a central pump, the heart, to every part of the body. The red blood cells are the *tank cars* shuttling on the bloodstream highway between tissue and the lung, where they load and unload oxygen, and carbon dioxide, by diffusion.

Hemoglobin molecules have 4 sites which accept one O_2 molecule each if the external O_2 concentration is high, and which release this O_2 molecule if the external oxygen concentration is low, Fig. 4.17. For humans, the hemoglobin molecules

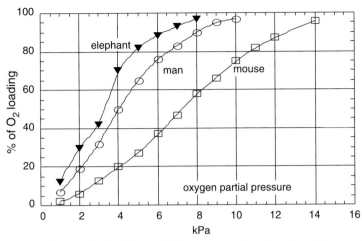

Fig. 4.17. Oxygen loading in the blood of different animals as function of the partial pressure of oxygen

charge up fully if the partial pressure of oxygen reaches about 10 kPa, and they give off the oxygen if p_{O_2} falls below 2 kPa. The hemoglobin molecule also binds and releases carbon dioxide in a similar fashion.

On the *lung-end* of the oxygen convection system, the red blood cells are rapidly charged because the oxygen partial pressure in the lung is on average $p_{O_2,\text{lung}} \approx 13$ kPa. This is well above the 90% oxygen saturation pressure of the hemoglobin molecules. On the *tissue-end* the diffusion rate should slow down as more and more oxygen is flowing out of the red blood cells. These *tank cars* have, however, a convenient property, known as the "Bohr effect": red blood cells also readily absorb CO_2, which increases the partial pressure of the O_2 component, just as if a heavier fluid had been pumped into the *tank car* so that the O_2 would float up and increase its pressure head [see for instance Schmidt Nielsen 1993]. Typically three oxygen atoms ($M_O = 16 \cdot 1.6 \cdot 10^{-27}$ kg) are needed for each carbon molecule ($M_C = 12$) with its 2 hydrogens attached ($M_H = 1$) in the hydrocarbon chain to produce one carbon dioxide molecule, and one water molecule. The sketch below shows this reaction symbolically and gives the molecular weight of the constituents.

$$3O + \overset{\text{H}}{\underset{\text{H}}{C}} = H_2O + CO_2$$

$$3 \cdot 16 + 12 + 2 = 2 + 16 + 12 + 32$$

Therefore, the mass ratio of the oxygen and fuel molecules is $M_{\text{oxyg}}/M_{\text{fuel}} = 48/14 = 3.4$, or $M_{\text{oxyg}} = 3.4 \, M_{\text{fuel}}$, and the oxygen consumption (flow rate) is $J_{\text{ox}} = 3.4 \, J_{\text{fuel}} = (3.4/\Delta h) \, I'$ kg/s. For the further calculations we set h = 30,000 J/g, and convert the mass flow rate J kg/s into a particle number flow rate N' by dividing J_{ox} by the mass of the oxygen molecule: $M_{O_2} = 32 \cdot 1.6 \cdot 10^{-27}$ kg. This yields

$N' = 2.21 \cdot 10^{18} \cdot b\, 3.6 M^{\frac{3}{4}} = 7.92 \cdot 10^{18}\, b\, M^{\frac{3}{4}}$ oxygen molecules/sec, where b is the activity factor.

The oxygen storage in the blood is so efficient that one cubic centimeter of blood can carry about as many oxygen molecules as one cubic centimeter of air at standard pressure and temperature, namely $n_{O_2} = 5 \cdot 10^{24}\, O_2$ molecules per m^3. Therefore, the volume flow of oxygenated air or blood is given as

$$\varphi = \frac{N'}{n_{O_2}} = 1.5 \cdot 10^6 \cdot b \cdot M^{\frac{3}{4}} \cdot m^3 \text{ blood/s, or } m^3 \text{ air/s.} \qquad (4.21)$$

For a person with $M = 70$ kg the volume flow for the resting body ($b = 1$) amounts to 37 cm³/s, or 2.2 liter/min. Actually the inhaled air flow is larger than φ in order to keep the oxygen concentration in the alveoli sufficiently high to enhance the diffusion rate. A typical person breathes about 6 liters/minute. This flow rate sets several allometric relations for the blood system, as we will see shortly.

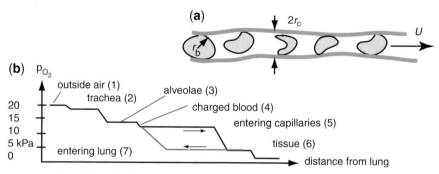

Fig. 4.18. (a) Red blood cells squeezing through a capillary, $r_c \approx r_b \approx 10\ \mu m$, length $l_c \approx 10$ mm. (b) partial pressure of O_2 molecules at various points shown in Fig. 4.16

All cells of the body must be bathed in oxygen and must get rid of carbon dioxide. This is accomplished by red blood cells drifting in a vast network of arteries, veins, and capillaries, which provide the highways and byways for the blood. The capillaries of big and small animals have about the same diameter, namely $d_c \approx 10\ \mu m$, which is equal to the diameter of the red blood cells d_{rbc}. That both diameters are equal is easy to understand. The red blood cells are then as close as possible to the walls of the capillaries. This maximizes the diffusion flux and it also helps to polish the capillary walls.

4.3.4 How Many Capillaries?

The number of capillaries can be evaluated if the diameter d_c of the capillaries, and their approximate length l_c are known. Diffusion carries the oxygen into the tissue, and diffusion only works over short distances. Therefore, the distance x from the wall of each capillary must be small. We assume that the maximum distance should be $\Delta x_{max} \leq d_c$, and that a typical capillary has the length $l_c \approx 5$ mm. Then we assign to each capillary a small volume element ΔV_c, which has the radius $r_s \approx 15 \ \mu m = 1.5 \ d_c$, and the volume $\Delta V_c = l_c \pi (1.5 \ d_c)^2 \approx 5 \cdot 10^{-3} \cdot \pi (15 \cdot 10^{-6})^2 = 3.5 \cdot 10^{-12} \ m^3$. The whole body of volume $V_{body} = M/\rho$ must be divided into such volume elements. Their number N_c is

$$N_c = V_{body}/\Delta V_c = M/\rho \ l_c \pi (1.5 \ d_c)^2 = 2.83 \cdot 10^8 M . \tag{4.22}$$

For simplicity we approximate the density of blood as $\rho \approx 1000 \ kg/m^3$, and set $b = 1$. Typically a person with $M = 70$ kg then has $N_c \approx 2 \cdot 10^{10}$ capillaries. The average flow velocity in these capillaries can now be calculated. The total fluid volume flux φ must be split up into N_c parallel channels, yielding a flux $\varphi_c = \varphi/N_c = 1.5 \cdot 10^{-6} \ M^{3/4}/2.83 \cdot 10^8 \ M = 6.3 \cdot 10^{-15} \ M^{-1/4}$ per channel. This volume flux must pass through the area $A_c = \pi (0.5 \ d_c)^2$. Therefore, the blood in the capillaries has the average velocity $u_c = \varphi_c A_c = 6.3 \cdot 10^{-15} \ M^{-1/4}/\pi (5 \cdot 10^{-6})^2 = 8 \cdot 10^{-5} M^{-1/4}$ m/s. This yields $u_c = 28 \ \mu m/s$ for a person of $M = 70$ kg. The blood is slowly creeping through the capillaries allowing for the complete unloading of oxygen from the red blood cells.

4.3.5 Laminar Flow in the Aorta

Any pipe system with the volume flux $\varphi m^3/s$, with the average flow velocity u m/s and diameter d m can be characterized by its Reynolds number $Re = d \cdot u/v$. The kinematic viscosity of blood is typically $v_{blood} = 4 \cdot 10^{-6} \ m^2/s$. For a given flow rate $\varphi = \pi (d/2)^2 u$, the radius $r = d/2$ can be given as function of Re and φ:

$$r = \frac{2\varphi}{\pi \cdot Re \cdot v} , \text{ or conversely } \ Re = \frac{2\varphi}{\pi \cdot v \cdot r} . \tag{4.23}$$

Thus the flow velocity u can be given as function of r, v, and φ, namely

$$u = \frac{\pi}{4} \cdot \frac{Re^2 \cdot v^2}{\varphi} . \tag{4.24}$$

Equation (4.23) shows that for the same flow rate a narrow pipe has a higher Reynolds number than a pipe with larger radius. Equations (4.23) and (4.24) can be used to characterize the flow in the aorta through which all the entire flow volume φ must pass. If the flow were to remain laminar at a Reynolds number of say

$Re \leq 2000$, the radius of the pipe could not be smaller than $r_{min} \approx \varphi / 1000 \, \pi \cdot v$. With this assumption the aorta would have radius, area, and flow velocity

$$r_{aorta} \geq 1.2 \cdot 10^{-4} M^{3/4} \text{ m}, A_{aorta} = 4.5 \cdot 10^{-8} M^{3/2}, \text{ and } u_{aorta} \leq 31.6 \, M^{-3/4} \text{ m/s}. \quad (4.25)$$

Radius and velocity are lower limits because it is known [McKay 2002] that some big vessels actually contain turbulent flow. Further, the calculations were based on an activity factor $b = 1$, but the ventilation system must accommodate metabolic rates that might exceed the resting state by a factor $b \approx 10$. It must also be acknowledged that blood is a fluid in which the viscous forces are not exactly proportional to the velocity gradient. Such liquids are called "non-Newtonian" fluids. The particles tend to bunch together in the middle of a channel, flowing at a speeds that does not change with radius, while large velocity gradients are generated near the wall. The Reynolds number based scaling calculated here may be considered a lower limit [e.g. see Clark 1927] for experimental results quoted by Schmidt-Nielsen as $A_{aorta} \approx 9.4 \cdot 10^{-6} M^{0.82}$. Obviously there are some other biological demands that influence the aorta size.

4.3.6 The Power and Frequency of the Heart

A blood system is a necessity for all higher animals. What is the energetic cost, namely the power of the heart driving this convection system? Power is the product of volume flux φ m³/s, and pressure drop Δp N/m².

$$P = \varphi \cdot \Delta p \quad (4.26)$$

The flow volume φ is fixed by the metabolic rate. The average pressure head generated by the heart is approximately constant for all mammals [Schmidt-Nielsen 1993], namely $\Delta p_{heart} \approx 13$ kPa [Tenny and Remmer 1963]. Therefore, the required power of the heart is

$$P = 13 \text{ kPa} \cdot 1.5 \cdot 10^{-6} b \, M^{3/4} \text{ m}^3/\text{s} = 1.95 \cdot 10^{-2} b \, M^{3/4} \text{ W}. \quad (4.27)$$

Typically for a man the power of the heart is $P_{man} = 0.5$ W. This is clearly only a small fraction of the total metabolic rate at rest $\Gamma = 3.6 \cdot 70^{0.75} = 87$ W. It is a small price to pay for the benefit of keeping the body supplied with oxygen and nutrients.

The relation $P = \varphi \, \Delta p$ can also be used to derive the allometric relation for the heart frequency. The heart can be thought of as a pump with a displacement ΔV, and a frequency f. The volume flux φ must be equal to $f \Delta V$. Therefore $P = f \Delta V \Delta p$. As stated above the pressure head of the pump is independent of body mass; however, the mass of the heart and hence its volume, and by inference the displacement ΔV, scale as the body mass M. Hence $P = \text{const} \, f_{heart} M$. This result can be combined with Eq. (4.27) which scales as $P = \text{const} \, M^{3/4}$ to yield

$$fM \text{ const} = \text{const } M^{3/4}, \text{ or } f = \text{const } M^{-1/4}. \tag{4.28}$$

This relation shows that large animals have a slower heart beat than small ones. Equation (4.28) can be used to estimate the pulse rate of a mouse knowing that a person typically has 72 beats per second $f_{man} \approx 72/60 \text{ s} = 1.2 \text{ Hz}$. We write the Eq. (4.28) for mouse and man and take the ratio $f_{mouse}/f_{man} = (M_{mouse}/M_{man})^{-1/4}$ then $f_{mouse} = f_{man} \cdot (M_{man}/M_{mouse})^{1/4} = 1.2 \cdot 7.7 = 9.2 \text{ Hz}$.

The mechanical energy per stroke cycle of the heart can also be derived from a pressure-volume diagram, Fig. 4.19. In an example given by Gordon [2001] the volume varies between 35 cm³ and 100 cm³, while the pressure varies between 3 and 120 torr. The cycle is executed in counterclockwise direction, requiring an energy input of 0.9 J.

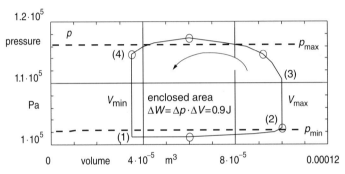

Fig. 4.19. p–V diagram of a heart according to Gordon [2001]. Stations of the cycle: (1) Inlet valve opens, (1) $\rightarrow$ (2) volume expands admitting blood. (2) Inlet valve closes. (3) Outlet valve opens, (3) $\rightarrow$ (4) volume is reduced, and blood flows into the aorta. (4) Outlet valve closed

The work ΔW done by the heart is the area enclosed by the curve, namely the integral of the pressure as function of the volume $\Delta W = \Delta p \, dV$. This quantity can be estimated from the area of the rectangle defined by the average maximum and minimum values of pressure and volume: $\Delta W \approx (p_{max} - p_{min}) \cdot (V_{max} - V_{min}) = (1.16 \cdot 10^5 - 1.02 \cdot 10^5) \text{ N/m}^2 \cdot (1.0 \cdot 10^{-4} - 3.5 \cdot 10^{-5}) \text{ m}^3 = 0.91 \text{ J}$. If this heart beats 72 times per minute it will consume the power $P = 0.91 \text{ J} \cdot (72/60 \text{ s}) = 1.09 \text{ W}$. This is the same order of magnitude as calculated under (4.27) with the metabolic power and a constant pressure drop Δp_{heart}.

4.3.7 The Ventilation System

The main function of the breathing system is to permit the diffusion of oxygen into the blood stream. The ventilation system consists of the windpipe, the bronchial tree and the alveoli which are mounted like the leaves of a tree, at the very ends of the finest branches. The alveoli are tiny spherical shells with very thin surfaces which separate the blood and air, but permit the passage of oxygen and CO_2 by diffusion. The total surface area of all alveoli together is quite large. The flow of oxygen particles N'_{diff} is given by the diffusion equation, discussed in Sect. 4.1.1

$$N'_{diff} = A \cdot D \cdot \frac{dn}{dx} = A \cdot D \cdot \frac{\Delta n}{\Delta x} \text{ molecules/s}. \qquad (4.29)$$

The oxygen must diffuse over a length of typically $\Delta x \approx 10$ μm. Diffusion constants have typical values of $D \approx 10^{-6}$ cm²/s for oxygen molecules in an aqueous environment, and the number density difference is of the order of $\Delta n \approx 10^{18}$ cm⁻³. A certain area of lung surface $A = A_{lung}$ is needed in order to pass the particle density flux $N' = 7.9 \cdot 10^{18} b M^{3/4}$ (O_2 particles/s) that is required for the metabolic processes. It can be calculated by solving (4.29) for the area, namely:

$$A_{lung} = \frac{\Delta x}{D \cdot \Delta n} \cdot 7.9 \cdot 10^{18} \cdot b \cdot M \quad . \qquad (4.30)$$

For a person with $M = 80$ kg this area amounts to $A \approx 10$ m². Actually our lungs are generously designed[3] with a typical area of 80 m². This surface area of the lung is about 40 times larger than the total skin area of a human body, and it is squeezed into the lung volume which only occupies about 6% of the body Volume $V = M/\rho$. Measurements of the lung volume, also called the vital capacity, have established [Tenny and Remmer 1963] that it scales as

$$V_l = 5.7 \cdot 10^{-5} M^{1.03} \approx 0.057 \, V \text{m}^3. \qquad (4.31)$$

How can such a large surface area be squeezed into such a small fraction of the body? The trick is to divide the surface into a large number N_{av} of small sub units, the alveoli, or air sacks, which are approximately spherical cavities.

Neglecting the space between the spheres, one has $V_l = N_{av} \cdot V_{av} = N_{av} (4\pi/3)$ $(d_{av}/2)^3$. The diameter d_{av} of these cavities is $d_{av} = 2(3V_l/4\pi N_{av})^{1/3}$. All of them together have a total surface area

$$A_l = N_{av} \pi d_{av}^2 = 4.84 \, N_{av}^{1/3} \, V_l^{2/3}. \qquad (4.32)$$

The area increases with the number of sub units N raised to the power 1/3. Typically for man with $M = 80$ kg the surface area is 80 m². It is known from measurements that the human alveoli have a diameter of about 200 μm [Astrand and Rohdahl 1970]. Their surface area is therefore $A_{av} = 4\pi (10^{-4} \text{m})^2 = 1.3 \cdot 10^{-7}$ m², yielding the number of alveoli $N_{av} = A_l/A_{av} = 80$ m²$/3 \cdot 10^{-7}$ m² $\approx 7 \cdot 10^8$. This prediction is slightly larger than the measured value $N_{av} \approx 2 \cdot 10^8$, probably because a significant fraction of the lung volume is occupied by the bronchial tree.

---

[3] The additional capacity may be needed when the body is working hard.

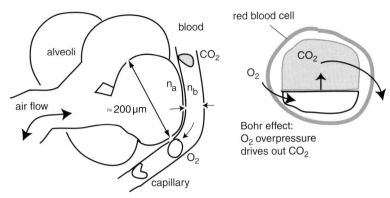

Fig. 4.20. Allveoli, the air sacks of the lung, where O_2 and CO_2 are exchanged

The allometric scaling of the lung area with body mass is obtained by substituting (4.31) into Eq. (4.32)

$$A_1 \propto N^{0.33} M^{0.68}. \tag{4.33}$$

In fact the total number of alveoli is not a constant, but it will vary from one individuum to the other according to the average oxygen demand. On average, soccer stars have more alveoli than physics professors.

4.3.8 Breathing

Air is inhaled in gulps (tidal volume) with each heaving of the lung. One can easily determine this volume since the tissue compressibility of the lung [McKay 2002], namely the ratio of the periodically expelled volume ΔV_1 and the total lung volume V_1 is constant: $\Delta V_1 / V_1 = 13.6\,\%$ [Schmidt-Nielsen 1993, p. 102]. Therefore the tidal volume is

$$\Delta V_1 = 0.136\,V_1 \approx 7.7 \cdot 10^{-6}\,M\ \mathrm{m}^3. \tag{4.34}$$

In order to generate the volume flow $\varphi = 1.5 \cdot 10^{-6} M^{0.75}$, the tidal volume ΔV_1 must be supplied at the frequency f_1 so that $\varphi = f_1 V_1$, or

$$f_1 = \frac{\varphi}{\Delta V_1} = \frac{1.5 \cdot 10^{-6} \cdot M^{3/4}}{7.7 \cdot 10^{-6} \cdot M^{1.04}} = 0.19 \cdot M^{-0.29}. \tag{4.35}$$

The ventilation system is generously appointed since the breathing apparatus must have room for suddenly increased demand, and since it is also used to activate the voice.

While the heart rhythm is regulated on a subconscious level, and cannot be changed at will, the breathing frequency can be varied arbitrarily. If these additional demands were not imposed, the breathing frequency $f = 1/T$ could be optimized for the diffusion physics at the lung surface A_1. Diffusion depletes the oxygen number density $n_a = N_a/V_1$ in the alveoli, and increases the oxygen number density in the blood $n_b = N_b/V_b$. The blood in the capillaries of the lung (with the total volume V_b) acquires oxygen from the air and carries away an amount N'_{conv} to power the metabolic body functions. In the absence of diffusion, the oxygen number density N_a would fluctuate in the lung, somewhat like shown in Fig. 4.21.

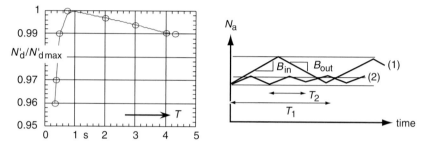

Fig. 4.21. (a) Normalized diffusion flux, as function of time. (b) Number of oxygen molecules in the lung, (1) slow breathing, (2) faster breathing $B_{in} = $ constant

To model this process we approximate the gradual variation of volume by a saw tooth function, and describe the waxing and waning of oxygen numbers in the lung by the in-breathing rate $B_{in}(t)$, and the removal rate $B_{out}(t)$, both measured in number of O_2 molecules/s. The coupling of these processes can be described by two differential equations

$$N'_a(t) = B_{in}(t) - B_{out}(t) - N'_{diff},$$ (4.36)

$$N'_b(t) = N'_{diff} - N'_{conv}.$$ (4.37)

The differential Eqs. (4.36) and (4.37) are coupled by the diffusion term derived from Eq. (4.30)

$$N'_{diff}(t) = A \cdot D \cdot \frac{n_a - n_b}{\Delta x}.$$ (4.38)

This system of coupled differential equations has been analyzed under the following conditions: $N'_{conv} = N' = 6 \cdot 10^{19}$ O_2/s; $A = 70$ m², $D = 18 \cdot 10^{-6}$ 18 cm²/s, $\Delta x = 10$ μm. The rates B_{in} and B_{out} where set at the maximum value for which the volume flow rate of the air would remain laminar in the windpipe. The period $T = 1/f$ was left as a variable parameter in order to see if the net diffusion rate N'_{diff} for oxygen would depend significantly on the breathing frequency.

Figure 4.21 gives the following result: the diffusion flux N'_{diff} has a shallow maximum at $T_{max} \approx 0.75$ s. For shorter periods the diffusion flux drops off rapidly, because the oxygen does not fill into the lung quickly enough so that n_a stays too small. For larger periods N'_{diff} drops off because the oxygen concentration in the alveoli is depleted by diffusion. However, the decrease is not very drastic and there is no significant penalty for breathing slowly. This has presented an opportunity for the evolution.

The breathing apparatus, which was initially designed as a gas exchange system was eventually turned to other uses. (i) The air in the lung could also absorb water vapor to assist in temperature control of the body. (ii) The air could also be squeezed through vocal cords to produce sounds. Our singing voice and our social interactions through speech could not have evolved without spare capacity in the lungs.

When interpreting these results, one must keep in mind that most organs of the body have multiple applications and that certain functions of the body are not optimized for a single use. For instance the blood system not only transfers oxygen but also conveys fuel (like fat), and moves heat to the extremities of the body.

4.3.9 Blood Circulation Time

The blood is pumped through the body by the heart at a mass flow rate J_h. The total blood mass m_b is pumped through the heart within the circulation time T given by $T = m_b / J$.

Taking typical values $m_b \approx 5.0$ liter=5 kg and $J \approx 0.06$ kg/s one finds the blood circulation time $T = 5$ kg/0.06 kg/s ≈ 83 s. Most of this time is spent while the blood "oozes" through the capillaries. At the capillary velocity $u_c = 28$ μm/s and a capillary length of $l \approx 2.0$ mm the capillaries are traversed in the time $\Delta t = 2.0 \cdot 10^{-3}$ m/$2.8 \cdot 10^{-5}$ m/s = 71 s.

4.4 From Digestion to Propagation

Fluid motion is needed to feed the body. Diffusion is a natural way to acquire external resources. It works as long as the living organism keeps its molecular pumps going to maintain concentration gradients. However, concentration gradients do not have a large reach. In order to feed large bodies, convection systems with pumps, valves, and pressure regulation had to be invented.

When animals get larger, they rapidly deplete the local resources and then must move to another resource rich area. They have to be able to walk, swim or fly. Mobility is achieved with larger body sizes. A whole new set of physical phenomena appears, when animals become large and venture into the arena of long range locomotion.

It is only a small step from oozing some body fluids into the surrounding to achieve jet propulsion. It is only a small step from having a wiggly tail to at achieve

small-scale random motion. However it is a larger achievement to swim aerody-
namically with flukes and flippers. It is yet another step to walk on land, and raise
the body up to trot and run. But the crowning achievement of propagation is to
take to the wing and fly through the air. As we will see in a later section this can
only be done by animals in a limited size range.

Table 4.5. Frequently used variables of Chap. 4

variable	name	units	name of units
c_s	sound velocity	m/s	meter/second
D	diffusion constant	m/s	meter/second
J	mass flow rate	kg/s	kilogram/second
l_c	length of capillary	m	meter
n	number of molecules	m^{-3}	number/cubic meter
Re	Reynolds number	1	
u	velocity	m/s	meter/second
N'	number of atoms per s	particles/s	meter/second
γ	adiabatic exponent	1	
φ	volume flow rate	m^3/s	cubic meter/second
λ_{HP}	flow resistance	1/ms	1/meter & second

Problems and Hints for Solutions

P 4.1 Why Breathing is Complicated for Birds

A certain diving bird winters near Vancouver and nests in the Arctic tundra to the
east of the Rocky Mountains. At ground level the pressure is $p_0 = 1$ atm, which corresponds to 10^5 N/m^2. The bird routinely dives to a depth of 15 m, where the pressure is much higher than at the surface of the water. On its migration, the bird
crosses the Rocky Mountains at an altitude of $y = 4000$ m. Determine the oxygen
concentration n_{O_2} (number of O_2 molecules/cm^3) in the air in the alveoli (lungs) of
the bird a) on the ground, b) if the bird is swimming at a depth of $d = 15$ m under
water, and c) at the highest point of the migratory flight.

P 4.2 Breathing at High and Low Pressures

How do animals cope with high altitudes where the oxygen concentration is very
low? Describe the differences in the lungs of mammals and birds. How do sperm
whales deal with the large change in external pressure when diving to large
depths?

P 4.3 Balloon Ghost Animals

Suppose a balloon-like simple animal has learned to generate H$_2$ gas, so that it can
float in the air. Assume such an animal weighed 3 kg, and 2/3 of the body mass was

needed to build the balloon for holding the gas. Calculate the radius R of the balloon, assumed to be spherical, and the skin thickness d. Discuss the problems which such an animal might encounter.

P 4.4 Diffusion and Metabolic Rates of a Nudibranch

A certain nudibranch, $M = 0.005$ kg, living in the water carries the gills (mass m) outside its body. Assume that (i) the gills are 10% of the total body mass, (ii) the gills are tree-like structures that have branches with an average diameter of $D = 500$ μm, filled with fine capillary vessels of $d = 10$ μm diameter right under the skin of the gills. a) Determine the surface area of the gills. b) Calculate the flow rate of oxygen N'_{O_2} into the gills. c) Use the relation between N', and Γ, which is derived in Sect. 3.3.2 to determine the metabolic rate of the nudibranch. e) Determine the metabolic constant a assuming that for these animals one can write $\Gamma = a\,M^{3/4}$.

P 4.5 The Primary System

Assume that the flow velocity in the arteries of any animal is safely below the onset of turbulence. Take a Reynolds number of $\mathrm{Re} = 2000$, so that the following scaling relations for the primary system apply: $u_a = 8 \cdot 10^{-3}/d_a$ m/s, $J = 2\pi d_a$ kg/s, $\Delta p_a = 10^{-3}\,L/d_a^3$ N/m², $N = (d_a/d_c)^4 \cdot (\Delta p_a/\Delta p_c) \cdot (L_c/L_a)$, where $u_a =$ average velocity in artery, $d_a =$ diameter of artery, J kg/s is the blood flow rate, Δp_a is the total pressure drop across the heart, Δp_c is the pressure drop across the capillaries, d_c is the diameter of capillaries, and N is the number of capillaries. With this scaling calculate u_a, J, Δp_a, for:

a)   a big warm blooded animal (horse, giraffe...),
b)   a cold blooded animal (crocodile, lizard...),
c)   a small animal (rat, bird...),
d)   a grade one student and a well trained adult athlete.

Comment on the numbers which you get. Do they make sense?

P 4.6 Careless Operation

In an open chest operation the aorta of a patient has been accidentally punctured by a round hole of 2.0 mm diameter. a) What is the average pressure inside the artery close to the heart? b) Why does blood escape out of the hole? c) What is the average velocity of the blood flowing out of the hole? d) How much blood will escape within the first 5 seconds? e) Could you use the same calculation method to accurately predict the blood loss in the first minute? (e) If a bypass of $L = 0.30$ m length and radius $R = 0.002$ m was inserted across the heart, and the pressure in the heart did not change during this procedure, how much blood would flow through the bypass in 5 seconds?

P 4.7 The Aorta, Your Second Heart

The aorta of a human cadaver has an inner radius of $R_0 = 9.5$ mm, and a wall thickness of $\Delta R_0 = 2.8$ mm. It is $L = 0.5$ m long. In the living body the aorta is expanded due to the pumping of the heart. In the diastolic state (d) the inner radius of the aorta is found to be $R_d = 13$ mm. When the heart valve opens in the systolic phase (s) the blood volume $\Delta V = 70$ cm^3 is ejected. Half of this blood is initially stored in the aorta expanding its wall to the inner radius R_s. This temporarily stored blood maintains the flow after the heart valve has closed. Assume a diastolic pressure and systolic pressure of $p_d = 80$ mmHg and $p_s = 120$ mmHg respectively, and a heart rate of 72 beats per minute.

a) How large is the stress σ in the wall of the aorta in the states (s) and (d)? Compare the stress with the muscle stress $f_o \approx 2 \cdot 10^5$ N/m^2 of the unloaded muscle. b) How much energy is stored in the elastic wall of the aorta? c) Using the information from either a) or b) determine Young's modulus of elasticity Y of the aorta wall. d) What is the average Reynolds number in the aorta? Is the flow laminar or turbulent? e) From the average discharge rate and the average aorta radius determine the pressure drop over the length of the aorta. f) Suppose a stenosis blocks 70% of the aorta cross section area but the heart produced the same volume flow rate, $\varphi = 70 \cdot 10^{-3}$ m^3/stroke. What would be the flow velocity and the Reynolds number at the constriction? g) Comment on the pressure and work generated by the heart with the blockage in the aorta.

P 4.8 The Neck of the Giraffe

Giraffes with their long necks (length L) are able to browse on tall trees, which are out of the reach of most other animals. This definite advantage comes at a price. The long neck makes breathing difficult. The metabolism $\Gamma = b \cdot 3.6\,M^{\frac{3}{4}}$ requires a certain intake of fresh air. The windpipe volume $V_w = \pi R^2 L$ must be filled first when exhaling, before any air volume V_{ex} can be exchanged and reloaded with oxygen. The flow should not be turbulent in the windpipe. The total volume ΔV displaced during the exhaling process $\Delta V = V_w + V_{ex} = C V_{lung}$ is a certain fraction C of the lung volume.

Assume an allometric relation for the breathing frequency $f = 0.9\,M^{-\frac{1}{4}} s^{-1}$, and the lung volume scaling $V_{lung} = 5.7 \cdot 10^{-5} \cdot M$, and calculate the maximum windpipe length L as function of the activity factor b.

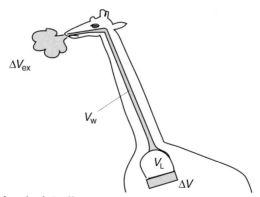

Fig. 4.22. Lung and neck of giraffe

Hints and Sample Solutions

S 4.1 Breathing for Birds

The pressure drops approximately by a factor 2 for every increase in height of 5500 m. In mathematical terms $p = p_0 e^{-Bh}$, where $B = mg/k_B T$, m = 0.029 kg/mol is the average molecular weight of the air molecules, $k_B = 1.38 \cdot 10^{-23}$ J/mol · K is Boltzmann's constant and T the temperature of the atmosphere. The formula holds only if the temperature T is the same at every height. You may take a value $B = 1.26 \cdot 10^{-4}$ m^{-1} corresponding to $T = 273$ K. Remember that air consists of 1/5 oxygen and 4/5 nitrogen, and 1 cm^3 of air at $T = 273$ K contains $n = 2.7 \cdot 10^{19}$ molecules.

S 4.2 Counter Flow Systems

Mammals breathe in-and out. Birds have a counter flow system, which transfers oxygen more efficiently, similar as counter flow heat exchangers transfer heat more efficiently. Sperm whales store oxygen in their muscles, the lungs collapse, since the water pressure is 100 atm at a depth of 1 km.

S 4.6 Careless Operation

a) With the help of Fig. 4.15 determine the pressure of the blood in the aorta as $p_0 \approx 10^5 + 1.3 \cdot 10^4$ N/m^2 = 1.13 · 10^5 N/m^2. b) The blood squirts out because of the overpressure $\Delta p = 1.3 \cdot 10^4$ N/m^2. The outside pressure is $p \approx 10^5$ N/m^2. c) The blood escapes in radial direction at some velocity u_r. Within the vessel there is no radial velocity, hence $u_0 \approx 0$. The density of the blood $\rho \approx 1000$ kg/m^3 is approximately like the density of water. Then by Bernoulli's equation one has $p_0 + 0.5 \rho u_0^2 = p_0 = p + 0.5 \rho u_r^2$, $u_r^2 = 2 (p_0 - p)/\rho = 2 \cdot 1.3 \cdot 10^4/ 10^3 = 26$ m^2, and $u_r = \sqrt{26} = 5.1$ m/s. d) The mass flow rate through the hole of 2 mm diameter, area $\pi (10^{-3})^2$ is $J = \rho A u_r = 0.016$ kg/s equivalent to 16 ml of blood. In the first 5 seconds the patient would loose $\Delta V = 5$ s · 16 ml/s = 80 ml of blood. e) At this mass loss rate the patient would loose $\Delta V = 60 \cdot 0.016$ l = 0.96 l of blood in one minute. Such a loss of blood would no doubt lead to a collapse of the circulation system and the heart would not be able to maintain the excess pressure Δp. f) The flow in the bypass can be described by Hagen Poiseulle

$$J = \frac{\pi \cdot r^4}{8 \cdot v} \cdot \frac{\Delta p}{L} = \frac{\pi \cdot (5 \cdot 10^{-3})^4 \cdot 1.3 \cdot 10^4}{8 \cdot 4 \cdot 10^{-6} \cdot 0.3} = 2.66 \text{ kg/s} .$$

In $\Delta t = 5$ s the amount $\Delta m = J \cdot \Delta t = 2.66$ kg/s · 5 s = 13.29 kg = 13.3 liters of blood would flow through the bypass.

S 4.7 Aorta

First find the aorta volume, length of the circumference C_o, C_d, and C_s and thickness as the wall is stretched. Next determine the tensile stress σ (force/area) in the wall for the states (s) and (d) in a similar way as in Fig. 3.24, where the tensile forces in a caterpillar skin are discussed. Young's modulus Y and the modulus of elongation k are discussed in Sect. 3.3.3. $T=\sigma A$, is defined as the tension force, where A is the cross section area of elastic material. For calculating the elastic energy stored in the wall $W=\int T dl$ assume a linear relation between the tensile and extension $\Delta L=(L_{sys}-L_{dias})$ between the diastolic and the systolic state $T=T_d+k\Delta L$ so that $W_{ds}=\frac{1}{2}k\Delta L^2$. The discharged volume flow is φ m³/s $= u_{av}A$, where u_{av} is the average velocity, and A is the pipe cross section area.

5. Animals in Motion

Im Anfang war die Kraft. In the beginning was the Force.
                    Goethe

Moving the Limbs and the Body

Motion, the displacement $\Delta s = s_2 - s_1$ of an object from a starting point s_1 to an end point s_2, is caused by forces in a roundabout way. Muscle forces do not make displacements directly, they generate accelerations. Accelerations change the velocity of an object, for instance, from rest ($u = 0$) to some velocity u_1. When an animal maintains a velocity for some time interval Δt it moves through the displacement $\Delta s = u \cdot \Delta t$.

Motion of the whole body of an animal often involves the relative motion of many members and joints in the articulated appendages: legs, arms, and wings. Many muscles take part. Consider the walking process. The toe moves relative to the foot. The foot swings around the ankle, the ankle rotates about the knee, the knee swings relative to the hip, and the hip pushes the center of mass of the whole body. The kinematics of relative motion combining rotation and translation is discussed in Sect. 5.1. Kinematics reveals time and position of objects. For instance how long does it take for a cat to jump up onto a bird feeder that is a certain height above the ground? Does the bird have sufficient time to fly off?

Kinematics does not explain why motions occur. The cause of a motion is revealed through the dynamics, encapsulated in Newton's equation, which links forces and accelerations, Sect. 5.2. Examples are: Landing on your feet from a high jump, the terminal velocity of a small falling object, and rocket propulsion of a squid.

Energy must be expended for any motion. Often one can predict parameters of motion using only the principle of conservation of energy. Animals know how to move with the least possible effort. They have had millions of years to optimize energy consumption on internal and external levels. For instance, they learned to minimize the power required for the limb motion by choosing suitable geometry, by developing light and strong materials, and by operating periodic processes at certain resonances. Such energy considerations are presented in Sect. 5.3.

5.1 Kinematics of the Motion

Kinematics describes the position and time of objects. Kinematics keeps track of an entire object or its many parts: the whole animal or it's individual limbs. Kinematics deals with translational motion, and rotational motion. Translation is analyzed in terms of position s, velocity $u = ds/dt$, and acceleration $a = du/dt$. Rotation is described by the equivalent quantities: Angle ø, angular velocity ω, and angular acceleration α.

A moving object may consist of several parts, like a leg with upper leg, lower leg, foot and toes. Then one must look at the velocities of every part, and distinguish absolute velocity measured in the laboratory reference frame from the relative velocity of its parts. Thus, different symbols are required to denote velocities: the letters u and v, in the upper and lower cases are occasionally used to indicate velocities. In particular, where the symbol for velocity could be confused with the kinematic viscosity, Greek letter v, or the volume, V the letters u, and U are used, respectively, to denote velocities.

5.1.1 Translational and Rotational Motion

The two forms of motion important for animals are translational motion (like the motion of the center of gravity of an animal), and rotational motion (like the swinging of a leg). Most animals have flexible bodies and articulated arms, and legs connected by joints to the body. In order to calculate the speed of a foot one must considers first the motion of the center of mass of the animal, second, the motion of the leg about the hip joint, third, the motion of the lower leg about the knee joint, and finally the motion of the foot relative to the lower leg. This is relative motion.

When animals move in a certain direction while swimming, flying, or running, their center of mass travels along a straight line. This is called translational motion. The average propagation speed $u_{av} = \Delta s / \Delta t$ is related to the distance Δs covered during the travel time Δt. When an animal experiences an average acceleration a_{av}, it must change its speed by the amount $\Delta u = a_{av} \Delta t$. Often one deals with the instantaneous values $u = ds/dt$ and $a = du/dt$.

All organisms with joints also perform angular motion, characterized by the angular position ø (measured in radian), the angular velocity $\omega = d\text{ø}/dt$ (measured in radian/sec), and the angular acceleration $\alpha = d\omega/dt$ (measured in radian/sec^2).

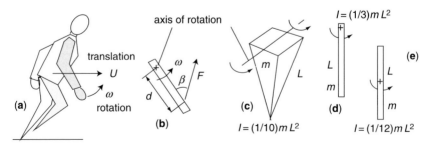

Fig. 5.1. (a) Translation and rotation. (b) Rotation parameter. (c) Rotating cone. (d) Rod hinged at one end. (e) Rod hinged in middle

The angular velocity ω can be estimated if one knows how long it takes to move through a given arc $\emptyset$. Remember that the 360° motion of the second hand of a clock performed in 1 minute corresponds to the angle $\emptyset = 2\pi$ when measured in radians. The second hand moves at the angular velocity $\omega = 2\pi/60\,\text{s} = 0.1047$ radians/s. A force which generates angular motion is called a torque τ, or moment. The moment depends on the moment arm d, namely the distance between the point where the force is applied and the rotational axis, and on the angle β between the moment arm d and the force F and the rotating member, see Fig. 5.1b.

$$\tau = F \cdot d \sin\beta \tag{5.1}$$

Note that $d\sin\beta$ is the perpendicular distance between the line of force and the axis of rotation, Fig. 5.1b. All the moments $\Sigma\tau$ acting onto an object of mass moment of inertia I generate the angular acceleration α according to

$$\Sigma\tau = I\alpha. \tag{5.2}$$

The mass moment of inertia I depends on the location of the rotation axis. In general I has the smallest value if the rotation axis goes right through the center of mass. I gets progressively larger as the rotation axis is moved away from the center of mass. In first approximation an arm or a leg may be approximated as a slender rod of length L and mass, hinged at one end. Its moment of inertia is

$$I = (1/3)ML^2. \tag{5.3}$$

If the mass M of a leg is concentrated close to the hip, and if the foot is very small and light (like a chicken leg) the limb may be approximated as a cone, which has $I = (1/10)ML^2$. Some values for different bodies are shown in Fig. 5.1.

In problems dealing with translational and rotational motion one often knows three parameters out of the four: position, velocity, acceleration, and time. Then one can use the equations of kinematics to find the missing variable. Table 5.1 shows the relevant relations.

Table 5.1. Relations of kinematics

linear motion (for constant a)		angular motion (for constant α)	
	missing variable	missing variable	
$u = u_0 + at$	x position	$\emptyset$ angle	$\emptyset = \emptyset_0 + \alpha t$
$x = u_0 t + \frac{1}{2}at^2$	$u = dx/dt$ velocity	ω angular velocity	$\emptyset = \omega_0 t + \frac{1}{2}\alpha t^2$
$u^2 = u_0^2 + 2ax$	t time	t time	$\omega^2 = \omega_0^2 + 2\alpha\emptyset$
$x = \frac{1}{2}(u_0 + u)t$	$a = du/dt$ acceleration	$\alpha = d\omega/dt$ angular acceleration	$\emptyset = \frac{1}{2}(\omega_0 + \omega)t$
$x = ut - \frac{1}{2}at^2$	u_0 initial velocity	$\emptyset_0$ initial ang. vel.	$\emptyset = \omega t - \frac{1}{2}\alpha t^2$

The use of these relations is illustrated by an example. A person, walking with stiff legs, typically rotates his leg ($L = 0.95$ m) through 1/6 of a turn ($2\pi/6$) radians in 0.8 s. The average angular velocity is $\omega = (2\pi/6)$radians/0.8 s $= 1.38$ radians/s. The average speed of the foot is $u = \omega L = 1.24$ m/s. The horizontal speed of the foot relative to the ground is zero at the instant when the foot leaves the ground. The foot is accelerated by gravity, and by the action of the leg muscles. The combined torque of these two forces accelerates the leg and gives it an angular acceleration, so that the angular velocity of the leg will grow. More will be said later about this process.

5.1.2 How to Manipulate Rotational Motion

Consider a small ball of mass M, which is rolled up an inclined plane by a previously compressed spring. The ball acquires an initial velocity u_1 with the horizontal and vertical components u_{1x} and u_{1y}, and an angular velocity ω_1. Neglecting the air resistance, the ball in free flight is only subjected to gravity which continuously changes the vertical velocity u_y, but leaves the horizontal velocity unchanged. Positions of the center of mass of the ball at equidistant time intervals Δt are shown in Fig. 5.2a. The ball traverses equal distance ds_x during equal time intervals Δt. However, in the vertical direction the spatial distance segments ds_y change continuously. They first get smaller, because the velocity decreases until the time (2) and then they get larger, because the vertical velocity increases.

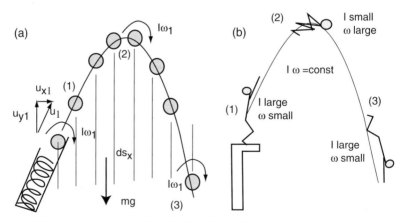

Fig. 5.2. (a) Translation and rotation of ball in flight. (b) Diver changing her mass moment of inertia in flight

As the ball rolls up the ramp, driven by the spring, its angular velocity increases reaching the value ω_1 at the point (1) where the ball becomes airborne, Fig. 5.2a. After the ball has left the ramp its angular velocity ω_1 remains constant. While airborne the angular momentum $I\omega$ does not change. The free flight motion of this ball equals the trajectory of a basket ball thrown with a spin.

Many animals control their mass moment of inertia during free flight. A cat uses its tail when thrown into the air in order to land on its feet. Birds do this when they twist and turn their wings. A gymnast changes her shape when she wants to make a somersault. The diver shown in Fig. 5.2b has the same intent.

The center of mass of the diver still follows the parabolic arc of the sphere shown in Fig. 5.2a. The diver starts with the angular momentum $I\omega_1$. This angular momentum remains constant while the diver is airborne. However, the diver changes her mass moment of inertia I. By stretching out she increases I so that ω must get smaller, and by pulling her arms and legs towards the body the diver decreases her mass moment of inertia, so that she must spin faster.

Sometimes animals change their mass moment of inertia in mid-flight to turn around in the air, sometimes they manipulate it in order to assist weak muscles that can only generate a limited torque. A case in point is the heron as it takes off from the ground.

5.1.3 How the Heron Starts Flying

The angular acceleration of an object depends on its mass moment of inertia I which in turn depends on the location of the axis of rotation, see Fig. 5.1.

A flat plate of mass m and length l, rotating around an axis at one edge (like a wing rotated about the shoulder joint) has $I=I_e=(1/3)\,m\,l^2$. In contrast, if the plate is rotated about a central axis, namely an axis through the center of mass, the mass moment of inertia shrinks to the value $I=I_c=(1/12)\,m\,l^2$. Herons instinctively know that. They have long wings. They wobble their body up and down with their legs as they start to fly, thereby rotating their wings about a central axis, Fig. 5.3, so that the mass moment of inertia I is much smaller as if they would keep the body still. Thus they only need a smaller force to flap their wings.

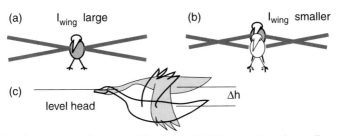

Fig. 5.3. (a) Wing rotation about shoulder joint. (b) Wing rotation about elbow position. (c) Body moves up and down in rhythm of wing beat, but head remains level

Keeping the body on a level while flapping the wings requires more torque (rotational force) than wobbling the body up and down. This motion could be irritating for the eyes. However, the head is held level relative to the ground as it moves up and down relative to the body at the wing beat frequency, Fig. 5.3c.

Pelicans and storks use a similar technique in full flight. The head is held level while body and wings oscillate up and down out of phase: the body moves down while the wings move up.

5.1.4 Linear Motion: Predators Fast Food

Kinematics is not only useful to describe the motion of individual animals and their limb motions – it can also be used to relate an animal to its environment. Think of the kinematics of attack and of escape strategies.

An animal can accelerate to reach a terminal speed u_{max}, which it may maintain for only a limited time T, or in a limited range respectively. A predator pursuing its food must know its own speed and range, and it must know the speed and range of its prey. Assume that a predator and its meal move in the same direction, say the x-direction as seen in Fig. 5.4, where the predator is faster than its prey, and has a larger range R_p, but the prey has more endurance $T_p < T_f$.

The thinner lines a, b, and c represents the prey. In order to make a catch the predator's space-time trace must intersect its prey's escape curve. The predator (thick line) starts running at time $t=0$. The prey starts running at the time τ. The reaction time τ consisting of (i) the time required to see, hear or smell, (ii) to recognize the danger, and (iii) to start moving. A catch occurs with certainty for line (a). The largest distance at which the predator can hope to make a catch is D. The predator just catches up to the prey (line b).

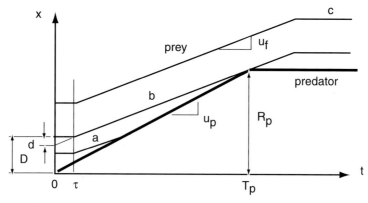

Fig. 5.4. Predator and its fast food. Simplified x–t diagram for calculating the maximum catch distance D

The best strategy for the predator is to come close to its victim. The best defense strategy for the prey is to be aware of any danger, and to move early. Both depend on the *distant senses* developed for attack and defense. The position x_p of the predator is given by $x_p = u_p t$. The range R_p of the predator is the distance reached at its endurance time T_p, namely $R_p = u_p T_p$, .

A catch occurs if $x_f = x_p$. The maximum distance D at which the predator can catch its meal is found from the relation $x_f = R_p = u_p T_p = (D-d) + u_f T_p, = (D - u_f \tau) + u_f T_p$, which can be solved for D:

$$D = T_p, (u_p - u_f) + u_f \tau . \tag{5.4}$$

If the predator and its potential meal have about the same speed, the endurance time T_f of the prey will make the difference. The reaction time τ of the prey is an extra handicap; it decreases the "safe distance" for the prey by the amount $d = u_f \tau$. The lost distance d can be reduced if the prey has senses that scan the environment over a long distance. It also helps the hunted if the hunted runs fast. (Faster hunters get better service in Nature's fast food restaurant.)

Speeding up from rest is a considerable energy expense. For instance, a zebra of $M = 200$ kg that speeds from 0 to 36 km/h ($u = 10$ m/s) must expend the mechanical energy $E = \frac{1}{2} M u^2 = \frac{1}{2} 200 (10)^2 = 10$ kJ. If the animal has an overall fuel efficiency of $\eta = 25\%$, this represents a food energy loss $\Delta Q = 10$ kJ$/\eta = 40$ kJ, which is more than one gram of body fat.

Animals are able to perform such calculations "instinctively" because they will not attack if the prey is too small, or if the chances of success of a hunt are too slim. Typically a lioness makes nine unsuccessful attempts before she manages to catch a meal. Speed and sharp senses are not the only defense strategies; twisting and turning also helps.

5.1.5 Connection of Angular and Linear Velocities

A person of leg length $L = 0.95$ m can walk 6 km in one hour with steps length of $S = 0.86$ m. These are $N = 6000/0.86 = 6980$ steps in 3600 seconds at the step frequency $f = N/3600 = 1.94$ steps/s, and the step period $T = 1/f = 1/1.94 = 0.52$ Hz. If ø is the maximum deflection angle of the leg the tip of the toe moves through an angle 2ø. The deflection angle can be found from a triangle with the sides L and $S/2 = 0.43$ m, namely sin ø $= 0.43/0.95$. To express the angle in radian solve ø $= \sin^{-1}(0.43/0.95) = 26.9°$, and convert to radian $26.9 (\pi/180) = 0.47$ rad. The angular velocity of the toe is $\omega = 2ø/T = 2 \cdot 0.47$ rad$/0.52$ s $= 1.80$ rad/s. The toe moves on an arc at the linear velocity of $u_{toe} = L\omega = 0.95$ m $\cdot 1.8$ radians/s $= 1.71$ m/s.

Suppose a soccer player kicks a ball using only the motion in his lower leg, Fig. 5.5. What is the velocity of the ball when the ball separates from the boot? To answer this question of relative motion one must know the forces that cause acceleration, and employ Newton's equation. We assume that the kicker can momentarily exert a force of $T = 1500$ N in the tendon below the kneecap with the moment arm $d_\perp = 0.06$ m. This generates a torque $\tau = 0.06 \cdot 1500 = 90$ Nm. The problem can be approached in various approximations.

First assume that the ball is gently accelerated by the foot, and the ball flies off when the foot has rotated through the angle Ø $= \pi/2$. Treat the lower leg like a thin rod, of $m = 7$ kg and length $L = 0.5$ m, which is hinged at one end with the mass moment of inertia $I=(1/3)\,ML^2=(1/3)\,7.0\cdot0.5^2=0.58$ kg m². From Newton's law $\tau=I\alpha$ calculate the average angular acceleration $\alpha=\tau/I=90$ Nm/0.58 kg m² $=154$ rad/s². Now find the time t that it takes the lower leg to rotate through an arc of $\emptyset=90°=\pi/2$ rad. This time is found from $\emptyset=\frac{1}{2}\alpha t^2$ namely, $t=(2\emptyset/\alpha)^{1/2}=\sqrt{(\pi/154)}=0.142$ s. Since the foot is steadily accelerated the angular velocity increases steadily. At the instant $t=0.142$ s the angular velocity has grown to $\omega(t=0.142)=\alpha\cdot t=154$ rad/s²·0.142 s$=22$ rad/s. Then the foot has the tangential velocity $u_f=L\cdot\omega=0.5$ m · 22 rad/s $=11.0$ m/s $=11.0\,(10^{-3}$ km/{1/3600 h}) $=39.6$ km/h. If the soccer player simultaneously rotates his upper leg, the lower leg velocity is added to the velocity at which the knee moves. If the ball flies off at this instant (with the same speed as the foot itself) then the foot would keep moving in the forward direction, retaining most of its kinetic energy.

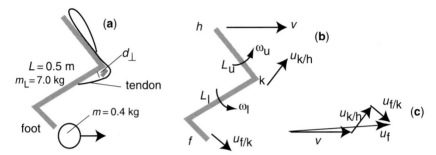

Fig. 5.5. Relative motion. (a) Leg dimensions. (b) Linear and angular velocities. (c) Velocity vector diagram, soccer ball velocity u_f

However, if the foot would stop dead after hitting the ball, having transferred to the ball *all* its kinetic energy, the ball would take off with a much larger velocity. This velocity can be found from the energy equation using values for the kinetic energies of the leg, and the ball. Assume the ball has the mass $m_b=0.4$ kg. Conservation of energy requires $\frac{1}{2}I\omega^2=\frac{1}{2}\,m_{ball}\,u_b^2$. Using real values: $0.5\cdot0.58$ kg m²· $(22$ rad/s)² $=140$ J $=0.5\cdot0.4\,u_b^2$, it results in $u_b=26.5$ m/s $=95$ km/h. This result is consistent with the fact that a good soccer player can kick the ball to reach speeds of ≈100 km/h.

If the ball was hit in mid swing, and a good fraction of the kinetic energy was converted into deformation work W_d, one can apply the relation of conservation of momentum for the foot $(M_f u_f+M_b u_b)_{initial}=(M_{f+b}\,u_{f+b})_{during\ contact}=(M_f u_f+M_b u_b)_{final}$.

One could also use the conservation of energy: $(0.5\,I\omega^2)_{before\ impact}=(0.5\,I\cdot\omega^2+0.5\,m_b\cdot u_b^2)_{after\ impact}+W_d$. However, to find the final velocity from energy consideration one would have to know how much energy is converted into lost deformation work W_d during the impact.

5.1.6 Relative Motion

The soccer ball example shows that motion analysis can be quite complicated. That example does not, however, describe the motion sequence of a free kick. When a soccer player executes a free kick he runs a few meters, increasing the speed of his center of mass, and while running he swings his leg around the hip joint and the lower leg around the knee joint. To find the absolute velocity of the foot one has to add all the relative velocities together.

To model a free kick assume first that the hip joint moves at the same speed u and has the same linear acceleration a as the center of mass of the player just before the foot hits the ball, when the leg is in the vertical position. Second, we assume that the upper leg, length L_u, swings at the angular velocity ω_u and has the angular acceleration α_u. Third, we assume that the lower leg of length L_l, rotates at ω_l about the knee. Then the knee moves at the speed $u_{k/h} = \omega_u L_u$ relative to the hip, and it travels at the speed $u_k = u + u_{k/h}$ relative to the ground. In turn the foot moves relative to the knee at the speed $u_{f/k} = \omega_l L_l$, while the knee moves at its own speed. Therefore, the foot has the absolute velocity

$$u_f = u_k + u_{f/k} = u + u_{k/h} + u_{f/k} = u + \boldsymbol{\omega_u} \cdot L_u + \boldsymbol{\omega_l} \cdot L_l. \tag{5.5}$$

These relative velocities have been written with *bold letters* to indicate vector quantities. Vectors can be represented by arrows, and the resultant vector can be determined by vector addition, see Fig. 5.5c. The vectors $\boldsymbol{\omega_u} \cdot L_u$ and $\boldsymbol{\omega_l} \cdot L_l$ always point into the direction of the tangent of the motion. Associated with such relative motions in rotating reference frames are angular and centrifugal accelerations. For instance, the knee has the tangential acceleration $a_{k,t} = a + L_u \alpha_{u,t}$, and a radial acceleration along the upper leg $a_{k,r} = L_u \cdot \omega_u^2$. The detailed discussion of the accelerations for this relative motion goes beyond the scope of this book.

5.1.7 Lifetimes and Biological Periods

All animals encounter various periodic processes and time scales. Most significant is the lifetime t_L, which varies from species to species. Generally large animals live longer. All higher animals with hearts have a characteristic heart beat period $t_H = 1/f_H$ or heart frequency $f_H = 1/t_H$. Life time and heart frequency can be stated as allometric relations of the body mass, discussed in Sec. 1.2. Locomotion frequencies, associated with the wing beat of birds, the flipping of tail fins of fish, or the leg frequency of land animals, also tend to scale with the body mass.

The allometric relations for lifetime and heart frequency have a curious consequence. All animals have about the same number of heart-beats: the heart frequency of mammals is quoted as $f_H = 241\, M^{-0.25}\,\text{min}^{-1}$, and the allometric relation for the lifetime derived with the parameters of Table 1.2 is $t_L = 11.8\, M^{+0.2}$ years. The total number of heartbeats Z_m during the lifetime of a mammal can be found by multiplying its lifetime and its heart frequency. Note that a year has

$60 \cdot 24 \cdot 365 = 5.26 \cdot 10^5$ minutes. $Z_m = f_H \cdot t_L = 1.5 \cdot 10^9 \, M^{-0.05} \approx$ billion beats. This number is nearly independent of body mass due to the weak mass dependence, because $M^{-0.05} \approx 1$. If the heart was built as a physical resonator one would expect its frequency to scale with the square root of the heart mass M_H. On the other hand the allometric relation for the heart mass is $M_H \approx$ const $\cdot M$ so that one would expect $f_H =$ const $M^{-0.5}$. Instead the allometric relation for the heart frequency is $f_{h,al} \approx$ const $M^{-0.25}$. Obviously the heart does not act as an elastic resonator. There is something still to be explained.

Table 5.2. Biological periods

time in s	10^{-6} =1 μs	10^{-4}	10^{-2}	1 s	10^{+2}	10^{+4} ≈ day	10^{+6} months	10^{+8} years
population								—
lifetime							———	
ovary cycle							———	
gland activities						———		
blood circulation					———			
breathing				———				
heart beat			———					
nerve			——					
brain functions	————————							

Animals encounter biological periods with a very wide range of time intervals, see Table 5.2. Some of these *periods* are only loosely defined. Some periods are tied to the seasons, where the timing device is the rotation of the earth around the sun, others are tied to the times of the day and hence directed by the daily rotation of the earth around its axis. Periodic processes could serve as *movements* for physiological clocks.

In addition, animals encounter and *use* various time scales, or frequencies. Time scales can be produced in clockworks that utilize (i) constant speed devices, (ii) energy exchange systems, or (iii) *dissipative* flow systems.

Constant speed systems generate characteristic frequency for instance, when an object is moved at constant speed over equidistant ridges (legs of grasshoppers), and (ii) when a sound wave bounces back and forth in an acoustic cavity like a flute or a vocal tract.

Time scales can also be associated with the process of energy flow and the conversion of energy from one form to another. Table 5.3 summarizes various time scales and energy flow processes. Notice that the power flux q' scales proportional to the mass M for free fall, pendulum, and muscle, and it scales allometrically for spring, cooling, and metabolism. The clockwork of energy exchange includes all oscillating simple harmonic motion devices (like the leg pendulum), periodically vibrating elastic structures (like ligaments, membranes, bones, air columns), and vortex shedding structures. Clockworks based on dissipative flow are connected to the decay, or recharging of certain quantities. For instance a starving warm blooded animal will lose a fraction $\Delta M/M$, of its body mass in a characteristic time Δt_M. When exposed to a very cold environment its body temperature T will drop by a certain fraction $\Delta T/T$ in the time Δt_T. A nerve will restore its action potential in a characteristic recharging time Δt_n.

Do animals have a sense of timing? If animals can measure time intervals they could in principle determine their velocity u, and their position x, in conjunction with their sense of balance which measures acceleration a. With an internal clock they can in principle calculate their speed $u = \int a\,dt$ and then know their position $x = \int u\,dt$ even with closed eyes.

Table 5.3. Process time scales, and heat flow rates

Notice that the power flux scales as $q' \propto M$ for free fall, pendulum, and muscle, and it scales as $q' \propto M^\alpha$, $\alpha \neq 1$ for spring, cooling, and metabolism. Reversible processes are marked <rev>

process	energy	converted to	time scale	power flux
(1) free fall from height h <rev>	potential (gravitation) $E_p = M g h$	kinetic $\frac{1}{2}Mu^2$	$h = \frac{1}{2}g\,t^2$ $\tau = \sqrt{(2h/g)}$	$q' = E_p/\tau$ $q' = 2^{-\frac{1}{2}} g^{\frac{3}{2}} h^{\frac{1}{2}} M$
(2) pendulum <rev> $y = L \sin \emptyset$, $y = L(1-\cos \emptyset) \approx L\emptyset^2/2$	kinetic $E_k = \frac{1}{2}Mu^2$	potential (gravit.) $E_p = Mgy \approx$ $\frac{1}{2}MgL\emptyset^2$	$\tau = T/4 =$ $(\pi/2)\sqrt{(L/g)}$	$q' = E_{kin}/\tau = E_p/\tau =$ $(1/\pi)\,\emptyset^2 L^{\frac{1}{2}} g^{\frac{3}{2}} M$
(3) elastic vibration <rev>	kinetic $E_k = \frac{1}{2}Mu^2$	potential (elastic) $E_e = \frac{1}{2}k\,x^2$ spring constant k	$\tau = T/4 =$ $(\pi/2)\sqrt{(k/M)}$	$q' = E_{kin}/\tau$ $q' = (1/\pi)k^{\frac{1}{2}} x^2 M^{\frac{1}{2}}$
(4) conduction area A; temp. diff. ΔT	internal energy	heat (random thermal energy)	time to loose 1 kJ, $\tau_{kJ} =$ $1\,kJ \cdot \Delta x/(Ak \cdot \Delta T)$	$q' = kA(T_h - T_o)/\Delta x$
(5) cooling by thermal conduction; $E_{int} = CM(T_h - T_o)$ object mass M; specific heat C,	internal energy	heat (random thermal energy)	$\tau = E_{int}/q'$ e-folding time $\tau = (CM\Delta x)/(kA)$	decreasing as $q' = q'_o \exp\{-t/\tau\}$ where $q'_o = kA(T_h - T_o)/\Delta x$
(6) vortex street of bar, length L, diameter D Strouhal number St,	free flow laminar kinetic $E_k = \frac{1}{2}m\rho u^2$ flow velocity u	$M_e = \pi\rho LR^2$; $R \approx D/2$ vortex energy $E_{eddy} = \frac{1}{2}M_e u^2$ $\approx (\pi/8)\rho L D^2 u^2$	$\tau = 1/f = D/(St\,u)$ eddy radius	$q' = E_{eddy}/\tau$ $\approx \frac{1}{2} \cdot St \cdot u^3 M_e/D$ $\propto M^{2/3}$
(7) metabolism	chemical energy	metabolic energy; finally → heat	time to metabolize 1 kJ, $\tau_{kJ} = 1\,kJ/\Gamma_o = 0$ $= 1\,kJ \cdot 0.28 \cdot M^{-3/4}$	basic metabolic rate $\Gamma_o = 3.6 \cdot M^{3/4}$ $= fALv = vM(f/\rho)$
(8) muscle, mass $M = AL\rho$ force $F = Af$, where $f \approx 0.5\,f_o = 10^5\,N/m^2$ (at max power); intrinsic velocity v_o	chemical energy humans $v_{o,m} \approx 2\,s^{-1}$ humming birds $v_{o,hb} \approx 20\,s^{-1}$ $f/\rho = 100\,J/kg$	mechanical work $W = F \cdot \Delta L$ typical maximum contraction $\varepsilon = \Delta L/L \approx 0.1$ $v = \varepsilon/\tau = \varepsilon/(L\tau)$	$\tau = \varepsilon/v$; at max. power $v \approx 0.3\,v_o$ humans: $\tau_{man} =$; 0.17 s; humming birds: $\tau_{hb} = 0.017$ s	$q' = P = Fu$ at max power humans $v_{man} = 0.6\,s^{-1}$ $P_{max,man} = 60\,W/kg$ $P_{max,hb} = 600\,W/kg$

5.2 Dynamics of the Moving Animal

Motion kinematics is only one part of a full analysis of locomotion. If the sum of all forces F, and the sum of all moments τ acting onto an object is zero, the linear momentum $m \cdot u$ and the angular momentum $I \cdot \omega$ of an object is constant. Changes of this state of motion are caused by unbalanced resultant forces ΣF, and by unbalanced resultant moments $\Sigma \tau$ respectively. Forces and moments generate accelerations, and accelerations change the linear velocity u or then angular velocity ω. The fundamental tool used to analyze these processes is Newton's equation, which is examined now for the linear motion. The linear momentum mu is changed by the net force ΣF according to

$$\sum F = \frac{d(m \cdot u)}{dt} = m \cdot c. \tag{5.6}$$

If the mass m is constant, and if one knows the net force ΣF acting onto the center of mass of an object, dead or alive, one can find the acceleration $a = dv/dt = \Sigma F/m$. Then one can determine how the animal gathers speed u, and how it changes its position x, namely $u = at$, and $x = \frac{1}{2}at^2$. In curvilinear motion, where an object is forced to travel along a segment of a circular path of local radius R, the velocity vector must continuously change its direction. This acceleration was already discussed in Sect. 5.2.1, it requires a force $F_n = mu^2/R$ that must act at right angles to the path. F_n points to the center of rotation. Monkeys must have a firm handhold when they swing from branch to branch in order to overcome this centrifugal force.

Objects may lose mass. This also represents a change of momentum. Then Newton's equation is written as

$$\sum F = \frac{d(m(t) \cdot u(t))}{dt} = m \frac{\partial u}{\partial t} + u \cdot \frac{dm}{dt}. \tag{5.7}$$

Table 5.4. Relations of dynamics

linear motion		angular motion	
position	x	angular position	ϕ
velocity [m/s]	$u = dx/dt$	angular velocity	$\omega = d\phi/dt$
acceleration [m/s²]	$a = du/dt$	angular acceleration	$\alpha = d\omega/dt$
on circular path	$|a| = u^2/R$		
translational inertia, mass [kg]	m	rotational inertia [kg m²]	I
Newton's IInd Law, force [N]	$\Sigma F = ma$	torque τ [Nm]	$\Sigma \tau = I\alpha$
momentum [kg m/s]	mu		$I\omega$
work [J=Nm]	$W = \int f \, dx$		$W = \int \tau \, d\phi$
kinetic energy [J]	$K = \frac{1}{2}mu^2$		$K = \frac{1}{2}I\omega^2$
power [J/s=W]	$P = Fu$		$P = \tau\omega$
linear momentum $(m_1u_1 + m_2u_2 + \ldots)_{initial} = ($ $)_{final}$		angular momentum	$I_1\omega_1 = I_2\omega_2$
work – kin. energy theorem	$W = \Delta KE$	if no friction loss	$W = \Delta KE$
rocket force [mass flow J]	$F = uJ$	centrifugal force	$F_n = mu^2/r$

For instance a squid or an octopus which is at rest may suddenly eject a jet of water at some rate $J=-dm/dt$. The minus sign indicates that the animal is losing mass. There is no net force initially, and its body speed is $v=0$. Hence the motion starts according to $0=m\partial v/\partial t-uJ$. The ejection of the mass flow generates a propulsion force $u\cdot J$.

The acting forces $\Sigma F=F_p-F_r$ can be devided into propulsion forces F_p and resistive forces F_r. *Propulsive forces* come from muscles, ejection of mass, gravity, or wind; they are exerted at the soles of the feet of land animals, at the tails or flippers of aquatic animals, at the wings of flying animals, or at the orifices of jets in squids. *Resistive forces* include the drag in air or water and friction. Relations of translational and rotational dynamics are summarized in Table 5.4.

5.2.1 How Animals Get Going – the Resultant Force

If one knows that an object can only move when a (net) force is acting onto its center of mass, one can easily see how floating or airborne animals get around: They are pushed by wind and ocean currents. However, the ground under the feet of land animals does not move. How can they travel at all? The answer is that land animals are indeed pushed by the ground because forces come in pairs. When a foot pushes against the ground the ground pushes back with an equal and opposite force. While the foot rests on the ground the center of mass of the animal may be lifted up and propelled forward.

If an animal just stands motionless on two legs, the soles of its feet exerts onto the ground the force Mg. The ground presses back with a normal force F_n equal and opposite to Mg. If the animal steps forcefully onto the ground with some vertical force F_y (generated by the muscles in its leg), it exerts the total vertical force $Mg+F_y$, and the ground reacts with the normal force $F_n=-Mg-F_y$, shown in Fig. 5.6. The vertical reaction force of the force $F_{ry}=-F_y=F_n-Mg$ accelerates the organism upwards. The horizontal reaction force $F_{rx}-F_x$ accelerates the animal forwards.

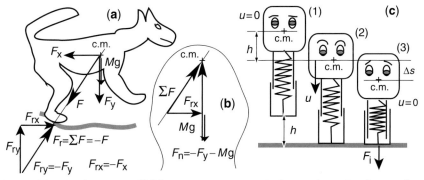

Fig. 5.6. (a) Animal jumping off. **(b)** Force vectors: F_n Vertical ground reaction force $F_{r,x}$ horizontal ground reaction force, ΣF resultant ground reaction force, **(c)** landing on one foot

When animals run the muscles in each leg generate a torque τ. If the leg has a length L this torque will produce a propulsion force F_h that presses the foot against the ground in horizontal direction

$$F_a = \tau / L .\tag{5.8}$$

If the ground is firm the foot will experience the reaction force $F_r = -F$. The ground exerts the total reaction force $F_r = F_{rx} + F_{ry} + Mg$.

If the line of action of this force goes through the center of mass cm the animal will be accelerated forward and upward without experiencing a torque. The horizontal force produces the horizontal acceleration

$$a_x = F_{rx} / M .\tag{5.9}$$

For the further motion analysis one may use the relations of kinematics, summarized in Table 5.4. Suppose the motion starts from rest, and the acceleration acts during the time interval Δt. Then the center of mass acquires the speed

$$u_x = a_x \cdot \Delta t = \Delta t \cdot F_{rx} / M .\tag{5.10}$$

If the acceleration remains constant throughout the distance Δs, the body will reach the speed

$$u_x = \sqrt{\frac{2 \cdot \Delta s \cdot F_x}{M}} .\tag{5.11}$$

The vertical motion can be analyzed in a similar fashion. If the animal is motionless the impact force opposing the ground reaction force F_N is equal to the weight Mg. Lift-up from the ground requires $F_N > Mg$. A running animal may produce $N \approx 3\,Mg$.

5.2.2 Landing on Your Feet

When an animal jumps down from some height h its feet and legs must absorb the impact force. To analyze this motion one first determines the velocity u shortly before impact

$$u_{imp} = \sqrt{2gh} .\tag{5.12}$$

To illustrate the principle we model the animal as a stiff body of mass M with a spindly leg that acts like a massless, heavily damped spring, Fig. 5.6c.

When the body comes to rest the spring is compressed by the distance Δs. Hence, the deceleration of the center of mass takes place over the distance Δs, while the body slows down from the impact velocity u_{imp} to the final velocity $u_{final}=0$. The average deceleration during this time is

$$|a|=u^2/2\Delta s=2gh/2\Delta s=gh/\Delta s.\tag{5.13}$$

The average impact force acting onto the center of mass is therefore

$$F=Ma=Mgh/\Delta s.\tag{5.14}$$

This force *shakes* up the stiff animal body. F is equal and opposite to the impact force F_i with which the foot of this model animal pushes onto the ground while coming to rest.

Note that F depends inversely on the deceleration distance Δs. Humans black out with accelerations exceeding a few g. Landing with stiff legs is a bad strategy. Cats always land with very soft legs.

5.2.3 The Jumping Flea

A jumping flea is capable of enormous accelerations when it jumps off. The insect accelerates for about $\Delta t \approx 0.7$ ms with an acceleration of $a \approx 2000$ m/s^2. This huge acceleration is achieved by releasing elastic potential energy.

The flea is modeled as a sphere of $R=1$ mm radius and a density of water. Its mass is $M=(4\pi/3)r^3\rho=(4\pi/3)(10^{-3})^3$ m^3 1000 kg/m^3 = 0.0042 g. The force that propels the flea during the acceleration phase is $F=Ma=4.2\cdot 10^{-6}$ kg $\cdot$ 2000 m/s^2 = $8.4\cdot 10^{-3}$ N. The starting velocity is found from the kinematics of the motion $u_0=a\Delta t=2000$ m/s$^2\cdot 7\cdot 10^{-4}$ s = 1.4 m/s.

When falling downward like a descending raindrop, the flea might have a terminal velocity of $u_t=110$ m/s as calculated in Sect. 5.2.6. However, the legs of the flea will increase the drag over and above the drag of a smooth sphere so that its terminal velocity is likely to be much smaller than 4.5 m/s. Since the starting velocity of the flea is much smaller than its terminal velocity, the drag force will be negligible as it jumps up. Then the maximum height of the jumping flea can be found from the energy equation, assuming that the energy loss due to air friction may be neglected. This yields $\frac{1}{2}Mu_0^2=Mgh$, or $h=u_0^2/2$ g $=(1.4$ m/s$)^2/$ 19.6 m/s^2 ≈ 0.1 m.

Before the flea started to jump all the mechanical energy, which later exploded into motion, was stored in a "compressed spring", namely a pad of resilin, an elastic substance in the leg system. The energy analysis of the starting process $\frac{1}{2}mu_0^2=\frac{1}{2}kx_0^2$ yields a relation for the spring constant $k=AY/L$ and the extensions x of the spring. In this calculation Y is Young's modulus, A is the cross section of the spring, and L is its unstretched length.

5.2.4 Forces in Angular Motion

Animals twist, turn, and rotate their appendages during their normal activities. Monkeys swing from branch to branch; birds, and mosquitoes flap their wings; dogs and sprinters rotate their legs. Sometimes the whole body performs rotational motion. Such motion sequences generate centrifugal forces, and therefore animals must intuitively master motion that includes rotational dynamics. Now we will discuss a few examples.

A small animal can turn in a small circle. Turning and acceleration requires good traction between the soles of the feet and the ground, characterized by a friction force F_f. The friction force F_f needed to turn on a circle of radius r is

$$F_f = M \cdot u^2/r .\tag{5.15}$$

If an animal can muster the force F_f, it is able to turn on a circle of radius $r = u^2 M/F_f$.

Consider a big animal chasing a smaller one and suppose both can generate the same friction force per unit mass, Fig. 5.7a, then M/F_f is the same for both, and the turning radii of both are related as

$$\frac{r_1}{u_1^2} = \frac{r_2}{u_2^2}, \text{ or } r_2 = r_1 \frac{u_2^2}{u_1^2} .\tag{5.16}$$

After the turn, the tighter turning radius of the smaller animal has put the distance $\Delta y = 2(r_2 - r_1)$ between it and the pursuer. It gets away.

The centrifugal force also affects monkeys swinging from branch to branch [Preuschoft 1996], as illustrated in Fig. 5.7b. The maximum force at the fingers is $\Sigma F = Mg + M(u_{max}^2/R)$, where $u_{max}^2 = 2gh$, and $h = R - R\cos\phi$. For the maximum angle $\phi = \pi/2$, ($\cos\phi = 0, h = R$) the force becomes

$$F_m = Mg + 2MgR/R = 3Mg .\tag{5.17}$$

Since the forces in the flexors of the fingers do not grow proportional with the body mass, big apes do not swing through big angles ϕ.

The sudden deceleration of a swinging appendage, like an arm or a leg, can generate a vertical force, Fig. 5.7c. Suppose the arm of length L, mass m, and mass moment of inertia I has been accelerated by applying the torque τ_1. Therefore the arm swings through an angle φ_1 in the time interval t_a. The arm reaches the angular velocity $\omega_1 = \alpha_1 t_a$ in the angular position (1). At this instant a different set of muscles is suddenly applied that deliver the force F. Acting with the radius of gyration R the force produces the torque $\tau_d = F \cdot R$, which decelerates the arm. The radius of gyration R is defined with $I = mR^2$, where I is the mass moment of inertia and m the mass of the limb. If the angular motion is brought to a complete stop in the time interval t_d when the arm is in the horizontal position one has $0 - \omega_1 = -(\tau_a/I)t_a = \alpha_2 t_a = (\tau_d/I)t_d = (F \cdot R/I)t_d$. This relation is solved for the decelerating force

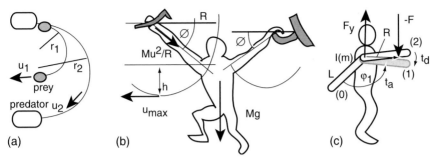

Fig. 5.7. (a) The fox and the hare. (b) Gibbon swinging from branch to branch. (c) High jumper gathering vertical momentum by swinging and stopping an arm

$$F = -(\tau_1/R)(t_a/t_d) . \tag{5.18}$$

An equal and opposite swing lift force $F_y = -F$ acts onto the hinge of the swinging arm while the deceleration occurs. If the acceleration time t_a is much longer than the deceleration time t_d, a substantial vertical force can be generated. High jumpers make use of this effect by launching the jump from one leg and using the other leg and arms for the generation of swing lift [Chimes 2002].

5.2.5 Moving Through Fluids

It is not a big step from modeling the internal motion of fluids in a body to discussing the propagation of small objects through fluids. Walls of arteries, and veins slow down the motion in pipe flow, due to the action of viscosity. This is described by Hagen-Poiseulle's relation, encountered in Sect. 4.2. At larger velocities the flow turns turbulent and one needs to consider the hydrodynamic forces discussed in Sect. 3.4.3. Free swimming organisms like water fleas, coho salmon, or squid are slowed down by viscous forces, *and* hydrodynamic forces.

Each organism of lateral dimension d can be characterized by a Reynolds number $Re = Ud/(\eta/\rho)$, where U is its velocity, and $v = (\eta/\rho)$ is the kinematic viscosity of water. The Reynolds number is the ratio of viscous forces and hydrodynamic forces. When Re is small, laminar flow is encountered. Then one only needs to consider the viscous forces, and can neglect hydrodynamic forces. When Re is very large turbulent flow is encountered, where the motion is dominated by hydrodynamic forces.

The critical Reynolds number separating the laminar and the turbulent regime is $Re_{cr,ex} \approx 100$ for objects like water fleas or killer whales moving through the fluid (external flow). When fluid moves through pipes (internal flow) it turns turbulent at the much larger value $Re_{cr,in} \approx 2300$.

One can also defined turbulent flow as an *excited phase* of the fluid state of matter in which eddies store, exchange, and gradually dissipate rotational kinetic energy and coherence energy. In this model the Reynolds number is the ratio of

Table 5.5. Forces in laminar and turbulent flow

Re $\ll$ 1 laminar flow	Re $\gg$ 1 turbulent flow	jet propulsion
Stokes friction	form drag	$F = J u$
on sphere, radius R	(hydrodynamic force)	J: mass flow rate
(viscose force)	$F = \frac{1}{2} C_D A \rho u^2$	u: jet velocity
$F_{\text{Stokes}} = 6 \pi R \eta u$	A: area obstructing the flow	

production and dissipation rates of eddies. The dissipation rate of eddies is relatively larger in internal flow than in external flow. Therefore a larger Reynolds number is required to reach the turbulent phase in pipe flow. The resistance force encountered by an object that swims through a fluid can always be interpreted as the energy per meter deposited by the object in the surrounding fluid. The resistance force is larger when the flow becomes turbulent, because the eddies "suck up" quite a bit of energy.

When encountering a flow situation one must first find out whether the flow is turbulent or laminar, so that one can use the proper resistance law. This will first be demonstrated for the motion of a small insect falling through still air and then for the motion of a squid.

5.2.6 Terminal Velocity of a Small Insect Falling in Still Air

Small objects that fall through still air settle down to a terminal velocity of decent, which is actually equal to the maximum speed that the organism can acquire. The velocity of any object is constant only if all forces that act on it are in balance: $\Sigma F = 0$. For a free falling object one must consider gravity, and Stokes friction. The analysis holds for the motion of a little insect without wings, or a raindrop as it starts to fall. Its mass is $M = (4\pi/3) r^3 \rho$. The motion is described by Newton's equation, where a is the acceleration.

$$a M = \Sigma F = M g - F_{\text{drag}}. \tag{5.19}$$

When the object starts falling, its velocity is practically zero so that $F_{\text{drag}} \approx 0$. Since the object is small, the Reynolds number must be small, so that the flow will be laminar – at least in the very beginning. Without drag the object is in free fall. However, as u increases F_{drag} gets larger, till finally $M g = F_{\text{drag}}$, and $\Sigma F = 0$. Then the acceleration disappears and u remains constant at the terminal velocity u_{term}, which is found from the force balance

$$6 \pi r \eta u_{\text{term}} = M g = (4\pi/3) r^3 \rho g. \tag{5.20}$$

Note that ρ is the density of the object (in many cases $\rho \approx \rho_{\text{water}}$), and η is the viscosity of the surrounding medium (air or water). In the absence of better knowledge we have used Stokes friction $F_{drag} = 6\pi\, r\, \eta\, u$, assuming that the flow remains laminar. We quickly check this assumption by solving for the terminal velocity

$$u_{term} = \frac{2}{9} \cdot \frac{g \cdot \rho \cdot r^2}{\eta} \ . \tag{5.21}$$

For a small raindrop of $r = 1\,$mm the terminal velocity calculated from (5.21) would be $u_{ter} = (2/9)\,10 \cdot 1000 \cdot 10^{-6}/1.6 \cdot 10^{-5} \approx 110\,$m/s. Then the Reynolds number would be of the order 10^4. Obviously the flow around the sphere is no longer laminar. Therefore one must use the hydrodynamic drag to get $0 = m\,g - F_{drag} = m\,g - \frac{1}{2}\,C_D\,A\,\rho_{air}\,u_{term}^2$, which is solved for u_{term}, namely

$$u_{term} = \sqrt{\frac{8 \cdot r \cdot \rho_{water} \cdot g}{3 \cdot \rho_{air} \cdot C_D}} = 4.5\,\text{m/s} \ . \tag{5.22}$$

The Reynolds number is then $\text{Re} = 4.5 \cdot 2 \cdot 10^{-3}/1.6 \cdot 10^{-5} \approx 560$. For this Reynolds number we find the drag of a sphere $C_D = 0.55$ from Fig. 3.18, and repeat the calculation of Eq. (5.22) to get $u_{term} = \sqrt{\{(8 \cdot 10^{-3} \cdot 10^3 \cdot 9.81)/(3 \cdot 1.29 \cdot 0.55)\}} = 6.1\,$m/s. Small things fall gently, like the rain in Vancouver. If a small insect is *falling* in an updraft of $u > u_{term}$ it will actually float up, and it may be carried over large distances.

5.2.7 Rocket Propulsion

Consider an object like the jelly fish polyorcas, Fig. 5.8, which gets around by expelling a stream of water at the mass flow rate $J = \rho\, u\, A$ kg/s with an average velocity u. This object will experience a thrust force $F_p = J\,u$.

The average outflow velocity $u = J/\rho A$ can be found if the mass flow rate J, the nozzle area A, and the density ρ of the expelled fluid is known. Therefore the object is propelled by the thrust force

$$F_p = \frac{J^2}{\rho \cdot A} \ . \tag{5.23}$$

Here we want to use this propulsion force to find the terminal velocity of a squid. Jellyfish and squid get their propulsion force $F_p = J\,u$ from rocket propulsion.

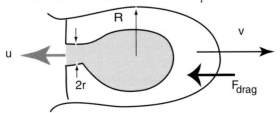

Fig. 5.8. Squid squirts

The mass flow rate $J = \rho\, u\, \pi\, r^2$ is related to the escape velocity of the fluid u, and the radius r of the rocket nozzle. It is known that the squid generates an internal pressure of about $\Delta p \approx 30$ kPa. For the flow escaping out of the orifice, Bernoulli's equation in the form $\Delta p = \frac{1}{2}\rho\, u^2$, must hold so that one can find the ejection velocity of the fluid.

$$u = \sqrt{\frac{2 \cdot \Delta p}{\rho}} = \sqrt{\frac{2 \cdot 30 \cdot 10^3}{10^3}} = 7.7 \text{ m/s} \tag{5.24}$$

With this information one can determine a terminal velocity u_{term} for the squid, which is reached when the acceleration is zero: $aM = 0 = F_p - F_{\text{drag}}$. Here we use the hydrodynamic drag for objects at large Reynolds number $Ju = F_{\text{drag}} = \frac{1}{2}\, C_D\, A\, \rho_{\text{air}}\, u_{\text{term}}^2$, where $A = \pi R^2$ is the obstruction area which the object presents to the flow. Solving for u_{term}^2 one gets

$$u_{\text{term}}^2 = \frac{2Ju}{C_D A \rho} = \frac{2\rho\pi r^2 u^2}{C_D \pi R^2 \rho}\ . \tag{5.25}$$

This leads to the terminal velocity u_{term}

$$u_{\text{term}} = u\,\frac{r}{R}\sqrt{\frac{2}{C_D}}\ . \tag{5.26}$$

Since C_D is only known from tables as function of the Reynolds number, one has to make an educated guess such as $C_D = 0.3$, since the squid has a very slender streamlined shape. Assuming $r/R = 0.1$ and taking the jet velocity $u = 7.7$ m/s we find $u_{\text{term}} = 7.7 \cdot (1/10) \cdot \sqrt{(2/0.3)} \approx 2$ m/s. This is indeed a typical empirical velocity.

5.2.8 Masters of Acceleration

Some animals are capable of enormous acceleration. They do not faint as humans do when subjected to accelerations of only a few g. They would be the ideal passengers to be launched on a spaceship, Table 5.6 shows jumping performance data of animals reported by various authors.

Table 5.6. Jumping performance of animals of body mass M. Height h, distance Δs, acceleration a, and peak power

animal	M [kg]	height of jump h [m]	distance of acc. Δs [m]	time of acc. Δt [s]	a [g]	peak power output [W/kg]	source of data
man standing jump	70	0.6	0.4	0.23	1.5		Bennet-Clark
Leopard antelope		2.5	1.5	0.43	1.6	115	Hill
lesser galago		2.25	0.16	0.047	14	915	Hall and Craggs
locust adult		0.45	0.04	0.026	11	330	Bennet-Clark
squid tentacle					33		Bennett-Clark
locust 1 instar		0.17	$7\cdot10^{-3}$	0.008	24	430	Bennet-Clark
flea	$5\cdot10^{-7}$	0.20	$7.5\cdot10^{-3}$ $\approx$300 L	$7\cdot10^{-4}$	200	2,500	Schmidt-Nielsen
click beetle	$4\cdot10^{-7}$	0.3	$7.7\cdot10^{-3}$	$6\cdot10^{-4}$	382		Schmidt-Nielsen
trout				0.1–0.2	15		Harper & Blake
pike				0.1–0.2	25		Harper & Blake

5.3 Locomotion and Energy

Energy analysis describes the motion of the center of gravity of an animal, and permits stages in the motion to be predicted. In the preliminary energy analysis of Chap. 2 we used the fact that every object has kinetic, potential, internal, and possibly elastic energy, and that the sum of these energies plus any energy produced by muscle forces, or lost to friction remains constant. Here we continue this analysis paying more attention to the different forms of motion: translation, rotation, and relative motion. Further, we look at motion strategies that operate at a minimum of power consumption. There are several strategies to minimize power: (i) choose a particular geometry for the limbs, (ii) convert some unneeded kinetic energy temporarily into elastic energy of parts of the animal body, or into gravitational potential energy (resonance), and (iii) store energy temporarily in eddies in the flow around the body.

Periodic energy storage schemes of resonance only work if the cycle frequencies of locomotion (which are set by the contraction frequencies of the muscles) are tuned into the natural frequencies of the elastic, or gravitational oscillations, or are matched to vortex shedding frequencies in the surrounding flow. In short the muscles motion must be in resonance with natural frequencies of the temporary energy storage modes.

Animals contract their muscles in order to move. They generate the required mechanical power $P = \eta b \Gamma_0$ at some efficiency $\eta \approx 0.25$ from their metabolic power $\Gamma = b \Gamma_0$. The less power is needed to execute a desired motion, the better. Active animals have a metabolic rate that is larger than the resting rate Γ_0 by a factor b. The activity factor b has typical values for birds in the range 12–15.

Energy analysis does not reveal how long it takes to go from one position to another. If that information is needed, one has to use the kinematics and dynamics of the linear and angular motion.

5.3.1 Energy Analysis of Moving Objects

A simple way to analyze the motion of an object like a ball, or a diver jumping of a diving board, Fig. 5.2 is to consider all the energies of the system: the translational kinetic energy $\frac{1}{2}Mu^2$, rotational kinetic energy $\frac{1}{2}I\omega^2$, gravitational potential energy $Mg\,\Delta h$, gravitational potential energy $Mg\,\Delta h$, the elastic potential energy $\frac{1}{2}kx^2$, and energy consumed by friction forces $F_{fr}\Delta s$.

Energy analysis is particularly useful when dealing with simple mechanical systems, which may be represented as mass points. The equation of conservation of energy is written for the system at two points in time labeled with the subscripts 1 and 2.

$$\frac{1}{2}Mu_1^2 + \frac{1}{2}I\omega_1^2 + Mgh_1 + \frac{1}{2}kx_1^2 = \frac{1}{2}Mu_2^2 + \frac{1}{2}I\omega_2^2 + Mgh_2 + \frac{1}{2}kx_2^2 + F_{fr}\Delta s \quad (5.27)$$

M is the mass of the object, x_1 and x_2 are the initial and final positions of the spring, h_1 and h_2 are the elevations, u_1 and u_2 are the initial and the final velocities, k is the spring constant, and Δs is the distance through which the body was exposed to friction forces. This energy equation may for instance be used to predict the center of mass velocity of a heavy object such as a leopard jumping down from a tree, or a ion jumping over a fence.

Consider a ball that is launched with some spin at point (1) in Fig. 5.2a. One can give conditions for point (2), $\frac{1}{2}Mu_{1y}^2 = +Mg\,(h_1-h_2)$. The terms with the angular velocity drop out because $\omega_2=\omega_1=$const. The vertical velocity u_{2y} is zero at point (2). Similarly the conditions at point (3) are given by $\frac{1}{2}Mu_1^2 + Mgh_1 = \frac{1}{2}Mu_3^2 + Mgh_3$. The conditions at point (1) can be related to the energy of the cocked spring: $\frac{1}{2}Mu_1^2 + \frac{1}{2}I\omega_1^2 \approx \frac{1}{2}k\,\Delta x_1^2$, where Δx_1 is the distance through which the spring has been compressed before the ball was launched.

One can treat a jumping flea as a mass point that is launched by releasing a cocked spring. If the initial velocity u_1 is known, one can determine how much potential energy was stored in the spring. If one analyses the internal construction and measures the elongation of the stretched tissue, one can derive its spring constant k and its Young's modulus. In order to estimate how high the flea could jump one would have to include the air resistance ($F_{fr}\Delta s = \int F_D\,ds$) in the energy analysis (5.27). The calculation becomes complicated, because the air resistance, or the drag force F_D of a moving object depends on its speed, see Sect. 3.4.3.

5.3.2 Cost of Transport and Resistive Force

Animals expend energy and power in locomotion. Energy is needed to overcome the resistive force F_r. The energy expended per distance Δx is $\Delta E = F_r\Delta x$. Then the force $F_r = \Delta E/\Delta x = E_{/m}$ can be interpreted as the energy expended per meter of travel, $E_{/m}$. In order to compare the performance of animals of different mass M, one defines the cost of transport parameter E_{tr}, namely the energy expended per meter and per kg of body mass M.

$$E_{tr} = \frac{E_{/m}}{M} = \frac{F_r}{M} \; J/(kg \cdot m) \,. \tag{5.28}$$

The energy per meter E_{tr} can also be used to compare the three modes of transportation (walking, swimming, and flying) employed by organisms and technical devices.

The cost of transport can be obtained from energy data. Tennekes [1997] discusses the wagtail, a small bird, which crosses the Sahara desert nonstop to reach its winter habitat. During this 1600 km journey the bird burns up about $\Delta m = 12$ g of body fat from a starting weight of about 35 g. Assuming an enthalpy $\Delta h \approx 32$ kJ/g fat, the bird uses up $\Delta E = \Delta h \cdot \Delta m = 384$ kJ of energy. At an estimated mechanical efficiency of 25%, the mechanical energy $\Delta E m = 96$ kJ is needed for the travel. Then the energy per meter is $E_{/m} = 9.6 \cdot 10^4 / 1.6 \cdot 10^6 = 0.06$ J/m, equivalent to a resistive force $F_r = 0.06$ N. At the average body mass during the flight of $M = 35$ g $- 6$ g $= 29$ g the bird's cost of transport is $E_{tr} = E_{/m}/M = 2.09$ J/(kg m).

Animals moving at a speed u must generate the mechanical power $P = F \cdot u$ by metabolic processes operating at the mechanical efficiency η. (Note that the letter η describes the efficiency and not the dynamical viscosity.)

$$P = F \cdot u = \eta \, \Gamma = const \, M^{3/4} \,. \tag{5.29}$$

One can solve (5.28) for the resistive force, $F_r = \eta \, \Gamma / u$, and substitute into (5.27), so that the cost of transport becomes a function of the metabolic rate and the travel velocity u.

$$E_{tr} = \frac{E_{/m}}{M} = \frac{F_r}{M} = \frac{\eta \cdot \Gamma}{u \cdot M} = const \cdot \frac{M^{-1/4}}{u} \,. \tag{5.30}$$

Figure 5.9 shows empirical cost of transport data for swimming, flying and running. These data can be represented as allometric relations.

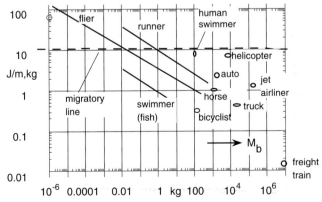

Fig. 5.9. The cost of transport decreases with mass; bigger is better. Data adapted from Schmidt Nielsen [1972]

$$E_{tr} = C \cdot M^{\beta} . \tag{5.31}$$

The empirical proportionality constant C differs for the various propulsion methods. They can be extracted from the empirical data, by measuring the value of the particular curve at the point where $M = 1$ kg. Running ($C_r \approx 7.5$) is more expensive than flying ($C_f \approx 2.3$), which again is not as efficient as swimming ($C_s \approx 0.60$). Swimming benefits from the fact that little energy is expended to keep the elevation, but of course the forward motion is hampered by drag. Birds encounter much less drag, but they must always work against gravity. The exponents for swimming, running, and flying are $\beta_s = -0.6$, $\beta_f = -0.46$, and $\beta_r = -0.59$.

One can use the power Eq. (5.29) and cost of transport relation (5.30) to derive the locomotion velocity as function of body mass:

$$u = \text{const} \cdot M^{-\frac{1}{4}} / E_{tr} = \text{const} \cdot M^{\beta - 0.25}. \tag{5.32}$$

Using the different empirical exponents β one obtains the locomotion scaling relations for swimming, flying, and running:

$$u_s = \text{const} \cdot M^{0.35}, \; u_f = \text{const} \cdot M^{0.21}, \; \text{and} \; u_r = \text{const} \cdot M^{0.34}. \tag{5.33}$$

The allometric exponents in these relations are larger than expected from other considerations (see Sect. 6.3.2, and Sect. 3.4.4) probably because the slopes β_s, β_f, and β_r in Fig. 5.9 are not known to sufficient accuracy.

5.3.3 Saving Mechanical Power by Slender Limbs

All limbs are articulated around joints, to carry out rotational motion. A muscle driving the particular motion of a limb of mass M applies a forces F at some moment arm d from the joint producing the torque $\tau = F \cdot d$. The limb has the mass moment of inertia I.

Newton's equation for rotational motion $\tau = \alpha I$ shows that the limb can be accelerated at the rate $\alpha = \tau / I$. The larger the acceleration, the faster the motion of the limb. Long and slender legs allow the feet to attain faster speeds. So to come to speed quickly, the limb should be long and have a small mass moment of inertia I. The magnitude of I depends on the mass distribution of the leg. If most of the leg mass sits near the foot, the mass moment of inertia is large, and then it is hard to rotate leg and foot.

However, if the leg mass is concentrated close to the hip joint, I is much smaller.[1] The mass moment of inertia can be calculated from the relation $I = \int r^2 dm$. A cylindrical member of mass M and length L, hinged at one end has $I_{cyl} = (1/3) ML^2$.

---

[1] A simple self experiment illustrates this point. Take a mass, say a full 2 liter juice bottle and swing it through a 90° arc holding it on the stretched out arm. It is not easy to do that quickly. However, if one bends the arm bringing the weight close to the body, then one can swing the weight much faster through 90°, achieving a much higher angular acceleration.

Fig. 5.10. Animal with legs (a) with small mass moment of inertia, (b) larger mass moment of inertia

A conical member of the same mass, hinged at the base, has only $I_A=(1/10)ML^2$, see Fig. 5.1. Consider two animals with legs of the same length, L and identical masses M, which are propelled by muscles generating the same torque τ, Fig. 5.10. Suppose the legs of one animal A are shaped like cylinders, and the legs of another animal B are shaped like cones hinged at the base. A will produce the angular acceleration $\alpha_A=\tau/0.333\,ML^2=3\tau/ML^2$. In contrast animal B can produce the much larger angular acceleration $\alpha_B=\tau/0.1\,ML^2=10\tau/ML^2$. If an acceleration α_1 is required to execute a certain motion animal A will need the torque $\tau_A=\alpha_1\cdot0.333\cdot ML^2$, whereas animal B will only need the $\tau_B=\alpha_1\cdot0.1\,ML^2$. Clearly the tender footed animal B is at great advantage. Just by having the mass of the limb concentrated near the joints, animal B can achieve the same acceleration with only 1/3 of the torque that animal A must apply.

All legs of fleet footed animals, like gazelles, or horses, mice, chicken, or robins have muscular upper legs, spindly lower legs and slim bony feet and fingers. Therefore, they are able to accelerate quickly.

5.3.4 Spring Loaded Animals

Many animals make use of elastic energy storage in their bodies.

Ants close their jaws, and fleas move their legs by the release of elastic energy. They activate cocked springs. Locusts move their wings with the flight muscles and simultaneously elastically deform their thorax shell and subsequently let the elastic forces of thorax pull the wings back into the opposite direction. Similarly, the cherry size jelly fish polyorcas pulls its wall radially inward with a single muscle [Megill 2002] in order to eject a water jet. In the *return stroke* the elastically deformed wall material opens the body cavity and sucks in the next charge of water. As shown by (3.15) the energy storage capacity $W=0.5\,LT^2/(AY)$ grows with the square of the applied tension force T. It is large when length L of the elastic member is large, and its cross section area A, and elasticity modulus Y are small. In some cases it is easy to calculate the forces and energies of elastic components. Elastic energy storage is also important for the generation of sounds: our vocal cords are periodically stretched as air rushes through the larynx.

Kangaroos have long tendons, which are stretched when the animal touches the ground in its hopping gait. Tendon material has a Young's modulus of $Y = 1.5 \cdot 10^9 \, \text{N/m}^2$, and it can be stretched by $\Delta L / L = 10\%$. Suppose a Kangaroo of $M = 25$ kg has tendons of $L = 0.3$ m length and $d = 10$ mm diameter (area $A = \pi (d/2)^2 = 7.85 \cdot 10^{-5} \, \text{m}^2$), which are stretched by maximally 8% so that $\Delta L = 0.08 \, L = 0.08 \cdot 0.3$ m $= 24$ mm. At maximum stretch the elastic potential energy $E_{el} = \frac{1}{2} k \Delta L^2$ is stored and the maximum force $F = k \Delta L$ is generated. The spring constant is therefore $k = Y A / L = 1.5 \cdot 10^9 \cdot 7.85 \cdot 10^{-5}/0.3 \, [\text{N/m}^2] = 3.93 \cdot 10^5 \, \text{N/m}$, and the stored elastic energy is $E_{el} = \frac{1}{2} \cdot 3.93 \cdot 10^5 \, \text{N/m} \cdot (2.4 \cdot 10^{-2} \, \text{m})^2 = 1.13 \cdot 10^2 \, \text{J}$ in each tendon. If the total elastic energy of both tendons $2 \, E_{el}$ is completely converted into kinetic energy in the next hop, the kangaroo bounces back, without using its muscles, to reach a vertical take-off velocity u_y found from $2 \, E_{el} = 0.5 \cdot M u_y^2$. Hence $u_y = \sqrt{(4 E_{el}/M)} = 4.3$ m/s.

Any body part held by elastic forces can perform oscillatory motion at a characteristic resonance frequency f_0. Mechanical energy is always saved if f_0 is tuned to other body frequencies, such as the leg, or the wing beat frequency.

5.3.5 Energy Storage in Elastic Body Components

Many structural body components have elastic properties, where an applied force F produces a displacement Δx or a lateral deflection $\Delta \phi$. In first approximation the elastic medium may be described by a linear response

$$F = k \Delta x. \tag{5.34}$$

For elastic components characterized by the modulus of elongation Y, cross section A, and length L the spring constant is $k = A Y / L$, see Sect. 3.3.3. Progressive springs, which become stiffer as the force is increased, may be characterized by a response $F = k \Delta x^2$, or even a higher power $F = k \Delta L^n$, or even a higher power $F = k \Delta L^n$, where $n > 2$. By compressing such springs by the distance ΔL elastic potential energy is stored. This energy can be found by integration. For a linear spring the stored energy is

$$E_{el} = \int_0^{\Delta L} F(x) dx = \int_0^{\Delta L} k x \, dx = \frac{1}{2} k \Delta L^2. \tag{5.35}$$

This energy storage does not occur instantaneously. It progresses with a time scale τ that depends on the elasticity k of the spring and the attached mass M. Potential energy moves in and out of storage at a rate set by the resonance frequency ω_{el} of the system. The characteristic energy transfer time τ is a quarter of the period $T = 2 \pi / \omega_{el}$. Animals can make use of elastic energy storage for reducing the cost of transport *provided* their locomotion time scale, which is set by the frequency of their feet, is *matched* to this energy transfer time

$$\tau = \pi / (2 \omega_{el}). \tag{5.36}$$

A condition for this time scale matching is now derived. The applied elastic force $F = k\,\Delta L$ acting on the oscillating mass M yields an oscillation frequency

$$f_s = (\tfrac{1}{2}\,\pi)\,\sqrt{(k/M)}. \qquad (5.37)$$

When k and ω_{el} are eliminated between (5.35), (5.36), and (5.37), a connection between the stored energy, and the energy storage time τ is established

$$\tau^2 E_{el} = \frac{\pi^2}{8} M \cdot \Delta L^2 . \qquad (5.38)$$

In order to achieve significant energy storage the resonance frequency must be tuned into the propagation frequency, or a multiple of it. If the animal changes its speed U it must also alter its elastic resonance frequency by changing the spring constant in order to fully utilize elastic energy storage.

An example of spring constant tuning was reported by Ferris et al. [1998]. While running, the stance leg undergoes a longitudinal elastic change in length ΔL; it acts like a compression spring with a spring constant k. Attached to it as oscillating mass is the upper body m_b. This system can oscillate at the resonance angular velocity $\omega_s = 2\pi f_s = \sqrt{(k/m_b)}$. The elastic resonance frequency f_s must be exactly twice as large as the propagation frequency of the leg f_1 in order to have the largest compression ΔL_{max} at the time t_1. This is when the stance leg passes through its vertical position, to reach the largest extension at the beginning and the end of each stride.

$$\omega_s/\omega_1 = f_{el}/f_1 = 2 \qquad (5.39)$$

Ferris et al. [1998] studied the stiffness of the leg for humans running at the constant speed $U = 5.0$ m/s on surfaces of different elasticity, and found that the combined spring constant of leg and ground was always the same, namely $k = 30$ kN/m $\pm 10\%$. The authors determined that 9% of the potential energy needed in each step could be stored in this elastic deformation. However, this energy would be largely wasted if the frequency of the periodic elastic force f_{el} were not tuned to the leg frequency f according to Eq. (5.39). The maximum stride angle and leg length were not reported, therefore we assumed typical values $\phi_o = 32°$, and $L = 0.85$ m. The propagation angular velocity is then $\omega_1 = U/(L \sin \phi_o) = 11.1$ rad/s. The average body mass of the runners was $m_b = 56.3$ kg $\pm 10\%$, yielding an elastic oscillation frequency $\omega_{el} = \sqrt{(k/M_b)} = 23$ rad/s $\pm 10\%$. Hence $\omega_{el}/\omega_1 = f_{el}/f_1 = 23/11.1 = 2.07$. Since this ratio is quite close to 2, one can conclude that the elastic vibration frequency of the leg is indeed tuned to the propagation frequency thus allowing maximum energy to be transferred into the leg spring and to be regained periodically in the propagation process.

It is suggestive to extend the concept of elastic energy storage and resonance tuning to other body components, such as the spine. Figure 5.38 shows the model animal *Bipedalus Physiciensis elegans*, a dinosaur type animal that walks on two strong hind legs, and has a flexible spine which vibrates rhythmically with every step. The center of mass stays approximately on a level path as the animal strides along. The spine with adjacent muscles (labeled as spine unit) may be considered as a solid elastic member that vibrates up and down. The motion becomes effortless if the elastic vibration is in resonance with the leg frequency. The hip joint happens to be at the highest point when one of the legs is in the vertical position. If the spine unit is stiffened by attached muscles, the compound stiffness k of the spine unit may be varied. Then the spine's resonance frequency can be kept in tune with the legs at any desired propagation speed.

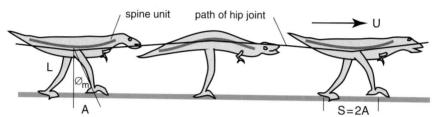

Fig. 5.11. The model animal *Bipedalus Physiciensis elegans* oscillates its spine with every step cycle

5.4 Continuous Motion

Animals have developed strategies in the water on the land and in the air to cover large distance at minimum energy expense in the pursuit of food, shelter, and mates. All forms of locomotion involve periodic motion of the locomotor extremities (legs, flippers, or wings). Features of simple harmonic motion and resonance can be detected in locomotion. Most animals are capable of different gaits, which change the internal resonance frequencies so that the animals can move with ease at slow or fast speeds. These strategies are discussed in the next chapter.

Table 5.7. Frequently used variables of Chap. 5

variable	name	units	name of units
f	frequency	1/s Hertz	
I	mass moment of inertia	kgm^2	
M	mass	kg	kilogram
T	period	s	second
U, u, v	velocity	m/s	meter/second
α	angular acceleration	radians/s²	radians/second squared

Problems and Hints for Solutions

P 5.1 Wind Speed and Rolling Resistance

A cyclist on an old bike, going east on level ground, with a rear wind, and a mechanical power of $P = 90$ W takes 12 minutes to cover the distance of $\Delta x = 4.0$ km. Returning home against the wind and applying the same power she takes 21 minutes. a) What is the wind velocity U_w, b) how much is the rolling resistance F_r? Assume a drag coefficient $C_D = 0.3$ and a cross section area of the cyclist $A = 0.7$ m².

P 5.2 Landing on the Feet

Two young people jump down from a height of $h = 2.0$ m onto the forest floor which has a mud hole and a rocky flat. The boy ($M = 70$ kg) lands with stiff legs on the muddy ground, which "gives" so that he depresses the ground by $\Delta y = 5$ cm when being decelerated from the impact velocity U_0 to 0. a) Determine the impact speed U_0. b) Calculate the deceleration (negative acceleration of the center of mass of the boy during landing), and determine the force on the soles of his shoes during impact. c) The girl ($M = 65$ kg) lands on the rock, but in order to reduce the impact force she lands with soft knees and moves her center of gravity relative to her feet by $\Delta y = 0.5$ m as she lands. What is the impact force on her soles of the feet during the landing?

P 5.3 The Energy Loss of a Hunt

a) Search the literature to collect data about a fast hunter such as wild dog, leopard, or lion. Find mass M, top speed U_t, length of leg L, length of step while foot is on the ground S, period $T = \omega/2\,\pi$ (either from $T = 1/n$ where n is the number of steps per second, or from the distance λ between two imprints of the feet on the ground: $T = \lambda/U_t$, duration τ of the hunt, or the range $R = U_t\tau$. Treat the motion of the rear leg like simple harmonic motion where the position of the foot relative to the vertical is $x = A \sin \omega t$, where $A = S/2$ is the amplitude, and $U = \omega A \cos(\omega t)$ is the velocity of the foot. The max foot velocity is $U_{max} = \omega A$, the instantaneous acceleration of the foot is $a = \omega^2 A \sin(\omega t)$, and the maximum foot acceleration is $a_{ma} = \omega^2 A$. b) Assuming that the maximum acceleration of the foot is equal to the maximum acceleration of the body, find the acceleration time t_1 needed to reach top speed $U_t = a t$. c) Determine the speed up distance s_1 (similar to the free fall distance $s = \frac{1}{2} g t^2$). d) Calculate the kinetic energy KE, which the animal attains at top speed. e) Assuming that KE was obtained by the application of an average force F acting during the time t_1 or over the distance s_1 the energy equation reads $KE = F \cdot s_1$ from which F can be found. f) Find the total energy expense of the hunt $E_{tot} = F \cdot s_1$ and express it in mass of body fat burned, taking into account a reasonable inefficiency of converting the body fat into muscle fuel.

P 5.4 Lifetime Number of Heartbeats in Birds and Mammals

The typical lifetime of a spark plug is about $N \approx 10^7$ times. Compare this number to the total number of heartbeats in the lifetime of birds and mammals.

P 5.5 Reaching Top Speed

Figure 5.12a shows a part of stiff legged bipedal animal of mass $M=20$ kg. The muscles are attached a distance $a=0.15$ m and $b=0.25$ cm away from the joint. The muscle generates a force of 800 N. The leg has a length of $L=0.75$ m.

a) Calculate the torque generated by the muscle. b) How much force $+F$ can be generated at the foot by this torque? Assume that the force F is constant, and that the foot does not slip, so that an equal and opposite force $-F$ appears at the center of mass. c) How long does it take and how far has it gone when the animal reaches a speed of (i) 10 km/h, (ii) 50 km/h?

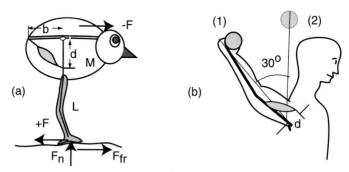

Fig. 5.12. (a) Bird. (b) Ball thrower

P 5.5 Wicket Cricket

A cricket player throws a 0.3 kg ball with stiff arm over his head, Fig. 5.12 b. Suppose he starts the throw at the position (1) and swings his arm through an arc of 30°, using only his deltoid muscle. How fast is the ball when he lets go in position (2)? To answer this question use the geometry of the muscles as shown, take the dimensions of your own arm, and assume for the calculation that the arm is a stiff member, mass M, length L, like a beam with constant cross section hinged at one end. $I=(1/3)ML^2$, specific muscle force $f=2\cdot10^5$ N/m². Explain all your assumptions, and approximations. a) Estimate the cross section of the Deltoid muscle and calculate its force. b) Calculate the average torque applied to rotate the arm. [Hint: you may simplify the problem by approximating the geometry so that the evaluation is more an estimate than a precise calculation; however, describe at the end if your method yields too small or too large a velocity]. c) Estimate the mass of the arm and calculate the mass moment of inertia for this rotation. d) Calculate the angular acceleration. e) Calculate the time it takes to rotate the arm from rest at position (1) to position (2). f) Calculate the angular velocity of the arm at position (2). g) Calculate the linear velocity of hand and ball at position (2). What is the kinetic energy of the ball and the arm at this instant? h) Discuss how the speed of the ball could be increased.

P 5.6 Gone by the Wind

A small insect of mass $M = 1$ mg $= 10^{-6}$ kg has climbed to the top of a free standing $h = 52$ m high Douglas fir, and is grabbed by a gentle horizontal wind of $u_x = 6$ m/s. To which distance S will it be carried in the horizontal direction?

P 5.7 Muscle Velocities

Make a list of the typical contraction speeds $v = \Delta L/(L \cdot \Delta t)$ of various muscles: wing muscles of locusts, humming birds, and pigeons, leg muscles of ants, mice, kangaroos, long distance runners, eyelids of a person, and plot v as function of L for all the different muscles.

Hints and Sample Solutions

S 5.1 Wind Speed and Rolling Resistance

a) Find the speed over ground U_1 and U_2. b) Find the mechanical energy $E_1 = P \cdot 12$ min $\cdot$ 60 s/min, and $E_2 = P \cdot 21$ min $\cdot$ 60 s/min that is expended on the trip to the east, and the return trip going to the west, and determine the total forces $F_1 = E_1/\Delta x$ and $F_2 = E_2/\Delta x$ that she has to overcome. These total forces $F_1 = F_{D1} + F_r$, and $F_2 = F_{D2} + F_r$ are composed of the rolling resistance F_r and the drag force e.g. $F_D = \frac{1}{2} C_D A \rho_{air} U^2$, where $U = v \pm v_{wind}$ is the relative velocity between cyclist and surrounding air. While F_D is different we assume that the rolling resistance F_r is the same going east an going west. c) You can write one equation for the total force going east and one equation for going west. This set of two equations has the two unknowns F_r and v_w. Find both.

S 5.4 Heart Beats in a Lifetime

According to the allometric relations big and small warm-blooded animals have all the same number of lifetime heartbeats. The heart frequency of mammals is $f_H = 241\, M^{-0.25}$ beats/min, and the lifetime is $t_L = 11.8 \cdot M^{+0.20}$ years $= 60$ min/h $\cdot$ 24h/day $\cdot$ 365 days/year $\cdot$ 11.8 $\cdot M^{0.20}$ years $= 6.2 \cdot 10^6\, M^{0.20}$ years. The total number of heartbeats during the lifetime of a mammal is found by multiplying both numbers $Z_m = t_L \cdot f_H = 1.5 \cdot 10^9\, M^{-0.05} \approx 1.5$ billion beats. This number is nearly independent of body mass due to the very weak mass dependency $M^{-0.05} \approx 1$. The allometric relations for birds are: $f_H = 155\, M^{-0.23}$ min^{-1}, $t_L = 28.3\, M^{0.19}$ years. These relations yield the total number of heartbeats $Z_b = t_L \cdot f_H = 2.3 \cdot 10^9\, M^{-0.04} \approx 2.3$ billion beats. It appears that the heart, of all animals is designed for a fixed number of beats. Birds have $2.3/1.5 \approx 1.5$ more heartbeats than land animals. One could say that birds on average live 1.5 times longer than land animals. This mirrors technical devices: airplanes in general live longer useful lives than cars. However, technical devices are generally made to stand up to much less use. A spark plug is built to last for typically 10^7 ignitions. The allometric scaling of the mass of the heart M_h was given in Chap. 1 as $M_h \propto M$. If the heart was a physical resonator one would then expect $f_h = $ const $M^{-0.5}$. Instead one finds $f_h \approx$ const $M^{-0.25}$. There is something still to be explained.

S 5.6 Gone by the Wind

One can give a limit for the minimum distance by assuming that the object falls all the way with the terminal velocity Eq. (5.22) $u_{y,\text{term}} = 4.5$ m/s. It will then descend at the angle ø given by $\tan ø = (\Delta y/\Delta t)/(\Delta x/\Delta t) = u_y/u_x = h/S$. Then $S = h\Delta x/\Delta y = h \cdot u_{y,\text{term}}/u_x = 52$ m $\cdot$ 6 m s^{-1}/4.5 ms^{-1} = 69.3 m. Of course the organism descends due to gravity, but is slowed down by Stokes friction. Assume the insect can be modeled as a sphere, then $F_s = 6\pi r \eta u_y$. This friction starts to become important as the animal gathers speed. Without friction it would have reached the terminal velocity within the time t_1 given by $u_{y,\text{term}} = g t_1$, or $t_1 = 4.5$ ms^{-1}/9.81 ms^{-2} = 0.45 s. At that time the animal has dropped the distance $\Delta y_1 = \frac{1}{2} g t_1^2 = 1.03$ m. More accurately one would have to solve Newton's equation: $Ma_y = \Sigma F = Mg - 3\pi d\eta u_y$, where d is the diameter of the animal modeled as a sphere. However, such an analysis overlooks the fact that the flow behind a tall freestanding tree is likely turbulent. Assume that the insect was sitting on a small branch of diameter $D = 1$ cm, then the flow has a Reynolds number Re $= Du_x/v = 0.01 \cdot 6/1.6 \cdot 10^{-5} = 3750$. This is turbulent flow with eddies that swirl around any little air traveler in an unpredictable manner.

6. Locomotion

The gaits of animals are tuned to resonate
With body parts that flex, or swing like pendulums
And presently absorb some energy,
Then let it go to help the locomotion.
The gear-shifts of the gaits are strategies
To change the length, or vertical acceleration
Or springy property of body parts that oscillate
In resonance at any speed with legs, feet, wings, or flippers.

Moving Around

Movement is a characteristic activity of all animals. The speed U of the movement is determined by a number of factors, both internal and external to the animal, and it is generally freely chosen.

All steady locomotion in air, water, and on land entails some forms of periodic motion. Often locomotion involves resonance, which is outlined in Sect. 6.1. Resonance helps to obtain maximum amplitudes at minimum input energy.

Locomotion in water is discussed in Sect. 6.2. Principles and limitations of flight are described in Sect. 6.3. Walking and running on land is outlined in Sect. 6.4. The motion in each medium has distinct gaits: steady swimming or escape-dashing in the water, stationary hovering, flapping-wing flying at low speed, and pseudo-gliding at high speeds in the air, and walking, trotting, or running on the land. Animals change their gaits, like shifting gears, in order to maintain resonance at higher speeds. In all three media, there are allometric relations between propagation speed and body mass of the form $u \propto M^{0.2}$.

6.1 Periodic Motion and Resonance

All steady locomotion involves some forms of periodic motion, which is generally accompanied by periodic *energy expenditure* due to acceleration and deceleration of limbs, and the raising and lowering of the body's center of mass. Such energy exchange occurs in swimming, in the flapping flight of insects and birds, in the brachiation of apes and monkeys, and in the various gaits of terrestrial locomotion. Much of this energetic cost of transport can be recovered through an exchange between surplus kinetic energy and gravitational potential energy in swinging appendages, like legs, arms, and tails [Alexander & Bennet-Clark 1976], or by an energy exchange with elastically deformed limbs and spines [Alexander 1988], or other body components. However, all this temporarily stored energy can *only* be recovered if the timing is right, namely if the fixed internal resonance frequencies f_{res} associated with pendulum or elastic oscillations, are tuned to the locomotion frequency f_L.

Periodic motion can be analyzed as a kinematics problem – resonance is a dynamic process.

6.1.1 Periodic Motion

Many motion sequences of animals are periodic with some cycle time T or frequency $f = 1/T$: wing beats of birds and insects, fish tail motion in steady swimming, motions of legs, and even in the vibration of sound producing organs. To study the physical parameters of such motion, consider a fish tail, of length L shown in Fig. 6.1.

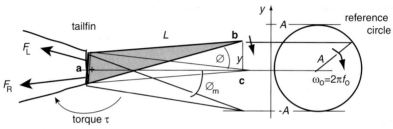

Fig. 6.1. Tail fin moving like a tuning fork

The tail angle Ø can be found from the lateral displacement y of the tip of the tail from the triangle abc as $y = L \sin$ Ø. This angle changes periodically with an angular velocity $\Omega = d$Ø$/dt$. The tangential velocity of the tail tip is therefore

$$u_{tan}(t) = L\Omega(t) . \tag{6.1}$$

The angular velocity Ω has a maximum value when the tail flips through its center position ($y = 0$), and Ω is zero when the tail tip has the largest deflection $y = \pm A$. One can find the instantaneous values of Ω and hence the angular acceleration $\alpha = d\Omega/dt$ by looking at the motion of the tail tip.

The mathematical description of the tail tip motion is easy if it may be approximated as simple harmonic motion: the tail tip swings to the left and the right like a sine wave with the lateral deflection $y = A\sin(2\pi f_o t) = A\sin(\omega_o t)$. Then the y-displacement is equal to the projection of uniform circular motion at the angular velocity $\omega_o = 2\pi f$ on the reference circle, shown in Fig. 6.1. The system has a natural frequency

$$f = \omega_o/2\pi = 1/T . \tag{6.2}$$

The velocity component in the y direction of the reference circle, u_y, is exactly equal to the y-component of the velocity of the tail tip:

$$u_y = dy/dt = \omega_o A \cos(\omega_o t) = L \sin \phi_m \cdot \cos(\omega_o t) . \tag{6.3}$$

The lateral acceleration of the tail tip is

$$a_y = du/dt = -(\omega_o)^2 L \sin \phi_m \cdot \sin(\omega_o t) . \tag{6.4}$$

A similar periodic motion arises when an appendage, like an arm or a leg, is loosely connected to a joint, and allowed to swing like a pendulum. Suppose the appendage is first pushed to one side to the deflection (amplitude) A, and then let go. Initially the pendulum will gather speed due to gravity, then swing through the rest position at maximum velocity, and thereafter decelerate as it ascends on the other side, to reach about the same lateral deflection A. Subsequently the motion is reversed, the pendulum swings back to the starting position, and continues oscillating back and forth. This pendulum motion is a continuous exchange of gravitational potential energy into kinetic energy, and back into potential energy. At the bottom of the swing the energy is all in the kinetic form. At maximum deflection of the swing all the kinetic energy has been converted into potential energy. The natural frequency of this motion is $f_o = C (g/L)^{1/2}$. The value of the constant C depends on the mass moment of inertia I of the appendage.

The tail of a fish is driven by muscles, which can only produce tension forces. Therefore, two opposing muscles are needed, represented by the top and the bottom curve in Fig. 6.2. They pull alternately in opposite directions.

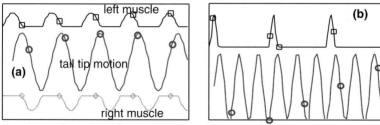

Fig. 6.2. (a) Muscle activation and tail motion for a compliant muscle. (b) Phase motion: limb motion activated in phase during every third oscillation period

For instance, the muscles pulling the tail to the left should be applied while the tail is swinging towards the left. The maximum force of the muscle on the right side of the body should be applied at the moment when the tail moves through its center position, Fig. 6.2a, so that the force is shifted in phase by $\frac{1}{4}$ of a period relative to the displacement, see Fig. 6.2a and b.

It is obvious that the muscle contraction must be carefully timed. This calls for sharp internal senses to determine the position of the moving limb. In principle the force could be applied at any arbitrary frequency f_F. However, if f_F is much higher than the natural frequency f of the system, the tail tip will be unable to follow – the tail just does not move at all. If the frequency of the applied force is small compared to f, the tail tip moves in phase with the force. This is not ideal because there should be a phase shift of ¼ period. The most efficient way is to make f_F equal to f. This condition is called resonance.

6.1.2 Resonance, a Principle to Reduce Energy Consumption

Any body part held by elastic forces can perform oscillatory motion at a characteristic resonance frequency f_o. Mechanical energy is always saved if f_o is tuned to other body frequencies, such as the leg or the wing beat frequency.

Some muscles of the body are contracted at regular intervals, exerting a time varying force $F_m(t)$ with a period T or frequency $f = 1/T$ in order to move a part of the body, so that the limb velocity $u(t)$ changes periodically with time. Members that are held by an elastic force (generated by ligaments, tendons, tissue, or cartilage), and limbs that are subject to gravity (like a pendulum) have their own characteristic mechanical resonance frequency f_o. If the frequency of the driving force f is tuned to the member frequency $f = f_o$, a small periodic force can maintain a large amplitude oscillation. This *amplification* process is of great technological and biological significance. Animals make use of it in walking, running, swimming, and flying, in hearing, voice production, and probably even in the process of detecting and emitting light and electric fields.

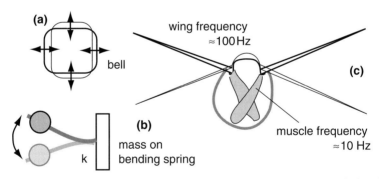

Fig. 6.3. Examples of periodic oscillations. (a) Elastic block or bell, (b) tuning fork or bending member with heavy mass m, (c) wings moved by elastic oscillation of thorax

It is not even necessary to energize the periodic motion in every cycle. Just as one can keep a swing going by pushing it only once in a while but precisely at the right phase, many technical and biological systems can be kept going at the resonance frequency f_o if the excitation occurs only every n^{th} cycle. Then the excitation frequency f_m is only f_o/n. However, careful tuning is required in such phased motion. Thus, high frequencies of wings or legs can be maintained if these extremities are tuned to internal resonances $f_{wing} = f_o$. In fact, some insects have a muscle frequency f_m which is a factor $n = 10$ lower than the wing beat frequency $f_{wing} = f_o$.

All systems with a restoring force (namely a force F that opposes a deflection s according to $F = -ks$) oscillate in simple harmonic motion (SHM) when displaced from the rest position by some distance Δs. A well known example of SHM is a mass m connected to a spring with the spring constant k, as shown in Fig. 6.3b. The mass oscillates with amplitude A and frequency f in vertical direction according to $y = A \sin 2\pi f t$, supported by the elastic force of the spring $F = -ky$. The angular frequency of this motion is $\omega_o = 2\pi f_s$. The acceleration calculated from (6.4) is $a_y = d^2 y/dt^2 = -\omega_o^2 A \sin \omega_o t$. Newton's equation $F = m\, a_y$ can therefore be written as $-ky = -kA \sin \omega_o t = m(-\omega_o^2 A \sin \omega_o t)$. The factor $-A \sin \omega_o t$ cancels leading to $k = m\omega_o^2$, so that one obtains the spring resonance frequency f_s

$$ f_s = \frac{1}{T_s} = \frac{1}{2\pi}\sqrt{\frac{k}{m}}. \tag{6.5} $$

A similar equation can be written for angular motion. In pendulum motion the spring force is replaced by gravity. For small deflection amplitudes a pendulum of length L oscillates at the pendulum resonance frequency $f_p = 2\pi\, \omega_p$

$$ f_p = \frac{C_m}{2\pi}\sqrt{\frac{g}{L}} \tag{6.6} $$

where the constant C_m is a function of the mass moment of inertia of the pendulum. Ideally, in the absence of friction, the system swings forever and ever.

Every organism tries to maintain its periodic motions of locomotion with the least possible energy input. In walking, swimming, and flying at some propagation velocity u the animal must overcome all external motion induced forces, symbolically represented by the net force F. Hence, locomotion requires the mechanical power $P = F \cdot u$. The animal accomplishes this task by pulling the limb with a periodic force $F_m(t)$ that is generated by the locomotion muscle(s). In inanimate mechanical systems power is only required while a locomotion force F moves an object at some velocity u. No motion, no power. In contrast, muscles consume metabolic power even if the limb is not moving at all. The force should act only when the object is moving in the direction of the force. Velocity $u(t)$ and force $F_m(t)$ must be carefully timed and applied at the right phase of the motion in order to obtain the most efficient energy transfer from muscle to moving limb.

Consider a harmonic deflection $y = A \cdot \sin \omega t$ which is associated with the velocity $u = \omega A \cos \omega t = \omega A \sin\{\omega t - \pi/2\}$, where $\omega = 2\pi f_o = 2\pi/T$ is the angular velocity on the reference circle. The phase of the velocity lags behind the phase of the displacement by exactly ¼ of a revolution, see Sect. 6.1.1. For the most efficient power input the driving force should also lag behind the displacement by $\delta = \pi/2$. This does not happen when the driving frequencies are very slow, because then the displacement will follow the driving force very closely ($\delta = 0$).

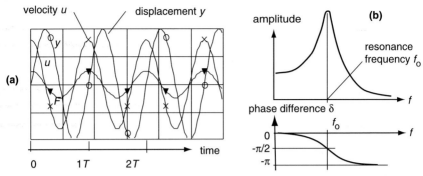

Fig. 6.4. (a) Displacement y, velocity, u, and driving force F for simple harmonic motion. (b) Amplitude A and phase difference δ as function of driving frequency f. Resonance occurs at $f=f_o$

In the other extreme, when the driving frequency is much larger than the resonance frequency f_o, the object is not at all able to follow the force. In this case the object is too slow to follow the force: just as the objects gets going in a certain direction the force has already reversed its direction. Therefore, the resulting amplitude remains very small. At high frequencies the object and the force are always going in opposite directions: $\delta \approx -\pi$ Detailed analysis of simple harmonic motion driven by a periodic force reveals that the phase angle δ changes smoothly from $\delta=0$ to $\delta=-\pi$ as the driving frequency f is increased, Fig. 6.4b. Exactly when the driving force is in resonance with the motion, at $f=f_o$, the phase angle has the optimum value $\delta = \pi/2$ for the best power transfer between the applied force and the resulting motion. Then very little energy is needed to maintain the motion, and the amplitude has its largest possible value. Figure 6.5 illustrates this point. In this graph the power (force · velocity) is given as a function of the normalized angular velocity ω/ω_0 for a pendulum that is made to swing at constant amplitude. In this example the motion has a damping constant $\gamma = \omega_0/5$. Note the rapid increase of power when ω/ω_0 goes beyond 1.

A nice illustration of a tuned resonance is the longitudinal oscillation of legs. During the running process the stance leg undergoes a small elastic change in length ΔL. The elastic force $k\Delta L$ acting on the body mass M yields an oscillation frequency $f_e=(\frac{1}{2}\pi)(k/M)^{\frac{1}{2}}$. The leg should have the largest compression ΔL_{max} at the time t_1, when the stance leg passes through its vertical position, and it should reach the largest extension at the beginning and the end of each stride. This is accomplished if $f_e=2\Omega_e$ is exactly twice as large as the propagation frequency f. Ferris et. al. [1998] studied the stiffness of the leg for humans running at the constant speed of 5.0 m/s on surfaces of different elasticity, and found that the combined spring constant of leg and ground was always the same, namely $k=3.0 \cdot 10^4$ N/m $\pm 10\%$. When the ground is soft the leg stiffens up, when the ground is hard, the leg becomes more elastic. The maximum stride angle and leg length were not reported therefore we assume typical values $\phi_0= 32°$ and $L=0.85$ m. The angular frequency of propagation is $\omega = 2\pi f= U/(L \sin \phi_0) = 11.1$ rad/s. The average

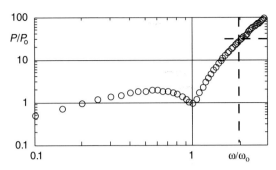

Fig. 6.5. Power P normalized to the power P_o at resonance needed to maintain a pendulum at a fixed amplitude as function of normalized frequency ω/ω_o

body mass of the runners was quoted as $M = 56.3$ kg $\pm 10\%$, yielding an elastic oscillation frequency $\omega_e = \sqrt{(k/M)} = 23$ rad/s $\pm 10\%$, which is about twice as large as ω. The leg stiffness is tuned, as expected, to produce resonance between the elastic vibration and the leg frequency.

The use of resonance in locomotion reduces the cost of transport. However, the energy savings come at a price – the motion is tied to the resonance frequency. Animals generally need to travel at various speeds. They therefore have developed various strategies to vary their resonance frequencies by tuning the parameters upon which the resonance depends: the spring constant k, the mass moment of inertia I, the length of the appendage L, or even the effective vertical acceleration of pendulum motion. Some examples will be discussed later.

6.2 Locomotion in the Water by Flippers and Tails

Nature has reinvented propagation mechanism for animals in air and water many times during the evolution of species, Fig. 6.6. Effective propagation requires first an optimal body shape and second efficient propulsion organs such as flippers or wings.

Locomotion in water, air, and on land can proceed in different gaits. Birds can glide and soar, or flap their wings rapidly for speedy advance. Land animals walk trot, or run, and fish can idle along or dart rapidly. The muscles must be applied to push backwards so that the body will move in the forward direction. In flying or swimming much of the retarding forces are associated with drag. The best strategy to reduce this drag is to avoid moving any limbs against the forward direction: generate propulsive forces by lateral motion. Flapping wings and flipping tails illustrate this strategy.

Since many body functions scale with body mass one might suspect that the empirical propagation speeds of various animals of different size do also exhibit allometric relations of the form

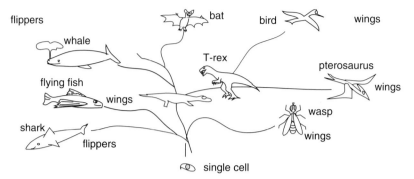

Fig. 6.6. Flippers and wings invented by insects, fish, birds, and mammals

$$U = a_u M^\alpha .\tag{6.7}$$

Indeed such a relation exists for flying animals, Sect. 6.3.2. The great flight diagram, Fig. 6.20, assembled by Tennekes [1997], yields the empirical flight velocity relation $U_{fl} \approx 15\,M^{1/6}$. The walking speed, Sect. 6.4.2 scales as $U_w \propto M^{1/6}$ and a similar relationship seems to exist for brachiation, Sect. 6.4.1, and for swimming, Sect. 6.2.1. In principle the propagation velocity can also be deducted from cost of transport data displayed in graphs like Fig. 5.9. Propulsion strategies are discussed in Sect. 6.2.2. The most energy effective strategies in swimming and flying are propulsive motions where the muscles apply forces at right angles to the swimming or flying direction. At very slow speeds the organism has only to contend with friction forces, Sect. 6.2.3. At high speeds animals extract forward momentum from vortex production, Sect. 6.2.4.

6.2.1 How Fast Are Swimmers?

Is there a systematic relation between body mass M of aquatic animals and their typical swimming speed U? Intuitively one would assume that small animals swim slower than large ones. Average velocities of aquatic animals can be found in the literature. Such values from various sources are plotted as function of body mass in Fig. 6.7.

The data can be approximated as an allometric swimming velocity function of body mass M

$$U_{sw} \approx 0.5\,M^{0.19} \approx a_{sw}\,M^{1/6}\,\text{m/s, where } a_{sw} \approx 0.5 .\tag{6.8}$$

To the best of the author's knowledge, no model has been proposed to explain this allometric velocity relation. The data suggest the scaling of body lengths $L \approx 1.7\,M^{1/3}$ m. However, for large fish with masses above 10 kg the scaling is better described by $L_{large} = 1.5\,M^{0.43}$ m.

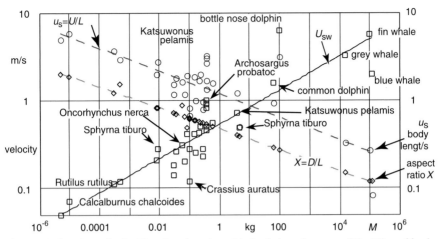

Fig. 6.7. Swimming velocity U_{sw} given in m/s and in body length per sec (U_{sw}/L); and body aspect ratios $X = D/L$ as function of body mass M. Data collected from various sources by M. Y. Yeung [2001]. The gray whale velocity was measured by Megill [2000]

From these data one can derive the specific swimming velocity $u_s = U/L$, which describes how many body lengths the animal advances per second.

$$u_s = U/L \approx 2M^{-0.1}\,\mathrm{Hz} \tag{6.9}$$

Also shown in Fig. 6.7 is the aspect ratio, defined as profile height D divided by body length L. The total drag or compound drag of a submerged object is composed of skin friction and hydrodynamic (vortex production) drag. The compound drag depends on the aspect ratio, as outlined in Sect. 3.34. It has a minimum for a certain aspect ratio. For mid-sized fish the drag minimum is found to be $X = D/L \approx 0.25$. This aspect ratio is indeed attained by fish in the body mass range up to 10^2 kg. However, for large aquatic animals the drag minimum is shifted to smaller aspect ratios. Large whales have aspect ratios approaching $X \approx 0.12$.

6.2.2 Propulsion Strategies at Higher Speeds

Flippers are built to generate forward forces. Think of an oar, it is pulled through the water at the velocity u in the backwards direction, so that the reaction force can drive the boat forward at the speed U. Swimming dogs cannot lift their legs out of the water. Instead they pull their legs forward with contracted claws and pulled in lower limbs to reduce the resistance, on the backward power stroke they extend their paws providing a larger area for increased resistance.

Flippers and wings are built to use a different principle: the main motion is not parallel or anti-parallel to the propagation direction, but rather at right angles to the body motion. Therefore, there are no phases where flippers or wings move op-

Fig. 6.8. (a) Oar and boat, (b) dog paddle

posite to their traveling direction. If the object moves with high velocity u through the fluid so that the Reynolds number $\mathrm{Re} = u\,d/v$ exceeds the value 100, eddies will appear. These vortices cause additional drag on the moving object. However, a skillful swimmer can make use of the eddies to improve the propulsion.

The main principle of generating thrust by lateral motion is first the creation of a rearward jet of mass flow J and velocity U, and second the judicious use of eddies that are produced in the surrounding fluid. These principles will be discussed in the following sections.

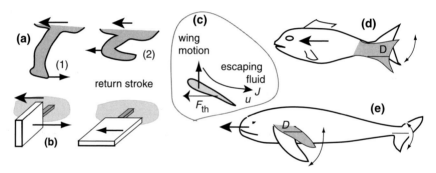

Fig. 6.9. Propulsion strategies: (a) flexing the limb, (b) rotating the limbs, (c) thrust production by wing, (d) lateral tail deflection, (e) flipping flippers or wings

6.2.3 Swimming at Slow Speeds

Any object pulled or pushed slowly through a fluid medium, like air or water, experiences skin friction drag forces. They arise because some fluid elements attach themselves to the body and are dragged along carrying with them adjacent layers of fluid in a region called the boundary layer, see Fig. 6.10 a. The object must continuously accelerate new layers of water as it slips through the fluid. Acceleration requires a force: the skin friction drag force.

Slow swimmers need only to fight the boundary layer drag. The propulsive force F can be generated by a flagella, connected to the body at the root R. It whips up and down, sideways or in a circle, like a cowboy may move his bull whip. The oscillating flagella looks like a transverse wave that emanates from the root and travels to the loose end, Fig. 6.10 b.

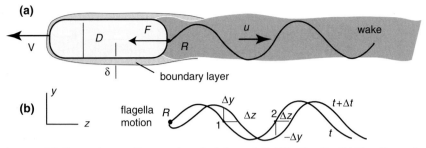

Fig. 6.10. (a) Slow swimmer (Re ≪ 100) pushed through fluid by flagella. (b) Flagella motion

Consider the motion after some time interval Δt. The wave has traveled rearwards through the distance Δz. Fluid elements attached to the flagella at points 1 and 2 must have been forced backwards through the distance Δz. There must be a net motion of fluid J kg/s in rearward direction streaming at the bulk velocity u, Fig. 6.10a. The body wave also has motion components in the y-direction. However, there is no net fluid motion in the y-direction because the lateral motion $+\Delta y$ of fluid elements at point 1 is annulled by an equal and opposite motion in $-\Delta y$ at point 2, half a wavelength downstream.

The rearward flow generated by the flagella acts similarly to the exhaust flow of a rocket. It generates a thrust force $F = Ju$ that propels the body and gives it a steady forward velocity V. In steady state the thrust force F is exactly equal to the drag force F_D that impedes the motion. For a spherical body of radius r propagating at slow speeds (Re ≪ 100) one may use Stoke's friction to relate the thrust force $F = F_{stokes}$, and the velocity V

$$F_{stokes} = 6\pi r\eta\, V. \tag{6.10}$$

For the calculation of this force refer to viscosity η values are given in Table 3.4.

Consider an organism of $r = 1.4$ µm swimming in blood plasma ($\eta_{bpl} = 1.6 \cdot 10^{-3}$ kg/ms) at 4 body length per sec. Its speed is $U = 4 \cdot 2\,r\,s^{-1} = 8 \cdot 1.4 \cdot 10^{-6}\,ms^{-1} = 11.2$ µm/s. It must generate a propulsive force $F = 6\pi\, 1.4 \cdot 10^{-6}\,m \cdot 1.6 \cdot 10^{-3}\,m \cdot 1.12 \cdot 10^{-5}$ kg/ms $= 4.7 \cdot 10^{-13}$ N. Not a big force indeed, but enough to propel the organism.

6.2.4 A Model for Fish Propulsion from Rest

Some time ago it was noted that the drag force acting on the body shape, such as a (dead) fish, is about 10 time larger than the muscle force which the fish could generate in steady swimming. Fish should not be able to swim at all [Gray 1955]. This contradiction is known as Gray's paradox. Dead fish create a flow field, which differs radically from the pattern that living fish generate in the water.

When objects move quickly through the water they produce eddies. It appears that swimming animals actually make use of these vortices for their forward

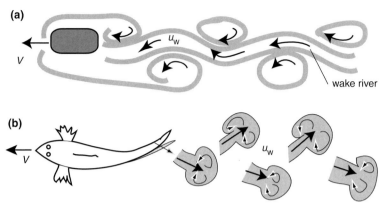

Fig. 6.11. (a) Von Karman vortex street of wakes of a dragged object (passive wake). (b) Active wake of fish

propulsion. A model for this motion is given below. There are fundamental differences between the flow field created by objects that are passively pushed through a fluid, Fig. 6.11a, and objects that actively propel themselves by pushing fluid, Fig. 6.11b. At Reynolds numbers exceeding Re ≈ 100 vortex pairs are ejected in a rearward direction.

Dragged objects produce a passive wake, the von Karman vortex street where the fluid in the wake river follows the object meandering between two rows of regularly spaced vortices, Fig. 6.11a. Drag and vortex shedding frequency are closely related [Ahlborn, Seto, and Noack 2002]. The vortices have the same sense of rotation as the eddies that are generated on the body itself – they mesh the forward motion of the object and its wake to the stagnant fluid sideways of the object. The energy required to establish the vortices and the wake-river, which meanders between them, causes the *hydrodynamic drag*, sometimes also called the form drag of the object.

In contrast, the fluid in the active wake of a self-propelled object moves in the opposite direction to the body and all the vortices of the active wake rotate opposite to the von Karman vortices.

The second difference of these two flow types is the manner in which the wake river is set into motion. In the passive von Karman vortex street, the wake fluid is dragged by tension. It is sucked forward by the reduced base pressure behind the object. The active wakes of the aquatic animals are compression driven. Fluid is pushed rearwards by forces generated by the tails or flippers. This process generally breaks up the wake flow into jet-like current sections, which are deflected alternately to the left and to the right, each carrying a vortex pair at their leading edge, Fig.6.11b. The flow field of the fore-body actually helps in the propulsion mechanism. The anterior part of a swimming fish behaves like a dragged object, producing "anterior" vortices that roll along the body with the rotational sense of a passive wake. The vorticity of the "wrong sign" is bought with an energy loss incurred when overcoming the form drag of the anterior part of the fish. The eddies set up by the fore-body have the same sense of rotation as the startup vortex in the

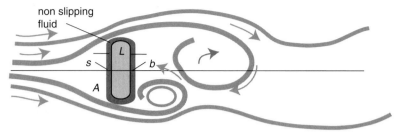

non slipping
fluid

Fig. 6.12. Body with non-slip boundary layer

fast start mode. They facilitate a structure in the water against which the animal can lean its tail to gain forward momentum. In slow and stealthy motion the fish might just annihilate the vorticity; however, for the generation of maximum thrust, the tail fin would invert the sense of rotation of the initial vortices and produce an active wake with vortices that turn in a direction opposite to the anterior body vortices.

The rate of production of vorticity ζ can be obtained by taking the curl of the Navier Stokes equation, see Batchelor [1967].

$$\zeta' = \partial \zeta / \partial t = -(u \cdot \nabla)\zeta + (\zeta \cdot \nabla)u - \nabla \times (1/\rho)\nabla p + v\nabla^2 \zeta \qquad (6.11)$$

Most important is the third, vorticity-generating term on the right hand side. This quantity is zero by the well known vector identity as long as the pressure field contains no discontinuities. However, $\nabla \times (1/\rho)\nabla p$ can be non zero when the potential field contains a jump at some contour line, where p/ρ attains different values if the contour line is approached from different directions. This happens for instance at the surface of a body. For further details see Ahlborn et al. [1997], and Ahlborn, Lefrançoise, and King [1998].

This substitution is valid because: 1) The inertia of the infinitely dense fluid prevents it from being moved by the fluid flow, thus making it effectively rigid. 2) The Navier Stokes equation ensures that the velocity is continuous across the interface between the two fluids, thereby establishing the no slip condition.

To see this most clearly, replace the rigid body and its awkward no-slip boundary condition with an equivalent volume of infinitely dense fluid. With suitable assumptions (6.11) can be integrated over an area to obtain the rate of change of circulation Γ' produced by the difference of the pressures p_s at the stagnation point in front and the base pressure p_b behind the object.

$$\Gamma' = \frac{\partial}{\partial t}\left\{\int_{area} \varsigma \, dA\right\} = (\Gamma_f - \Gamma_{in})/\Delta t = \frac{1}{\rho}(p_s - p_b) \qquad (6.12)$$

This equation may be interpreted in two ways: (i) when $p_s > p_b$ the right hand side is positive the final circulation Γ_{fin} is larger than the initial circulation Γ_{in}. This implies that vorticity is generated by the drag force.

$$F_{\text{drag}} = \int_{\text{fa}} p \, da - \int_{\text{fb}} p \, da \approx A(p_a - p_b) \tag{6.13}$$

where A is the cross section area of the object obstructing the flow. Integration is carried out over the front surface f_a, and the back surface f_b. (ii) If on the other hand vorticity is destroyed $\Gamma_{\text{fin}} < \Gamma_{\text{in}}$, the right hand side must be negative and p_b must be larger than p_s. Thus a propulsive force F_p is present. It is related to the rate of change of the circulation.

$$(\Gamma_{\text{in}} - \Gamma_f)/\Delta t = (1/\rho)(p_b - p_s) \approx F_p/\rho A \tag{6.14}$$

With the condition $\Gamma_{\text{fin}} = 0$, Eq. (6.14) describes the stealth motion of propulsion, arising from the destruction of some initially existing vorticity. The propulsive force increases if the initial and the final circulation Γ_{fin} and Γ_{in} have different signs, so that

$$(\Gamma_{\text{in}} - \Gamma_{\text{fin}}) = (\Gamma_{\text{in}} - (-\Gamma_{\text{fin}})) = (\Gamma_{\text{in}} + \Gamma_{\text{fin}}). \tag{6.15}$$

Hence, one can see that a large propulsion force can be generated through the reversal of the direction of the vorticity, and any initial vorticity of the dragged fore-body actually increases the propulsion force.

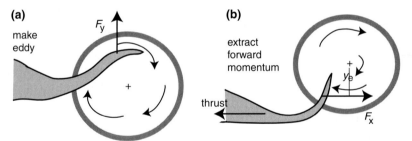

Fig. 6.13. Propulsion model. (a) Produce an eddy. (b) Extract forward momentum

The inspiration for this propulsion model came from experiments with flat plate tail models in a water tank. Flow visualization revealed the vortex structure Ahlborn et al. [1991]. The initial side flip of a tail generates a start up vortex, which has the "wrong" sense of rotation. The forward force is produced on the return flip where the tail pushes against this eddy and either stops it or even turns it in the opposite direction. Some of the energy seems to be stored in elastic deformation, because slightly flexible tail fins generate more forward momentum then either stiff or soft tail fins.

The first sideways flip generates an initial eddy with the energy $E_1 = \frac{1}{2} I \omega^2$ and the angular momentum $G_1 = I \omega = (2 E_1/\omega)$. Subsequently, the tail is bent to apply a push in the $-x$ direction, which changes the angular momentum of the eddy from

G_1 to G_2 by applying the force F_x with the moment arm y_e representing the torque τ. Let $\Delta t = T/4$ be the time that the force is applied $\tau = F_x y_e = dG/dt \approx \Delta G/\Delta t \approx (G_2 - G_1)/(T/4)$. Then a force spike F_x is created.

$$F_x = 4(G_2 - G_1)/(y_e T) \tag{6.16}$$

Measurements with fish tail models mounted on a force transducer platform revealed this force spike at the time when the vortex pair separates from the tail Ahlborn et al. [1997].

In *stealth* motion the fish may elect to leave no eddy behind making $G_2 = 0$. Then $F_x = -4 G_1/y_e T$. A larger force F_x can be extracted if G_2 is negative. One can show that the least energy is needed if the angular momentum is exactly inverted: $G_1 = G_2$. In inverting the direction of circulation, the fishtail turns the flow field of the fore-body (which would yield a passive wake) into an active wake.

As in so many other circumstances, nature has managed to turn an adverse event into an advantage. The eddies generated by the fore-body would mean extra drag. However, in reversing the direction of the circulation of these eddies the fish is able to extract additional forward thrust. These eddies act like stepping stones, and the total kinetic energy lost to the water stays at a minimum.

6.3 On the Wing

Air is less dense than water by the factor 1000. Therefore, animals who take to the wing, can never be neutrally buoyant. They must continually produce lift to support their body weight, and propulsive forces to move. This calls for different strategies than aquatic propulsion. The smaller density of the air leads to much smaller inertia forces residing in any eddies. Therefore, only insects and very small birds can lift their weight by shedding eddies in unsteady motion. Air glides smoothly over the wings, as seen in the flow field simulations (Figs. 6.14 and 6.15) showing a wing profile in a towing tank photographed by F. Ahlborn [1918].

The wing profile is dragged through the water in a tank at some speed U and it is photographed with time exposure $\Delta t = 0.025$ s by two different cameras. One camera is dragged with the wing (Fig. 6.14), showing the streamlines of the water that sweeps by. The other camera sits motionless besides the tank (Fig. 6.15). Notice how the wing is smeared out in this photo since it has moved some distance during the exposure time Δt. As usual in Fluid Dynamics, the propagation velocity is designated in this section by the upper and the lower case letter U. There should not be any confusion with the internal energy U since internal energy is not an important parameter of fluid flow.

Fig. 6.14. Stream lines around a wing profile as seen from the moving body. Photo Friedrich Ahlborn, Collection [1918]

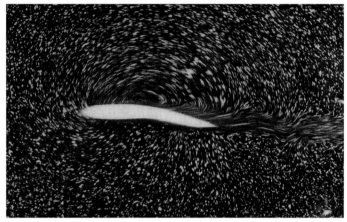

Fig. 6.15. Stream lines around wing profile. Reference frame fixed in the fluid. Photo Friedrich Ahlborn, Collection [1918]

6.3.1 Generation of Lift

Lift can be generated in steady and unsteady motion. The unsteady lift is associated with vortex shedding. Insects and humming birds make use of this continuously, and bigger birds use it during take off and landing. However, bigger flyers don't possess the muscle power to use unsteady lift for more than a few wing beats, and they could run afoul of the drag/lift crisis where only very small vortices are shed.

The steady lift force is generally given as function of speed U, wing surface area S, density of air ρ, and an empirical constant C_L the lift coefficient.

$$F_L = \tfrac{1}{2} C_L \rho S U^2 \tag{6.17}$$

Lift in steady motion can be explained (i) by the Bernoulli effect, and (ii) by the continuous deflection of air in downward direction. The air passes more quickly

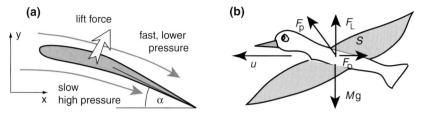

Fig. 6.16. (a) Bernoulli effect on wing. (b) Bird in steady flight

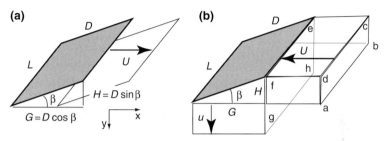

Fig. 6.17. (a) Wing moving relative to air. (b) Air flow relative to wing

over the top surface of the wing than over the bottom surface, generating a lower pressure above the wing. The air stream on the lower side is deflected downward by an airfoil with attitude angle β. This down wash acquires a velocity component $-u_y = u$, which gets larger with increasing β. The down draft represents a vertical momentum flux that generates a thrust force with vertical component $+F_y$ which lifts the wing. Both effects are combined in the lift coefficient, which one can formally write as $C_L = C_B + C_L$, where C_B describes the Bernoulli component of the lift and C_L the down wash component.

From empirical data as shown in Fig. 3.23 it is known that C_L grows with the attitude angle β. This dependency can be derived from first principles for the down wash thrust force F_y, using the conservation of mass. Consider a flat plate of length L and span D, inclined at the angle β, which moves through the air at the speed U. This model wing is shown in the reference frame of the air in Fig. 6.15. We neglect the frontal mass flow, which slips over the upper edge of the wing, and assume that all the mass is deflected downwards. In the reference frame of the wing, Fig. 6.17b the air is approaching at the speed U. Every second the mass of air, contained in a box of height H, width L, and length $\Delta x = U \cdot 1$ s, must be deflected downwards. The frontal cross section of the wing $L H = L D \sin \beta$ depends on the angle of attack β. The projected area of the wing as seen from below is $L G = L D \cos \beta$. The incoming mass flow $J_x = \rho H L U$ must be equal to the mass flow $J_y = \rho G L u$ departing from the wing in vertical direction, since no mass is lost in this process. Hence the air is deflected downwards at the average speed u

$$u = \frac{\rho \cdot H \cdot L \cdot U}{\rho \cdot G \cdot L} = \frac{H}{G} \cdot U = U \cdot \frac{\sin \beta}{\cos \beta} = U \cdot \tan \beta. \tag{6.18}$$

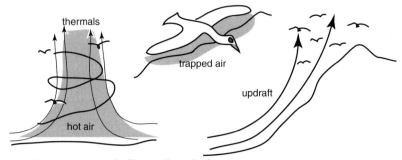

Fig. 6.18. Thermals, ground effect, and up-drafts

The vertical mass flow is $J_x = J_y = \rho\,GLu = \rho\,LuD\cos\beta$. Therefore, the vertical component of the lift force is:

$$F_y = J_y u = (\cos\beta \cdot \tan\beta)\rho\,DLu^2 = \tfrac{1}{2}\,C_L\,\rho\,DLU^2. \tag{6.19}$$

The lift force is a function of the wing surface area $S = DL$, the hydrodynamic impact pressure $\tfrac{1}{2}\rho U^2$, and down wash lift coefficient C_L. For both wings

$$C_L = 2\cos\beta \cdot \tan\beta = 2\sin\beta. \tag{6.20}$$

This coefficient represents one part of the lift coefficient $C_L = C_B + C_I$. The other part C_B is attributed to the Bernoulli effect. The Bernoulli component of the lift is the major effect particular for small attitude angles β. The derivation of Eq. (6.19) shows that the lift force is a function of the wing surface area DL, the density of the air ρ, the square of the velocity U, and the attitude angle β. In reality, lift coefficients must be obtained empirically. Typical values are $0.2 \leq C_L \leq 1.5$.

In addition to this steady lift many birds use other effects to stay aloft, Fig. 6.18. Seagulls, albatross, eagles, and vultures utilize updrafts and thermals to gain elevation. Many sea birds fly close to the water surface in order to use the ground effect, where air is trapped between wings and water surface. The birds then glide on a cushion of increased pressure.

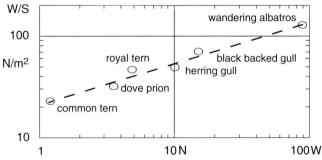

Fig. 6.19. Wing loading $W/S = Mg/S$ [N/m²] of sea birds. Data adapted from Tenekes [1997]. Note that the wing loading is generally less than 10^{-3} atm (1 atm $\approx 10^5$ N/m²)

The wings must carry the weight $W=Mg$ of the bird. A figure of merit is the wing loading W/S. This quantity increases with body mass, Fig. 6.19, and with speed U, see Eq. (6.21). Note that these wing loading values are not very large. The maximum $W/S \approx 100 \, \text{N/m}^2$ encountered by the wandering albatross is only 1/1000 of an atmosphere. When inflating a balloon by blowing into it, our lungs easily produce a pressure Δp that is hundred times as large ($\Delta p \approx 10^4 \, \text{N/m}^2$).

6.3.2 The Minimum Flight Velocity

Any flat surface that is exposed to a flow of air of speed U and is tilted at some angle of attack α will experience a lift force F_L and a drag force F_D, see Sect. 3.4.3 and Sect. 3.4.6. The lift force $F_L = \frac{1}{2} C_L S \rho_{air} U^2$ depends on the surface area S of the plate, the drag force $F_D = \frac{1}{2} C_D A \rho_{air} U^2$ is a function of the frontal area of attack $A = S \cdot \sin \beta$. The drag force is considered a loss, while the lift force keeps the flyer aloft as long as $F_L \geq Mg$. The lift coefficient C_L depends on the angle of attack β, see Fig. 6.16a. With a typical value $C_L = 0.6$ at $\beta \approx 6°$ taken from Fig. 3.23, one finds $\frac{1}{2} C_L \rho_{air} r \approx 0.3$, thus

$$W = Mg = 0.3 \, S \, U^2, \text{ or } W/S = 0.3 \, U^2. \tag{6.21}$$

This lift force increases with the velocity. The larger the velocity U the more weight W can be supported. With the help of (6.21) one can determine for any given weight Mg and surface area S the minimum speed U that is needed to exactly support the weight Mg by the lift of the wings. Let L be the typical lateral dimension of a flyer, then for flying objects that are similar $W \approx L^3$, or $L \approx W^{1/3}$. Then $S \approx L^2 \approx W^{2/3}$, so that

$$W/S = \text{const } W/W^{2/3} = \text{const } W^{1/3} = \text{const } (Mg)^{1/3}. \tag{6.22}$$

The quantity W/S can be eliminated between (6.21) and (6.22) in order to obtain the flight velocity U as a function of the 1/6th power of the body mass M

$$U = a_f M^{1/6}. \tag{6.23}$$

Experimental values of flight velocities as function of body masses are known from the great flight diagram by Tennekes [1997]. These empirical values are reproduced in Fig. 6.20. The flight velocities can be approximated by an allometric relation with the exponent $\frac{1}{6}$, shown as dotted line U_f. The empirical metabolic factor $a_f \approx 15$ can be found by reading the velocity from the dotted line at the mass $M = 1$ kg. The gentle slope of the flight velocity curve has great importance for the lifting heavy bodies. If the flying speed is increased by a factor 2, the weight of the flying body can be increased by a factor of $2^6 = 64$.

Figure 6.20 shows that this rule apparently applies to every actively flying structure. The graph covers over 12 orders of magnitude in weight, from the tiny

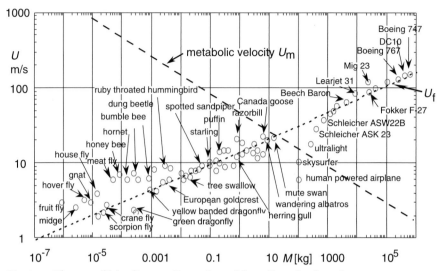

Fig. 6.20. The great flight diagram. Data adopted from Tennekes [1997]

scorpion fly to a Boeing 747. Sky surfers and ultra lights, do not fit in because they use updrafts to stay aloft.

The Gossamer Albatross, built by the team of McReady, crossed the English Channel to win the Kremer Prize. It was powered by a cyclist, who produced a steady 125 W of mechanical power. The ultralight flew very close to the water surface to utilize the ground effect. Halfway across the Channel it encountered a tanker, and in order to avoid a collision the aviator pulled the plane up and over the ship. This required so much extra effort that the cyclist did not have enough energy left to fly the plane up the beach to the waiting dignitaries. Instead he crash-landed the Gossamer Albatross into the surf of the French coast.

Birds have to overcome the drag resistance $F_D = \frac{1}{2} C_D A_f \rho U^2$, which depends on the front surface area A_f, the speed U, and the drag coefficient C_D. If they fly the distance d they will expend the energy

$$\Delta E = F_D d . \tag{6.24}$$

By solving this equation for $F_D = \Delta E / d$ one can see that the drag resistance may be interpreted as the energy expended by the object per distance of travel d. The energy ΔE has been imparted to the fluid. It resides initially in the induced fluid motion, in eddies, and wake rivers. Gradually this macroscopic kinetic energy will dissipate into heat. Note also that $\Delta E / (d \cdot M) = E_t$ is the cost of transport E_{tr}, which is empirically known from Fig. 5.9.

When flying for the time Δt at the velocity U a bird must apply the power

$$P = \Delta E / \Delta t = F_D U . \tag{6.25}$$

Hence, one can determine the drag resistance from the energy balance (6.24) or from the power balance (6.25).

$$F_D = \Delta E/d = M \cdot E_{tr} = P/U = \eta \, b \, a \, M^{3/4}/U \, . \tag{6.26}$$

One can also define a specific energy, namely the energy consumption per meter and per unit weight (mass).

$$E_{tr} = F \cdot d/Mg = P/(MgU) \, . \tag{6.27}$$

6.3.3 Why Big Birds Cannot Fly

Flight requires enough speed to become airborne. The model Eq. (6.23) suggests the relation $U_f = a_f \cdot M^{1/6}$. The great flight diagram yields the allometric constant $a_f \approx 15$ to give a limit for the velocity of level flight:

$$U_f \geq 15 \, M^{1/6} \, . \tag{6.28}$$

This speed must be generated by the body, which operates at the metabolic power $\Gamma = b \, a \, M^{3/4}$. A certain fraction η of this metabolic power will appear as mechanical power $P_{mech} \leq \eta \, \Gamma$. At the metabolic velocity U_m the mechanical power must overcome the drag force $F_D = F_L(A \, C_D/(S \, C_L)) = Mg(A \, C_D/(S \, C_L))$, where A is the frontal cross section, and S the wing surface area. The lift force F_L must offset the weight of the body $W = Mg$. Hence one has

$$P_{mech} = U_m F_D = U_m \, Mg \, A \, C_D/(S \, C_L) \leq (ab\eta) \, M^{3/4} \, . \tag{6.29}$$

We solve this equation for U_m in order to obtain the velocity that can be sustained by the metabolic power of a bird having wings with the lift to drag ratio C_L/C_D, namely

$$U_m \leq a_m M^{-1/4}, \text{ where } a_m = (S/A)(C_L/C_D) \cdot (ab\eta/g) \approx 40 \, . \tag{6.30}$$

This metabolic flight velocity decreases with body mass. Here we have set $C_L/C_D \approx 6$, $S/A \approx 12$, $\eta = 25\%$, $a\,b = 24$ ($b \approx 6$, $a \approx 3.6$), and have drawn U_m into the great flight diagram. Active flight is possible only for $U_m \geq U = 15 \, M^{1/6}$. For masses larger than about 15 kg the achievable metabolic velocity U_m falls below the required flight velocity $U_f \approx 15 \, M^{1/6}$, indicating that larger animals do not have the muscle power, to sustain the flight velocity necessary to stay aloft. The mute swan with $M \approx 15$ kg is the largest flying bird listed on this diagram. The human powered airplane is actually not an "honest flyer" since it uses the ground effect to generate extra lift. It seems that nature again has reached the limit of what is physically possible.

6.3.4 The Hovering Flight of Insects and Humming Birds

Insects and very small birds are able to hover motionless in midair. Their wings, length L, and span width D, swirl around at frequencies somewhere in the range $30\ \text{Hz} \leq f \leq 500\ \text{Hz}$, generating a down draft of air with some mass flow rate J. Their wing motion may be approximated as shown in side view in Fig. 6.21.

The wing tip follows a figure 8 path. Part of the motion can be modeled as steady flight, namely when the wing tip travels from the forward position 9, in Fig. 6.21, to the back leaning position 11, and on the return stroke, when the wing tip travels from position 3 to position 5. In these quasi steady motion phases the lift force F_L is given as usual by (6.17) where the surface area of each wing is proportional to DL, and $U = u_x \propto Lf$ is the average horizontal velocity. The lift coefficient C_L depends on the attitude angle α of the wing.

A major contribution of the lift is generated in the turn around phases illustrated in Fig. 6.21 by the positions 12–1–2 and 6–7–8. This lift arises from vortex shedding, similar to a fish generating an impulsive force when it whips its tail from one side to the other [see Ahlborn et al. 1997]. Without this non-steady lift contribution, hovering is apparently not feasible, [see for instance Dickinson 2001]. The turn around generates a vortex pair flanking a strong temporary downdraft with mass flow J and the average downward velocity u_y. This downdraft creates the unsteady lift force

$$F_{Lu} = J\,u_y \propto L\,D^3\,f^2. \tag{6.31}$$

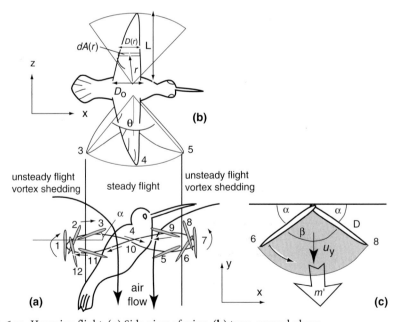

Fig. 6.21. Hovering flight. **(a)** Side view of wing, **(b)** turn-around phase

Hovering flight can be achieved only by animals within a certain body mass range. The limitations arise from several physical restrictions. 1) Vortices, which are needed to generate unsteady lift, are shed only in the Reynolds number range $Re \geq 50$. No vortices are produced below this range. 2) The metabolism of the animal must be able to generate the required mechanical power. 3) The produced lift force must exceed the weight. These conditions are now explained qualitatively. For more details see problems P 6.3 and P 6.4.

The typical length dimension of vortex shedding in the turn-around phases is the wing span width D. The typical velocity u of this motion is the speed of the lower edge of the wing. It scales with frequency f as $u \propto Df$. Therefore, the minimum Reynolds number condition $Re = uD/v = \text{const } D \cdot Df \geq 50$ sets the frequency range

$$f \geq /C_1 D^{-2}, \tag{6.32}$$

where C_1 is a constant. The lowest permitted frequency $f = C_1 D^{-2}$ is shown as the Reynolds number limit, line (1) in the log-log plot Fig. 6.22. The metabolic rate $\Gamma = \text{const } M^{4/3}$ of flying animals yield an upper bound of the hovering range. The mass scales as $M = \text{const } D^3$. The available muscle power $P_{met} = \eta \Gamma \propto M^{3/4}$ must exceed the needed mechanical power $P = Fu$. For a scaling estimate we set $F = F_L = \text{const } L D^3 f^2$ and $u = \text{const } Lf$, to get $P_{met} \propto \Gamma \propto M^{3/4} \propto D^{9/4} \geq P \propto D^5 f^3 D$. This relation can be solved for the metabolic power limit

$$F = C_2 D^{-11/12}, \tag{6.33}$$

which is schematically shown as curve (2) in Fig. 6.22.

The lower bound of the hovering range arises from the requirement that the lift force $F_L \propto D^4 f^2$ must be larger than the body weight $Mg \propto D^3$, resulting in:

$$f \geq C_3 D^{-\frac{1}{2}}. \tag{6.34}$$

The limiting mass-lift curve is labeled (3) in Fig. 6.22. These three limits enclose a small triangle in the $f - D$ plane, Fig. 6.22, where hovering is possible. Only

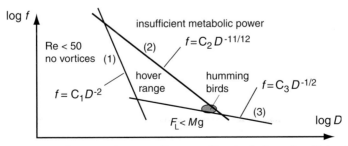

Fig. 6.22. Frequency f versus wing span D range of unsteady hovering flight. Curve (1) Reynolds number limit, (2) metabolic limit, (3) minimum lift limit

small animals can hover; the smaller the flyer, the larger the required wing beat frequency. Humming birds with $M \approx 0.01$ kg occupy a corner of this hovering range triangle. They are the largest hovering animals.

Hovering flight in essence involves *horizontal* wing motion. Larger birds can also "hang in the air" for a short while, beating their wings up and down. This flapping flight involves a different technique: the vertical air stream needed to generate upward thrust is made by a *vertical* motion component of the wings.

6.3.5 Flapping Flight

Flapping flight allows birds to generate forward thrust <u>and</u> lift to stay aloft at horizontal speeds U that are smaller than the minimum flight speed suggested by the great flight diagram $U \approx 15\,M^{1/6}$, because the relative velocity between wing and lift structure is increased by the flapping motion, Fig. 6.23a. The attitude angle α of the wings changes smoothly throughout the stroke cycle Fig. 6.23b. Thrust generation occurs during the downward strokes of the wings, Fig. 6.23c. Lift can be produced in the down-stroke as well as in the up-stroke phase.

Two distinctive gaits can be identified. At low speeds the wings can interact with the oncoming air to produce vortices that generate lift and propulsive force as they are shed from the trailing edge of each wing. In this mode vortices are periodically generated, which look like the rungs of a ladder, Fig. 6.23d. The vortex "rungs" are produced with every downbeat of the wings. They generate a spike in the propulsion force F_x, and in the lift. The vortices merge into a longitudinal vortex trail on either wing tip, which follows the flyer like two continuous bands. At higher speeds the interaction with the air generates a different gait. The wings leave the space faster than vortices can be formed locally. Then the flyer leaves behind only the two longitudinal wingtip vortices. The bird sails on the incoming air with angles of attack that generate lift in every wing position.

The transition velocity U_{tr} between these two gaits can be estimated as follows. In the slow gait the wings shed one vortex in each rapid motion part of the downward stroke, which typically takes ¼ of the period $T=1/f$. The drag associated with this vortex shedding process is experienced by the bird as lift. In order for the wing to interact with the vortex, the center of mass of the bird should move by less than the span width of the wing D. Hence, the bird's horizontal velocity U should be less than $D/(T/4) = 4Df$. Therefore, one can give the approximate transition velocity as

$$U_{tr} \approx 4Df. \qquad (6.35)$$

Crows with wing spans of $D \approx 0.15$ m typically flap their wings at $f \approx 4$ Hz, when they take to the air. Then they get vortex shedding lift for horizontal velocities of less than $U_{tr} = 4 \cdot 0.15 \cdot 4 = 2.4$ m/s.

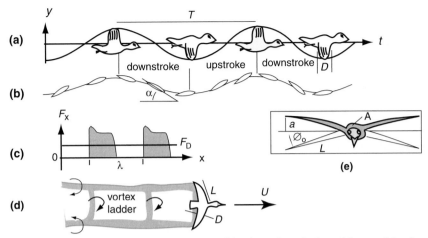

Fig. 6.23. Flappping flight: (a) wingtip motion, (b) orientation of wings, (c) propulsive force F_x and drag F_D. (d) Vortex pattern in slow flight, (e) wing amplitude a and deflection angle ϕ

Lift in flapping flight may be explained as a drag force produced by the downwards velocity of the wings. The flapping frequency of the wings f depends on the desired speed U. We model the wing tip oscillation as simple harmonic motion oscillating between the maximum deflection angles $+\phi_0$ and $-\phi_0$. The angular velocity on the reference circle is $\omega = 2\pi f$. The maximum deflection of a wing, width D, and length L, is $a = L \sin\phi_0$. Let u_y be the average downwards motion of the wings, namely ½ of the maximum speed of the wing tip

$$u_y = \tfrac{1}{2}\,\omega\,a = \pi f L \sin\phi_0. \tag{6.36}$$

As a wing moves downward at the speed u_y, it displaces air that must flow out in the rearward direction with a velocity u_x. The thrust $F_x = J_x u_x$ is generated by the rearward mass flow J_x, which grows proportional to u_x and the wing area DL. This thrust must overcome the drag force $F_D = \tfrac{1}{2} C_D \rho A U^2$ where $A = C_0 L^2$ is the frontal cross section area of the bird, and C_0 is a constant. In steady motion both forces are equal $F_x = J_x$, $u_x = C_1 DL\rho\,u_x^2 = C_2 DL\rho f^2 L^2 = F_D = \tfrac{1}{2} C_D \rho\,C_0 L^2\,U^2$. Various parameters can be cancelled and the constants can be combined to yield a relation between the wing beat frequency f, flight velocity U and body mass M.

$$f = C\frac{U}{\sqrt{DL}} = C_f\,M^{-1/6} \tag{6.37}$$

The constant C_f depends on the maximum deflection angle ϕ_0, the attitude angle of the wing α, the scaling factor C_0 between frontal area and length dimension, and the scaling between wing area DL and body mass M.

The birds expend mechanical power first for overcoming the fluid dynamical resistance $P_x = F_D U$ associated with their forward motion in x-direction and second in the vertical motion of beating their wings, which must support their body weight $P_y \approx M g \, u_y$. The total power expenditure scales as

$$P = P_x + P_y \approx C_4 \, L^2 \rho_{air} \, U^3 + C_5 \, M^{5/6} \, LU \, . \tag{6.38}$$

Flapping flight must be very energy consuming. Birds use it only for taking to the air, and for descending. Very big flying birds avoid flapping their wings. Albatross sail on the up-drifts above ocean waves, and vultures wait for thermals to develop in the morning sun to reach great heights.

6.4 Locomotion with Arms and Legs

Our ancestors living in the trees swinging by arm from branch to branch like pendulums, were *brachiating*. Monkeys have different gaits in this form of locomotion. They use internal linkages to increase the pendulum frequency when they want to move faster. Pendulum motion also plays an important role in terrestrial locomotion. Walkers swing their arms in the same rhythm as their legs. They angle the elbows to reduce the effective length of their arm pendulums (and thereby increase the pendulum frequency) when they start to run.

Large animals generally move faster than small animals. This could be attributed either to the fact that they have bigger legs, or that they have stronger muscles and therefore can swing their legs at higher angular velocities. Actually, the larger inertia of the legs of the bigger animals offsets much of the advantage from long legs. In fact small animals generate much higher angular velocities, yet have slower walking speeds. Running is an entirely different story, because the running motion always involves a vertical acceleration, caused by the impact force F_y which is considerably larger than the weight Mg of the runner.

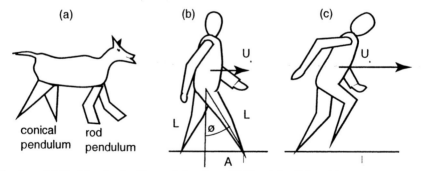

Fig. 6.24. (a) Model animal with pendulum legs. (b) Bipedal walking. (c) Running

6.4.1 The Arms Race of Tree Dwellers

Apes and monkeys face a different problem when swinging from tree to tree, Fig. 6.25. At first glance they resemble passive pendulum objects with amplitudes set by the distance between the pivot branch and the jump off branch [Preuschoft et al. 1985, 1996], and frequencies set by the length of the pendulum object. However, similar to land animals that can walk, trot, or gallop, tree dwellers have developed distinctive gaits [Bertram et al. 1999]. These include a slow mode called *continuous* brachiation and a faster mode called *ricochetal* brachiation. In the slow swing mode the monkey holds on to a support from the beginning of the swing until it grabs hold of the next branch at the end of the swing. The duration of the pendulum swing is $T_s/2 \approx 1.1$ s. In the ricochetal mode the swinging about a support point is followed by a free flight period that ends when the monkey grabs the next point of contact. The pendulum swinging part of this gait only takes about $T_r/2 \approx 0.55$ s. At first thought one would assume that the swing period should be identical, because the monkey's arm and body length is about the same in both gaits. However, on close inspection there are significant differences.

In the continuous brachiation the whole body acts like a pendulum of a total length L that is a function of arm length L_a and body length L_b. Body and arms are held stiff to yield a frequency f_b that scales with $\sqrt{g/L}$. Figure 6.25b is a simplified sketch of body and arm as described by Bertram et al. [2001]. The speed of continuous brachiation is then $U_b = \text{const} \cdot L \cdot f = \text{const} \sqrt{L}$. By inserting the scaling relation $L = \text{const} \cdot M^{1/3}$ one finds the allometric propagation relation:

$$U_b = \text{const} \cdot M^{1/6}. \tag{6.39}$$

In the ricochetal gait the body flexes about the shoulder joint. At the beginning of a swing the monkey quickly bends its arm relative to the body so that the shoulder joint is pushed forward, while the center of mass C of the body is held back. The arm thereby acquires a higher rotation speed. The arm moves somewhat like a shortened pendulum of length L_A; it has a reduced period. Subsequently, the body swings around the moving shoulder joint. The shoulder joint then receives the relative acceleration a_S, which has a normal and tangential component $a_S = a_{S,t} + a_{S,n} \propto L_A + U_S^2/L_a$. At the bottom of the swing only the normal component of the acceleration is present so that at this instant the vertical acceleration is increased to:

$$a_{S,v} = g + U_S^2/L_a. \tag{6.40}$$

This effective vertical acceleration affects the pendulum frequency, given by (6.6). Since $a_{S,v}$ is larger than g, the pendulum frequency of this arm pendulum is higher than for an ordinary pendulum with the same arm length. The locomotion is faster.

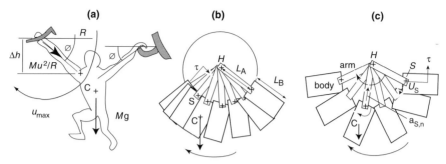

Fig. 6.25. (a) Articulated body of tree dweller. (b) Body held approximately stiff in continuous brachiation. (c) Body rotates about shoulder joint in ricochetal brachiation. In (b) and (c) body positions are shown in time intervals of $\Delta t = 0.1$ s

At other instances of the arm-pendulum-swing the shoulder joint has a different vertical acceleration. However, a_{Sv} is on average significantly larger than g, so that the frequency of the body pendulum is increased. This qualitative description is in good agreement with the time sequence of body positions, redrawn in Fig. 6.25 after Bertram. Through flexing the body at the shoulder joint and internal rotational motion, the pivot of the body pendulum acquires an additional vertical acceleration which substantially increases the body pendulum frequency. The gears of the ricochetal gait involve the use of muscles that articulate the arm motion relative to the body. Since muscle forces are involved that may be adjusted in strength, this fast gear likely enables a range of propagation speeds. Similarly, as for terrestrial locomotion, the gears of the faster gait of tree dwellers are muscle groups that give the pendulum hinge (here the shoulder joints) an additional vertical acceleration.

6.4.2 Walking

Physical models of a life function often paint the biological reality with a broad brush to derive general features, neglecting the finer details that are important to the zoologist. Here we model the motion of free swinging legs of a running or walking animal as a pendulum process, in order to derive propagation velocities, scaling with body mass, and energetic parameters. A pendulum oscillates at the frequency f

$$f = \frac{\omega_0}{2\pi} = \frac{C}{2\pi}\sqrt{\frac{g}{L}}, \tag{6.41}$$

where ω_0 is the angular velocity, g is the gravitational acceleration, and L is the length of the leg. An ideal pendulum, called a mathematical pendulum, which has all it's mass concentrated at the end point, features the value $C = 1$. The constant C describes the distribution of mass along the leg. If the leg can be modeled as a rod one finds $C_{rod} = \sqrt{(3/2)} = 1.22$, and if it is approximated as a cone hinged at the base,

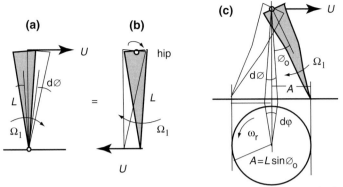

Fig. 6.26. (a) Leg motion in the laboratory reference frame. (b) Motion in the body reference frame. (c) Leg motion and reference circle

one finds $C_{cone}=\sqrt{(5/2)}=1.58$. The leg pendulum has the maximum deflection ϕ_0, which is typically 30° for many animals including humans [see for instance Dumont and Waltham 1997].

The stance foot stays on the ground during one half of the motion period, while the pendulum leg swings from $\phi=\phi_0$ to $\phi=-\phi_0$. Most authors focus the attention onto the stance leg and describe it as an inverted pendulum [see for instance Mochon, and MacMahon 1980]. The gravitational force Mg keeps the pendulum on the ground against the centrifugal force MU^2/L. This "ballistic walking" model is governed by gravity, similar to the regular pendulum. The time it takes an inverted rod pendulum to fall from the vertical position to the maximum deflection angle ϕ_0 is identical to the time it takes the free swinging pendulum to come back from its maximum deflection to the center position. The ballistic pendulum does not have an intrinsic period, because it needs the second leg to complete the cycle, whereas the free pendulum model has a well defined full period $T=1/f$. Since we want to study the frequency of the motion, we model the legs as free swinging pendulums. Then use can be made of standard pendulum theory to describe the processes of walking and running, and to define the transition point between both modes [Ahlborn and Blake 2002a].

Relative to the hip joint, the leg performs angular motion. We assume left- right symmetry. Then the angular velocity Ω_{free} of the free swinging leg (shown on Fig. 6.26b at the instance t_1 when the leg swings through its vertical position) is equal to the angular velocity of the stance leg, Ω_{stance}. Then the speed of the hip over the ground $U_{hip}=L\Omega_{stance}=L\Omega$ is equal to the speed of the foot relative to the hip $U_{foot}=L\Omega_{free}=L\Omega=U_{walk}$. The speed of the hip U_{walk} is the wanted quantity. The speed of the foot can be easily given by pendulum theory for the instant t_1, where the leg goes through its vertical position.

The leg oscillates in this model relative to the hip. The amplitude $A=L\sin\phi_0$ depends on the maximum stride angle ϕ_0. Typical stride angles are $\phi_0\approx30°$ leading to $A\approx0.5\,L$. Provided the toes do not touch the ground, the foot oscillates in x-direction as $x=A\cos\omega t$. It swings through its vertical position at the speed

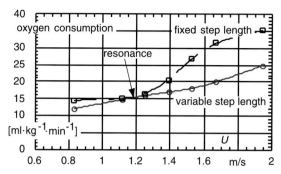

Fig. 6.27. Power in restrained (fixed step length) and unrestrained walking [Kryzka 1999]

$$U_{foot} = \omega\, A = \omega\, L \sin \phi_0 = 0.5\, L\, C\, \sqrt{(g/L)} \approx 0.79\, \sqrt{(Lg)} = U_{walk}\,. \tag{6.42}$$

In general, the leg length scales as $L = \text{const} \cdot M^{1/3}$, so that one finds for walking animals and humans the scaling relation

$$U_{walk} = a_w M^{1/6}\,. \tag{6.43}$$

Equation (6.42) shows that either the angular velocity ω or the step deflection angle ϕ_0 must be changed if the speed is to be altered. The speed may be easily raised by increasing ϕ_0. However, it does not bring much to increase ϕ_0 beyond $\approx 30°$. In fact animals maintain ϕ_0 at $27-30°$ over the range from normal walking to running speeds [see for instance Farley et al. 1993].

Beyond speeds where ϕ_0 has reached about 30°, an increase of velocity U means an increase in angular velocity ω. It is very energy consuming to move the legs above their pendulum resonance frequency ω_0. An increase of the metabolic energy expenditure shows up, as an increase in the oxygen uptake. Figure 5.36 displays data from a walker on a treadmill, moving at various speeds U. If the treadmill speeds up the walker increases his stride length A using the natural angular velocity $\omega_0 = \text{const}\,(g/L)^{1/2}$ of the leg. However, if the walker is forced to move at a constant stride length, then a change in speed means a change of the angular velocity away from the natural value ω_0. Indeed the oxygen consumption goes up, see Fig. 6.27. The lowest energy consumption is always at resonance. Therefore, runners should increase their resonance velocity ω_0 in order to minimize their energy consumption.

6.4.3 Running

Walking is often not fast enough. Animals run to catch a meal or to avoid becoming one. On first sight it appears that one can only increase the pendulum resonance frequency by shortening the length L, see Eq. (6.41). Unfortunately this does not increase the velocity U, since a shorter leg means a shorter horizontal dis-

Table 6.1. Mass M, impact force F_{max}, leg length L, and measured speeds U_{exp} reported by Farley et al. [1993] and derived parameters: resonance velocity U_r, vertical acceleration a_{vert}, metabolic endurance running constant C_r, and maximum tension leg length scaling constant C_2

animal	M [kg]	F_{max} [N]	L [m]	U_{exp} [m/s]	U_r [m/s]	A_{vert} [m/s²]	C_r	C_2
kangaroo rat	0.112	6.0	0.099	1.80	1.84	53.6	1.71	0.205
white rat	0.144	3.0	0.065	1.10	0.93	20.8	1.05	0.124
wallaby	6.86	300	0.330	3.00	3.04	43.7	3.14	0.174
dog	23.6	500	0.500	2.80	2.60	21.2	3.01	0.174
goat	25.1	500	0.480	2.80	2.47	19.9	3.02	0.164
red kangaroo	46.1	2000	0.580	3.80	4.01	43.4	4.15	0.162
horse	135	3000	0.750	2.90	3.27	22.2	3.27	0.146
					average		2.76	0.16
					standard dev.			±0.01

placement $S = 2A = 2L \sin \phi_0$. In fact the velocity U decreases when L is reduced, see Eq. (6.42). It was pointed out by Preuschoft [1999] that a vertical acceleration should increase the resonance frequency of the leg pendulums.

When a pendulum is speeding upwards at the vertical acceleration a_1, the gravitational acceleration g in Eq. (6.41) must be replaced by the total vertical acceleration $a_{vert} = a_1 + g$ of the pendulum hinge point. It is well known that a pendulum in an ascending elevator swings slower if the elevator is decelerating, and swings faster when accelerating. Animals accelerate their hip and shoulder joints in a vertical direction when they trot or run. By generating the vertical acceleration a_{vert}, an animal can shift the resonance frequencies of its leg-pendulums exactly to the value $\omega = \omega_{res} = U/L \sin \phi_0$, where the leg pendulum swings by itself exactly as fast as the horizontal motion requires. From first principles one can derive the relation for the resonance condition of a physical pendulum modeled as a cone ($C = \sqrt{5/2} = 1.58$) in an accelerated reference frame

$$\omega_0 = 1.58 \sqrt{(a_{vert}/L)} . \tag{6.44}$$

The velocity is given for the maximum stride angle $\phi_0 \approx 30°$, as $U = A\omega = \omega_0 L \sin \phi_0 \approx 0.79 \sqrt{(L a_{vert})}$. The vertical acceleration $a_{vert} = F/M$ can be calculated if the impact force F and the mass M of the animal is measured. The resonance locomotion velocity is

$$U_r = C \sin \phi_0 \sqrt{(LF/M)} \approx 0.79 \sqrt{(L a_{vert})} . \tag{6.45}$$

According to this model, the vertical acceleration is matched to the horizontal speed. Experimental support of this prediction comes from the comparison of velocities predicted with (6.45) and experimental values, Table 6.1.

The important results are summarized in Fig. 6.28. Over the reported mass range from kangaroo rats $M = 0.1$ kg to small horse $M = 135$ kg the predicted velocities U_r agree very well to the measured velocities U_{exp}. This leads to the conclusion that indeed the pendulum frequencies are tuned by vertical acceleration, as predicted by the model Eq. (6.45). This matching of vertical acceleration and speed is achieved by animals with body masses ranging over three orders of magnitude.

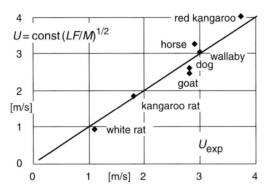

Fig. 6.28. Measured and predicted velocity

The leg length L of an animal is expected to scale with the cube root of the body mass. One can therefore define a proportionality constant C_2 by

$$L = C_2 M^{1/3}. \tag{6.46}$$

This constant is calculated for the 7 animals shown in Table 6.1. The average value of the constant is $C_2 = 0.16 \pm 0.01$. Then the constant of the walking speed relation (6.43) can be further specified:

$$U_{walk} = a_w M^{1/6}, \text{ where } a_w = 0.79 \sqrt{(C_2 g)} \approx 1.1. \tag{6.47}$$

For humans, the proportionality constant for the leg length is typically $C_{2,h} \approx 0.2$. Then, by using Eq. (6.47) the walking speed scales as $U_{walk} \approx 1.1 M^{1/6}$.

Since the model propagation velocities agree well with the experimental data, one can use the model and draw some further conclusions regarding the running velocities as a function of body mass. Vertical acceleration during the running process is the key to tuning the leg pendulums into resonance. In the galloping gait the hip attains an additional vertical acceleration due to the action of the rump muscles. The tuned resonance formalism suggests that galloping developed in the course of evolution as a means to increase the vertical acceleration, which in turn allowed these galloping animals to achieve higher velocities with the energy saving feature of leg pendulum resonance. From Eq. (6.45) one can derive the maximum impact force F as a function of the traveling speed U, in order to assess the muscle mass and bone strength needed to generate it.

$$F = \left(\frac{1}{C \cdot \sin\phi_o}\right)^2 \frac{M}{L} U^2. \tag{6.48}$$

In the running process the muscles generate the vertical force F_y and a vertical velocity u_y, which will be of the same order of magnitude as the horizontal velocity U at which the feet are swung. Hence we set $u_y = C_y U$. The mechanical power generated by the impact force $F_y u_y = \eta \Gamma = \eta b a M^{3/4}$ has to be provided by the metabolic activity of the animal. By squaring Eq. (6.45) and multiplying u_y into the numerator and denominator one obtains

$$U^2 = 0.79^2 \frac{L F_y u_y}{M u_y} = 0.62 \frac{C_{el} M^{0.20} \eta b a_o M^{0.75}}{M C_y U}. \tag{6.49}$$

On the right hand side of (6.49) the leg length L has been replaced by a body mass function. Running is a strenuous exercise. Therefore, the maximum tension scaling from Table 3.8 is used here, namely $L = C_{el} M^{0.2}$. After contracting the exponents a relation of the form $U^3 = \mathrm{const}\, M^{-0.06}$ is found, which defines a constant C_r and the metabolic endurance running velocity $U = U_{er}$

$$U_{er} = C_r M^{-0.023} \approx C_r. \tag{6.50}$$

The metabolic running constant C_r depends on the parameters C, ϕ_o, C_{el}, η, b, a, and C_y, which are thought not to depend on the mass. The exponent -0.02 is quite close to 0. Any product with a function x raised to the power 0 is independent of x. Therefore (6.49) predicts that the top running speed is approximately independent of the body mass M. To get an estimate of the metabolic running constant one can solve (6.49) for $C_r = U_{er}/M^{-0.023}$, and use the experimental data of body mass and (endurance) running speed from Table 6.1 to evaluate C_r. The average value is $C_r \approx 2.8$.

In contrast, the kinematic walking speed U_{walk}, given by (6.43), increases with body mass to the power $1/6$. Large animals walk faster than small ones. But medium-size, and large animals seem to run quite comfortably at similar speeds. Greyhounds can keep up with horses. However, at a certain body mass running becomes too large an effort since the specific muscle power P/M decreases with body mass as $P/M \propto M^{-1/4}$. Therefore very big animals walk rather than run. They lack the muscle power to continuously accelerate their body vertically as required for tuning their legs to running speeds. Of course, this general scaling rule does not take into account specific running oriented body designs. More will be said about this topic in Sect. 6.4.5.

6.4.4 The Transition from Walking to Running

In the pendulum model the walking speed U_w is given by Eq. (6.43). The transition from walking to running occurs when the traveling speed U exceeds U_w, so that the leg pendulum is forced to move above its resonance frequency $\omega_0/2\pi$. Then the walking/running threshold can be defined as

$$U_{w/r} \geq U_w = C \sin\phi_o \sqrt{Lg} \,. \tag{6.51}$$

If the stride angle ϕ_o is determined with an accuracy of $\pm 3°$ the threshold running velocity from Eq. (6.51) can be given with an uncertainty of $\pm 8\%$. The motion may be expressed in terms of a Froude number Fr, which compares gravitational and centrifugal forces

$$Fr = U_w^2 /gL \,. \tag{6.52}$$

The Froude number at the walking/running threshold can be found by entering $U_{w/r} = C \sin\phi_o \sqrt{gL}$ into (6.52). This yields

$$Fr_{r/w} = (C \sin\phi_o)^2 \,. \tag{6.53}$$

Typical stride angles ϕ_o for fast walking fall in the range from $31°$ to $34°$. The mass distribution constant C for a leg modeled as a rod is $C_{rod} = 1.22$ and modeled as a cone is $C_{cone} = 1.58$. With an average value between these two limits of $C \approx 1.4$, and a stride angle $\phi_o = 32°$, the transition Froude number is $Fr_{r/w} \approx 0.5$. This value is in fairly good agreement with the measurements by Kram et al. [1996], who found $Fr_{r/w} = 0.45$. In contrast the inverted pendulum model predicts the transition Froude number $Fr_{r/w} = 1$, based on the assumption that the centrifugal force MU^2/L is equal to the gravitational force Mg at the transition point. Equation (6.53) can be solved for the mass distribution constant C of a leg

$$C = \frac{\sqrt{Fr_{w/r}}}{\sin\phi_o} \,. \tag{6.54}$$

One could in principle use measured values of the run/walk transition Froude number $Fr_{r/w}$, and the stride angle ϕ_o to determine the mass distribution constant C from Eq. (6.53).

6.4.5 Why T-Rex Was No Endurance Runner

The walking speed (6.43) depends on M raised to the power 0.167. Walking velocities increases steadily with body mass. Animals with long legs walk faster than small ones. Big animals like giraffes, hippos, and elephants, which have body masses above 1 tonne, generally do not run, they amble about. The walking speed is shown as a line in the propagation velocity diagram, Fig. 6.29.

One can show that endurance runners have an upper body mass limit. Here we are not concerned with short time sprinting activities. For short times, an organism may go into overdrive in anaerobic motion where lactate builds up in the muscles. Such bursts of power, which may last typically for one minute, help to move the animal out of harm's way or catch the elusive meal.

The upper mass limit for steady runners arises from the fact that the walking speed increases steadily with body mass M, albeit with only a small power of M, while the steady running speed (6.50) is, in first approximation, independent of body mass. There is a terminal body mass M_t where both speeds become equal. At this body mass its metabolic power does not allow the animal to run continuously any faster than it can walk. The animal just does not have the muscle power to accelerate its body with every step in the vertical direction. For small animals running is faster than walking, but this advantage disappears at the mass M_t where both velocities are equal.

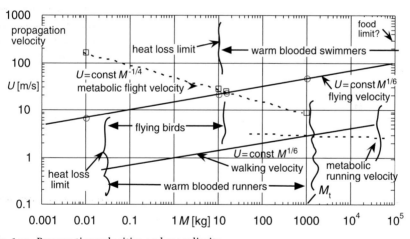

Fig. 6.29. Propagation velocities and mass limits

In steady running the muscles generate the vertical force F_y, and produce a vertical velocity u_y. The product $F_y u_y$ is the produced metabolic power. As shown by Eq. (6.50) the metabolically available endurance running velocity can be approximated as $U_{er} \approx C_r M^{-0.023} \approx C_r$. For an estimate, the empirical value $C_r \approx 3$ is used which was previously derived from the experimental data in Table 6.1. The endurance running velocity $U_{er} \approx 3$ m/s is shown as a dotted line in Fig. 6.29. The intersection of this line with the walking velocity curve defines the limiting body mass $M_t \approx 1000$ kg. An error of 20% of this constant would shift the mass limit by a factor $1.2^6 \approx 3$.

With this uncertainty in mind one may arbitrarily estimate an upper limit for long distance running at $M_t \approx 10^3$ kg. Therefore animals with body masses greater than 1 ton are unable or unwilling to continuously generate the vertical forces needed to raise their leg frequencies above the pendulum walking frequencies.

Based on this estimate T-rex, with its body mass of about 5 tons was probably not a long distance runner. Of course, over short periods of time, bigger animals could likely generate large bursts of power to speed up, but they would never run steadily over extended periods of time. Alexander [1991] comes to the same conclusion, based on the structure of fossil bones.

The locomotion speed for flying $U = 15\,M^{1/6}$ is also shown in Fig. 6.29, and the metabolic flight velocity $U \approx 15\,M^{-1/4}$, which sets an upper limit for the body mass of flying birds. Fig. 6.29 also includes results from energetic considerations for warm-blooded animals. Lower mass limits for homeotherm animals are set by the thermodynamic requirement to stay warm. In the range of body masses there is a certain window of opportunities for terrestrial locomotion and flight of warm-blooded organisms, which indeed many animals have found.

6.5 From Efficient Use of Energy to the Smarter Use of Information

Energy and materials make life tick. Early on animals have learned to be energy-wise through intimately understanding mechanics. They have optimized the energy consumption when swimming, running, or flying. Their bodies make best use of available materials. They do not squander calories in cold climates nor do they forgo means to keep cool in hot environments.

Much of the locomotion of large animals serves the purpose to access new food sources in a daily or seasonal rhythm. To make this migration energy-effective the animal should not use up more calories in getting there than what it will harvest during or following the travel. A figure of merit is the cost of transport E_{tr}, discussed in Sect. 5.3.2. All locomotion over extended distances uses periodic motion, the undulation of fins or flippers, the flapping of wings, or the pendulum action of legs. Much energy can be saved if the kinetic energy of the rapid motion of mid-swing can be transformed into elastic or into gravitational potential energy. Such energy conversion requires careful timing, because every energy storage process has its own time scale, or frequency. The secret of resonance in simple harmonic motion is the precise matching of the elastic or pendulum frequency to the desired driving frequency. In locomotion, the desired traveling speed $U = $ const$\cdot f_L$ sets the desired frequency f_L. Animals have mastered the art of tuning internal frequencies to the locomotion frequencies.

Suppose the transportation system of an animal reaches maximum efficiency, so that every joule of mechanical energy goes into overcoming drag and air resistance. Is that all that the animals can do to prosper on little sustenance? It is not, because in order to make rapid motion useful there is also a need for effective long distance senses, and a brain to recognize where the animal is going and respond to the quickly changing situations. The instruction to build bodies to perform such functions is stored in the genes.

The genes contain the blueprint for each organism. This information is the condition *sine qua non*, life's lifeline to its origin and its own uniqueness. However, other information is floating around which contains messages about opportunities or dangers arising in the surrounding. Events of nature or actions of other beings leave trails of data that may enable the organism to better survive – if it can decode the messages.

Data bits are transferred by sound, and light, and by electric or magnetic fields. Animals have acquired senses to read these signs, and have developed brains to analyze them. The meaning of these data streams must be learned. Deciphered information saves energy in many ways. If you see where the food is, you do not have to waste energy in random searches. If you spot your enemy from a safe distance, you can move out of harms way, and do not need strong armaments.

Evolution proceeds in revolutionary steps. In human activities we distinguish the "Gutenberg" revolution, where the printing press lead to an enormous dissemination of information, the industrial revolution where technical inventions made work and locomotion easy, and the information revolution where the internet has been shrinking distances and lead to knew activities. The animal kingdom experienced three similar steps. The training of young, where individuals benefit from experience of their parents, mirrors the Gutenberg revolution. The adoption of technologies such as jet propulsion in the water, the running on land, and flying through the air (which provides energy efficient transport) corresponds to the industrial revolution. Finally the information revolution was anticipated by the animal kingdom when long distance senses, and sonar appeared, combined with intelligent brains and language, so that animal group activities could evolve. We humans are just the leading edge of this evolution.

Table 6.2. Frequently used variables of Chap. 6

variable	name	units	name of unit
a	linear acceleration	m/s²	
a_{sw}, a_f, a_w	allometric constants of swimming, flying, and walking	–	
C_D, C_L	drag and lift coefficient	–	
E_{tr}	cost of transport	N/kgm	
f	frequency	1/s	Hertz
Fr	Froude number	–	
I	mass moment of inertia	kg m²	
J	mass flow	kg/s	
L	length of limb	m	meter
S	wing surface area	m²	
T	period	s	second
u, U	velocity	m/s	
W	weight	kg	kilogram
X	aspect ratio	–	
Γ	circulation	m²/s²	
α	angular acceleration	radian/s²	
ω	angular velocity	radian/s	
ζ	vorticity	m²/s	

Problems and Hints for Solutions

P 6.1 Drag of a Fish
Determine the drag coefficient for a fish of $M = 0.1$ kg using the cost of transport data E_{tr}. Determine its swimming speed from the great swim diagram, and work out its mechanical power consumption.

P 6.2 How Large are Gaia's Biggest Flying Birds?
A new small planet Gaia is discovered which has a gravitational constant is $g_G = 3.0$ m/s². This planet has a flora and fauna similar to earth, and an atmosphere with the same composition, pressure ($p = 1$ atm), temperature, and density $\rho = 1.29$ kg/m³ similar as the earth. Make a log-log sketch of the "great flight diagram" that you expect to find on this planet. Also show on your sketch the metabolic flight velocity curve for this planet, and then estimate the mass M of the largest flying birds on this planet.

P 6.3 Minimum Reynolds Number of a Hovering Insect

A certain flying insect, body mass $M=0.2$ g, wing beat period $T=1/f$, has wings with an average width (span) of $D=2$ mm. The wing tips follow a figure 8 motion, as shown in Fig. 6.21. The bug rotates the wings about the upper edge (without lateral motion) in $\Delta t = T/6 = 1/6f$, as it turns the wings from the forward to the backward stroke direction. Assume a wing angle of attack $\alpha = 25°$ for the forward and backward stroke steady phases of motion. a) Express the average velocity, u_e, of the lower edge of the wing as a function of wing span D, and frequency f. b) Calculate the wing frequency necessary to achieve vortex shedding at the edge (Re $=100$).

P 6.4 Vortex Shedding Lift of a Hovering Insect

The wings of a certain hovering insect, $M=0.08$ g, have a length of $L=14$ mm length and an average width (span) of $D=2$ mm. The wings turn around from the forward to the backward position in the time interval $\Delta t = T/6 = 1/6f$. Assume an angle of attack $\alpha = 25°$. a) Find the average vertical force F_v of both wings associated with the turning of the wings as function of the wing frequency. Assume that the force arises from the jet action of two wings $F_v = 2Ju_v$ of the air, which is pushed downwards as the wings flip over. b) Determine the minimum wing beat frequency needed for hovering, $F_v = 2Ju_v = Mg$.

P 6.5 Lift from Swinging Arms

Estimate the lift force that is generated by swinging one arm from the full back position, (where the hand is level with the shoulder) to the full forward position and stopping it there abruptly when the hand is level with the shoulder joint. Model the arm as a stiff rod of length $L=0.6$ m where all the mass $M=8$ kg is concentrated at the distance $0.75\,L$ from the shoulder. Estimate in a self-experiment how quickly you can swing your arm through ½ revolution – namely measure the time for 10 rotations of your arm and divide by 20. Assume that the arm is stopped within 10°, or 1/36 of a full rotation $v^2 \approx 2\,a\,\Delta s$.

P 6.6 On Walking and Leg Frequencies

Find the leg length of a small dog, a person, a cow, and a giraffe, and determine their walking speed. Estimate the impact force of a runner of $M=75$ kg, he is running at a speed of 6 m/s.

P 6.7 Why one Swings the Arms when Running

Explain how swinging the arms during running helps to lift the center of mass. Find the step frequency of a runner of average height $H=175$ cm, running 100 m in 11.5 s. Estimate his arm length and determine how much he should angle his elbows to bring his arm frequency in tune with the leg frequency.

Hints and Sample Solutions

S 6.3 Reynolds Number of Hoverin Insect

a) First determine the angular velocity ω, of the wing rotation. With reference to Fig. 6.21 assume a turnaround time $\Delta t = T/6 = 1/6f$. From Fig. 6.21 one has $\omega = \beta/\Delta t \,[\text{rad/s}] = 6(\pi - 2\alpha)f$. The speed of the lower edge of the wing relative to the center of rotation is $u_e = \omega D = 6fD(\pi - 2\alpha) = 6fD\beta$. b) Vortex shedding occurs when the Reynolds number exceeds a minimum value. $Re_{min} \approx 100 \le Re = u_e D/v = 6fD^2\beta/v$. Then $f \ge 100v/\{6D^2\beta\}$. For $D = 2$ mm, $v = 1.6 \cdot 10^{-5}$, $\alpha = 25° = 0.82$ rad, $\beta = 130° = 2.268$ rad. Then find $f_{min} \ge 1.176 \cdot 10^{-4}/D^2 = 29.3$ Hz.

S 6.4 Vortex Shedding Lift of Hovering Insect

a) First find the volume swept by the wings, assuming that the insect rotates the wings during the time interval $\Delta t \approx 1/6f$ about the upper edge (without lateral motion). If the wing is modeled as a rectangle the swept volume is $\Delta V = \pi L D^2 \beta/2$, where $\beta = 130° = 2.27$ rad is the angle through which the wing rotates at turn around. The Volume flow rate is then $V' = \Delta V/\Delta t = 3\pi fL D^2\beta = 1.2 \cdot 10^{-6}f\,\text{m}^3$, and the associated mass flow rate is $J = \rho \Delta V' = 1.54 \cdot 10^{-6} \cdot f\,[\text{kg}]$, where $\rho = 1.29$ kg/m³ is the density of air. The expelled air must flow vertically downwards. Its average velocity u_v can be derived from the conservation of mass, since it has to issue from an opening of length $L = 14$ mm, and width $\Delta x = 2D\sin(\beta/2) = 3.6 \cdot 10^{-3}$ m. The volume flow rate can be written as $\Delta V' = L\Delta x u_v$. Then the vertical velocity is $u_v = \Delta V'/L\Delta x = 1.2 \cdot 10^{-6} \cdot f\,\text{m}^3/(1.4 \cdot 10^{-2} \cdot 3.6 \cdot 10^{-3}) = 2.37 \cdot 10^{-2} \cdot f\text{m}$. Remember that f has the unit 1/s, therefore the right hand side has the unit m/s. a) The vertical force for both wings is then $F_v = 2 \cdot J \cdot u_v = 1.54 \cdot 10^{-6} \cdot f$ kg $\cdot 2.37 \cdot 10^{-2} \cdot f\text{m} = 7.3 \cdot 10^{-8}$ kg m $\cdot f^2$. Note that the units of kg m $\cdot f^2$ is N. b) The wing beat frequency f must be so large that the lift force must be equal to the weight of the animal $Mg = 8 \cdot 10^{-5}$ kg $\cdot 9.8$ m/s² $\approx 8 \cdot 10^{-4}$ [N]. Hence one has $f_{min}^2 = 8 \cdot 10^{-4} \cdot 8/7.3 \cdot 10^{-8}$ leading to $f_{min} = 105$ Hz. Actually, the lift force F_{st} during the steady flight phases 2–6 and 8–12, Fig. 6.21, is slightly smaller than the turn-around vortex shedding lift. One can define an average lift force $F_y = (1/T)\{F_v \Delta t_v + F_{st} \Delta t_{st}\}$, where Δt_v is the turn around time and Δt_{st} the steady flying time. For instance if $F_y = (2/3)F_v$ the minimum hovering frequency is increased by the factor $\sqrt{(3/2)} = 1.22$.

7. Waves, the Carriers of Information

In the beginning was the word
Gospel of John

Genetic Codes, and External Information

Life needs energy, matter, and information: matter and energy to act, and information to direct the action. Energy and matter continuously flow through the body of each organism; energy flows daily via food consumption, while matter flows at a much slower rate. Internal information flows as the genetic message from generation to generation, with minor evolutionary modifications. This internal information is expressed in the genetic code just as meaning and laws are encoded in written words. The genes carry meaning over countless generations of organisms just as the laws of Hammurabi, the wisdom of Genesis, or the gospel of St. John have been passed on by the words of the holy texts over the millennia.

Internal genetic information is the *conditio sine qua non*, the indispensable starting instruction of every animal, yet it is not enough to guide an organism through its life. All higher animals supplement this initial endowment with external information that is collected by senses, processed by the brain, and often temporarily stored in memory. This external information together with the internal information makes an organism a true individual. The specificity of the information fixes an organism at a unique place on the ladder of evolution, creating the *self* since this information treasure is really owned by the individual and will disappear at its death. External information improves survival and builds *personality*.

Organisms with appropriate collection equipment can obtain from the environment, practically free of charge, all the principal ingredients of life: matter, energy, and external information. Matter and energy is harvested with the food. External information is collected by senses. Good food feeds the body, good information expands the mind, and enhances survival. External information is useful for many reasons: 1) to distinguish food from poison, 2) to recognize opportunities, 3) to foresee dangers, 4) to help get by with less matter and less energy, 5) to evolve the self, 6) to hold together family communities, 7) to advance societies, and 8) to develop a world consciousness.

All higher animals actively exchange *external* information through language that may be carried by chemicals, expressed as body motions, or pronounced through optical, acoustical, or electrical means. Mobile and roaming animals need acute long distance senses. They use sight and sound. Eyes and ears are built to read small perturbations of radiation and sound fields. Such fields carry messages quickly and over long distances.

Chapter 7 begins with a description of the evolution of total information available to animals and humans, Sect. 7.1. The major part of this wealth of data can be captured by the long distance senses. A short review of the different senses and signal sensitivities is given in Sect. 7.2, with particular attention given to the human senses. Sound and light are the basis of the long distance senses. The wave properties of sound and light are discussed in Sect. 7.3 in order to understand how signals can be encoded into such wave fields, and how the signals are transmitted. The last area of this chapter, Sect. 7.4, describes physical effects that can alter wave fields. In short, Chap. 7 discusses the physical basis of the detection processes in ears and eyes.

7.1 External and Internal Information

Organisms perpetuate themselves by passing on the genetic information to their progenies. However, all higher animals supplement these internal instructions by external information. The steady *increase* of external data bits available to all animals is likely one of the driving forces of evolution. Therefore, a closer look at the physical effects that enable the collection of external information is an appropriate endeavour.

7.1.1 From Genes to Brain and Senses

Information must be stored. Animals store the data in the read-only-memory (ROM) of their genes and the rapid-access-memory (RAM) of their central processing unit, the brain. The members of the animal kingdom have experienced phenomenal increases of processing power during the course of evolution. Intelligent animals are often small. It appears that information can be substituted for body size and energy, and thus the quote "Brain is better than brawn" is literally relevant. However, how many grams of brain could substitute a kilogram of body mass is still unknown.

The first organisms stored information only in their genes. Genes are the instructions, which should be immutable for eternity. But mistakes are made occasionally in copying genetic information, and inaccuracies can cause the end of life at an early developmental stage. However, such copying errors can also lead to the incorporation of new and sometimes better features into the gene script. Nature has turned, what appeared to be a great handicap for life's survival on earth, into an enormous advantage: The ROM functions of the genes possess a minute mutation and selection capability, where *error is the chisel of evolution*. Genes roll the dice but the new organisms must pass the test of physics laws. These laws ultimately determine which throw should count.

Organisms that have senses which respond to signals from their surroundings, improve their survival chances. Stimuli from the surroundings may lead to a passive pre-programmed response, to an optimized intelligent reaction by each or-

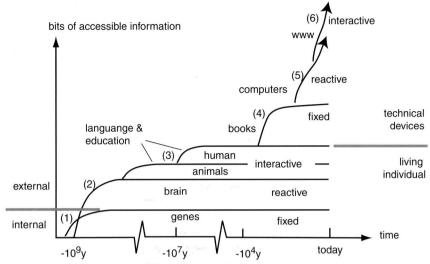

Fig. 7.1. Explosion of information available to organisms

ganism, or to a concerted reaction of a whole group that possesses an interactive language.

The use of information in the animal world evolved in three steps. Primitive organisms only use information stored in their genes, the passive generic data carriers. Higher life forms, which arrived later in the evolution of life on earth, carry brains that can evaluate the stimuli from senses and *stored data*.

The brain converts data into *information*, from which in turn *knowledge* may be abstracted to help the organism make intelligent choices. A German proverb states *Wissen ist Macht*: knowledge is power. Initially, the brain was only needed to recognize food opportunities, danger, readiness of mates to produce off-springs, or same-species territorial claims. These indications would be advertised through views, smells, gestures, or noise signals. The active and intentional displays likely turned into chemical, visual or acoustical language, which is the third layer of clues that an organism may receive from its surroundings. Language facilitates social interactions and education. It connects families and groups, so that the knowledge of the whole tribe may be passed on from generation to generation. Filtered *data* become information. In turn, different streams of *information* can be sorted and organized by a brain to extract *knowledge*, namely a pattern that guides the organism through foreseen or unforeseen situations. Knowledge is a vision of reality, the art of seeing structures beyond the superficial facts. Layers of different knowledge culminate in a category called *wisdom*.

7.1.2 Organic and Technical Evolution

Along the chain of evolution *homo sapiens* finally appeared. This species developed sophisticated languages, and invented tools that became more and more advanced thus mirroring the information evolution of the biosphere. First, the passive data carriers were invented: scripts, and libraries. Second, computers were developed to process data into information. Third the world wide web was established, which interactively links a person with society. Figure 7.1 depicts this explosion of data as a function of time, and Table 7.1 shows how the evolution of human tools mirrors the organic evolution from data to intelligence. Hopefully the increased information resource will lead to superior moral behavior.

Table 7.1. Information

level	living organisms	technical devices
passive data storage	genes	books
active optimized response	brain	computer
interactive exchange of knowledge	language groups	www

Human culture begins with language and collective memory of groups. However, civilization and major religions only seems to have evolved with the advent of writing. Writing can replace and reinforce the human memory and can carry wisdom over periods where the memory of individua might fail.

7.1.3 The Information and Material Hierarchies

In becoming modern man, the human species has experienced an explosive growth of available data bits. To make sense out of this flood of impressions, the human brain has grouped the data bits first into information, then condensed information streams into knowledge, and joined bodies of knowledge into wisdom. On rare occasions the human race has drawn great benefits from the inspiration of intellectual giants. Each higher level filters out unimportant details and concentrates the understanding of the lower levels by abstraction. Every step is a reduction of entropy of information, Fig. 7.2a.

The information pyramid bears some similarity with two other pyramids on which life rests: the pyramid of biomass on earth, Fig. 7.2b, and the pyramid of matter in the universe, Fig. 7.2c.

The sea of human culture is gradually augmented by the rivulets of wisdom that spring from the from the minds of intellectual giants like Moses, Plato, Buddha, Pythagoras, Shakespeare, Leonardo, Mozart, and Einstein. They add flashes of new intellectual awareness to the general knowledge.

The dowry of genetic information bestows upon every organism the selfish instinct to live against all odds, to *go forth and multiply*. This means fight for survival, one against the world. Me first, then my family, and let the rest of the world

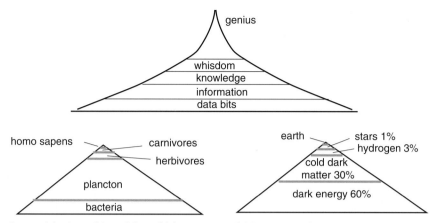

Fig. 7.2. (a) Hierarchies of data, (b) bio-mass, and (c) matter in the universe

look after themselves. Such instincts sanction war and destruction. Advanced levels of information add a new perspective: the common core of goodness and virtue in all cultures and religions is the respect for other humans and all forms of life. The concern for fellow man and all God's creatures is possibly not encoded in the genes.

7.2 Contact and Distant Senses

Every organism is woven into the biosphere by biology and physics. The biological connection arises from the food chain: every animal eats some living structure. The concern of each animal is not to become a meal too soon. But where is the food? Where are one's predators? Where are the dangers? Should you stay where you are to eat or move and hide?

The biological interdependence of the biosphere is complemented by a physical connection: every organism lives in a sea of wave fields and forces, which contain some information about the immediate and distant surrounding. The information is mediated by sound and light waves, by electric and magnetic fields, by gravity, and the temperature field. Some of these fields carry information over large distances while others have only a short range.

7.2.1 Signals and Sensor Sensitivities

Animals detect nearby and far away events with different sensors. The contact senses of smell and taste respond to individual atoms or molecules, pressure sensors responds to physical contact, and pain can be associated with all kinds of deviations from the normal state of the body, Fig. 7.3. Taste, which warns the organisms of poisonous foods, has the largest number of receptors, Table 7.2.

Table 7.2. Human contact senses

sense	pressure	pain	smell	taste
sensor location	everywhere	everywhere	nose	tongue
signal due to	surface force, deformation	deviation from some set norm	molecules	molecules
# of receptors	$5 \cdot 10^5$	$3 \cdot 10^6$	10^7	10^8
# of nerves	$\approx 10^4$		$2 \cdot 10^3$	$2 \cdot 10^3$
sensitivity			≈ 5 ppm	

Distant senses detect the magnetic field of the earth, electrostatic aura of organisms, and the wave fields of sound or radiation, which are caused or perturbed by objects in the surrounding. These field senses are optimized for the particular environment of every animal. From the detected quantities, organisms extract copious clues that tell about its surrounding. Organisms process these data bits not to overcome boredom, like a lazy TV watcher, but to improve their chance of survival. The signals that must be decoded are often quite small.

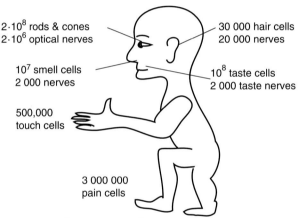

Fig. 7.3. Human sensors and nerve connections

For instance, the energy and power of signals carried by light and sound is typically six orders of magnitude smaller than energies involved in body motion. The information gleaned from small signals can save the organisms much energy. Think of a monkey observing that a distant tree has no ripe fruit. This animal does not have to travel there and need not climb up to find the dinner table empty. The information may often be essential for survival: if a rabbit can see a fox from a distance, the rabbit can get quietly out of harms way.

The evolution of sensors plays heavily into the predator-prey relationship: the predator must stack the odds in its favor by coming so close that the prey is safely within catching range. The prey must try to spot the danger well before the predator has penetrated into the danger zone. In a stable situation like that between wolf and reindeer, the race of sense development and intelligence reaches a stalemate: if the wolf kills too many deer, his off-springs will starve to death. In addition, the wolf mainly kills the sick and old reindeer and thereby keeps the herd healthy. This benign balance is disturbed when animals with superior senses and unspecialized taste invade a protected biosphere, like the dingo in Australia. The indigenous population often cannot evolve senses and intelligence fast enough to defend itself, and the predator just switches to other food once it has eliminated one particular species.

Biological detectors are very sensitive and they are economically built: sensor ranges are generally matched to the signal strengths that occur in the normal environmental niche of each animal. Humans do not hear ultra sound at 40 kHz, however, bats and some of their prey do perceive this high pitched sound.

Table 7.3. Mainly human field senses

field	gravity seismic	sound wave	light	heat, cold	magnetic	electric
sense	semicircular canals	ear	eye	skin pits	magnetic cell	electric cells
signal due to	inertia	fluctuation of air	photons $E \approx 3 \cdot 10^{-19}$ J	heat	magnetic moment	electric charges
# of receptors		$3 \cdot 10^4$	$2 \cdot 10^8$	$10^4 - 10^5$		
# of nerves		$2 \cdot 10^4$	$2 \cdot 10^6$			
bit rates		$4 \cdot 10^4$ b/s	$5 \cdot 10^7$			
typical back ground	$g = 9.81$ m/s^2	$I_b = 60$ dB	$I_o = 10$ W/m^2 10^{20} photons/m^2	$10°$ C		
typical signal	0.01 g	$(0.1 - 10)\, I_b$	10^{-10} W	± 10 C$°$		
sensitivity threshold	10^{-4} g	1 dB	single photons/s 10^{-18} W	relative $\pm 10°C$		feel ≈ 5 V on tongue, sharks detect 10^{-7} V/m
derived information	acceleration a_x, a_y, a_z	intensity frequency time delay	image on retina, location and distance of object	intensity direction, quality	direction	direction Faraday effect $u \times B$
perception	up, down direction of motion	direction size, 2D distance	objects in 3D	2D objects	north, up/down	own velocity position & size of organic matter, mates
perturbed by	too large signals	noise, walls obstacles	too much or too little light, opaque obstacles	too cold objects		too large distance, too high conductivity
active production		bat, whale, dolphin	fish, plankton, insects			electric fish

Some of the detected quantities are perceived as absolute values. Some people have an absolute ear – they know when a tone is 440 Hz. A painter can distinguish nuances of yellow and red to estimate the wavelength of light with little error. A seasoned electrician can touch a live wire and know that it is only 120 V and not 240 V, and a good seal skin sorter can distinguish the length of the hairs in a pelt simply by touching the pelt with his fingertips. Some detector sensitivities are truly remarkable: a shark can detect electrical fields as small as 0.1 μV/m. That is the field of a 2.0 mV battery connected to two metal plates across the straight of Dover, with one plate mounted in France and the other in England.

Other senses are of a more qualitative character. One can assign a qualitative value to hot water: one can easily say which of two pails of water is hotter. But water of intermediate temperature feels hot when one's hand was first placed in cold water and feels cold when the hand was first placed in warm water. Likewise, the length of time can be difficult to judge. For instance it is hard to tell when two hours have passed if you are sitting in a dark room. However, short time intervals can be estimated with an uncertainty of typically 20% in comparison to some periodic activities, for example counting 1001, 1002, 1003 to mark off seconds. In later chapters it will be shown that some senses have evolved to a resolutions which approaches the absolute limits set by physics.

How does nature go about inventing a sensor? Is it like a technical "request for proposal" issued to every body cell, or is it a steady evolution of the sense from rudimentary beginnings of different components? Many organs, like flippers and wings, have been independently invented by different animals. Is that also true for the senses?

The distant senses rely on data carried by sound and light waves. To understand how animals use these phenomena a few wave properties relevant to signal propagation are summarized first.

7.2.2 What Is Extracted from the Background?

Evolution has taught its creatures to read *information* from the tiny perturbations created in the wave fields by every object, dead or alive. Intuitively perceived physics helps animals to 1) identify an object and the object's intentions, 2) the distance of the object, 3) the direction in which the object is moving.

Every organism is bathed in the fields of radiation, sound, gravity, in electrical fields, and magnetic fields. When these fields are constant and steady, they carry no message regardless of their absolute intensity. A glaring white light, or a pitch-dark night, the absence of sound in a sound chamber, or a constant everlasting noise of surf and or wind, may cause discomfort, but these noises do not carry information. All useful data are extracted from contrasts. A picture may be made up from one million colored or black and white dots, arranged to show a face or a landscape. Sound consists of sound bits, namely sequences of high and low intensities emitted at many different frequencies. The frequency and intensity data must be sorted and analyzed for identification of the source and its location relative to the receiver.

The physical connection of animals to the biosphere starts with data bits, which are acquired by sensors. The sensors feed the data bits into nerves connected to the brain. The nerves conduct electrical signals, using physics principles that are already apparent in the first single-cell organisms. The brain processes the data together with information from internal monitors, to come up with an optimized response at every moment during the life of an organism.

The principal information carriers, light and sound, complement each other conveniently. Everything that moves makes sounds, because sound is produced by motion. Sound can be heard day and night. During the day and in moonlit nights, everything becomes visible even if it is motionless and does not produce any sound. A picture is worth a thousand words, but during the night all cats are gray.

Images require parallel processing of information, whereas sound signals are received in sequence. Sequential processing is the first step in a cause-effect training that leads to logical thinking. Viewing and vision on the other hand, is more akin to insight and inspiration. Thus, sight and sound are supporting different forms of higher intelligence.

Table 7.4. Information carried in light and sound

information	sound	light
who is there?	sound spectrum, tones, frequency f, characteristic intensity, identity of source (e.g. voice ...)	shape (image, geometrical optics) characteristic movements, characteristic size, color, wavelength λ
message received	sequential	simultaneous
how far?	compare detected intensity with characteristic intensity I, decibel β, sonar	2-eyes triangulation: compare apparent size to characteristic size or intensity of other relevant objects, image size
which direction?	2-ear delay time. speed v sonar	image within field of view
coming or going?	change of intensity, sonar (Doppler effect $\Delta f/f$)	compare size in time sequences of images

7.3 Wave Fields

For intelligent animals with sharp senses the distant surrounding is an immense target of opportunities. The wave fields of light and sound, which carry messages from remote objects, can be characterized by intensity, wavelength, frequency, and phase velocity.

Most objects are indirectly lit by ambient radiation (just as trees are lit in broad daylight). Some objects are self-luminous, like glow worms, or fires. Others give their location away by their own noises, like mosquitoes in flight or horses clinking their hooves. Some higher animals have learned to illuminate their targets by sonar beams or by bio-luminosity.

An object can only be seen if it appears to be the origin of optical or acoustical radiation of some intensity I_s. The source must generate a wave field that carries a detectable minimum intensity I_{ob} to the location of the observer. I_{ob} is always much smaller than the intensity I_s, because the wave intensity is diminished by spherical spreading, absorption, and scattering. An observer with a sensory organ of the collection area A intercepts the power $P_{ob} = I_{ob} A$.

Sounds are analyzed in the ears sequentially. Light waves can be analyzed simultaneously as images with lens or pinhole optics, or with parallel detector arrays.

7.3.1 Some Properties of Waves

Waves can carry signals from a source to a receiver. To understand the flow of information one must consider the source, the wave transfer process, and the reception process, as indicated in Table 7.5. The source may be self-luminous, indirectly illuminated, or actively scanned by a search beam. If the wavelength is small compared to the size of the sensory organ of the receiver, information can be extracted simultaneously as imaged by a multi-detector array. An example is the human eye with its 10^8 photosensitive cells in the retina of each eye. Light waves have wavelengths of $\lambda \approx 0.5$ μm, whereas eyes have entrance pupils in the range of millimeters.

Sound is received sequentially by most animals. Sound frequencies f in the range $5\ \mathrm{Hz} \le f \le 100{,}000\ \mathrm{Hz}$ can be resolved by biological detectors. Only the very highest frequencies in this range lend themselves to acoustic imaging.

Table 7.5. Parameters and physical processes in wave fields from source to receiver

source intensity I_s	wave field speed c, v, frequency f, wavelength λ	receiver intensity I_{ob}
1) self luminous	spreading	entrance pupil
2) indirectly illuminated by scattered wave field	refraction reflection	simultaneous image or
3) actively illuminated by search beam or sonar	absorption scattering by small objects	sequential records

All waves have a wavelength λ, a frequency f, and a phase velocity v. These quantities are related as

$$v = f\lambda .$$
(7.1)

A surface wave on water, Fig. 7.4, serves as example. The height of the displacement at the maxima is called the amplitude ξ_0, and is a measure of the intensity of the wave, namely $I = \mathrm{const}\ \xi^2$ W/m². One can assign an intensity value to every point in a wave field. There are transverse waves, and longitudinal waves. In transverse waves the medium oscillates at right angles to the propagation direction. Examples include surface waves on water, and light waves. In longitudinal waves, such as sound waves, the medium oscillates in the direction of wave propagation.

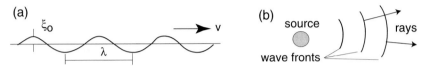

Fig. 7.4. (a) Surface wave on water, (b) wavefronts and rays

Waves spread into space, Fig. 7.4b. Their direction of travel can be described by rays, which are the wave front *normals*. As a wave travels away from a source its intensity is reduced by two effects: (i) by the spreading of the wave fronts with distance, and (ii) by absorption due to the intervening medium. In addition, the wave field may be diluted by diffraction, or scattering. Waves may change their propagation direction by reflection or refraction. Reflection and refraction are important for the design of eyes, and these effects can also lead to some natural channeling of wave energy. While scattering diminishes the intensity of a wave, the scattered light is not lost, it just travels into a different direction. It ceases to be a true messenger from the original source, but the scattered light may bathe some non-luminous objects, and make them visible to an observer. Ideal seeing requires plenty of light, which must be scattered off the object. Both, the amount of light which reaches the eye, and the receiver sensitivity determine from which distance an object can be seen.

Wavelength, frequency, and intensity play a role in seeing and hearing. It will be interesting to find out how animals perceive sound in a frequency range that stretches over four orders of magnitude, from about 10 Hz to 100 kHz. Compare to that the narrow range of light frequencies from $4.6 \cdot 10^{14}$ Hz to $6.3 \cdot 10^{14}$ Hz (corresponding wavelengths from $4.8 \cdot 10^{-7}$ m to $6.5 \cdot 10^{-7}$ m) that are perceived by our eyes, which are connected to nerves that conduct signals only at the low bit rates of typically 100 bits per sec.

7.3.2 Amplitudes, Wavelength, and How Things Move in Waves

The displacement ξ in any wave traveling in the z-direction at the phase velocity $v = \omega / k = \lambda f$ can be described by the wave function $\xi = \xi_0 e^{-i\{\omega t - kz\}}$ or by the simpler form

$$\xi = \xi_0 \sin\{2\pi\, ft - (2\pi/\lambda)\,z\} = \xi_0 \sin\{\omega t - k z\}\,, \tag{7.2}$$

where ξ_0 is the amplitude, f is the frequency; $\omega = 2\pi f$ the angular velocity, λ is the wavelength, and $k = 2\pi / \lambda$ the wave number. The oscillating quantity ξ stands for the pressure p in sound waves, for the electric field E in a light wave, for the lateral deflection y in a water surface wave, and so on. It is useful to recall the typical wavelength and amplitudes of sound and light waves. Light waves that are visible to humans fall into the narrow range $4.8 \cdot 10^{-7} \leq \lambda_{light} \leq 6.5 \cdot 10^{-7}$ m. Red light has a longer wavelength than blue light. Sound is an oscillation of the molecules of matter through which the sound travels. No sound can propagate in a vacuum. Audible sound waves have the range $2 \cdot 10^{-2}$ m $\leq$ sound ≤ 20 m. A low pitched (low frequency) sound wave has a longer wavelength than a high pitched sound. Elephants can hear lower sounds than humans. Dogs and bats can hear much higher frequencies. The wavelength of light and the displacement amplitude of sound are both much smaller than the width of a human hair. Figure 7.5 compares wavlength ranges of electro magnetic radiation to the displacment amplitude of sound waves and to the size of objects.

The quantity oscillating in a wave has the velocity u

$$u = \delta\xi / \delta t = \omega\, \xi_0 \cos\{\omega\, t - k\, z\} \tag{7.3}$$

and the acceleration a

$$a = \delta^2\xi / \delta t^2 = -\omega^2 \xi_0 \sin\{\omega\, t - k\, z\} . \tag{7.4}$$

Equation (7.2) is a solution for the general wave equation

$$\frac{\partial^2 \xi}{\partial t^2} = v^2 \frac{\partial^2 \xi}{\partial z^2} . \tag{7.5}$$

A physical quantity for which one can derive such a second order differential must have perturbations that travel as waves at the phase velocity v. For periodic waves of frequency f and wavelength λ one has $v = \lambda f$. Heinrich Hertz discovered that he could derive a wave equation starting with Maxwell's equations. Therefore, he predicted that there should be electromagnetic (EM) waves that travel through

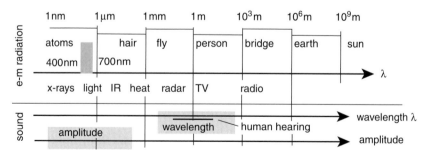

Fig. 7.5. Wavelengths and amplitudes of light and sound

vacuum at the velocity $v = c_0 = 3 \cdot 10^8$ m/s. And then he proceeded to generated EM waves. Not long thereafter Marconi transmitted signals across the Atlantic. Modern technology would be impossible without radio, TV, radar, x-rays, and other special frequency ranges of the electro-magnetic spectrum, Fig. 7.5. The phase velocity for sound waves depends on the compressibility $(\delta\rho/\delta p)$ of the medium through which the sound travels. In soft media like air, which have a high compressibility, the sound travels slowly. In less compressible media, like water, the sound speed is higher.

7.3.3 The Inverse Square Law

Intensity is one parameter that helps to estimate the distance of an object. The energy carried by a wave through a surface area of 1.0 m² is called the intensity I. Recall from Chap. 2 that all objects of temperature T emit (electromagnetic) radiation of intensity $I_s = P = A_s \varepsilon \sigma T^4$. The emissivity ε indicates how much the object's radiation differs from the emission of a perfect black body. $\sigma = 5.67 \cdot 10^{-8}$ W/m² K⁴ is Stefan's constant, and A_s is the surface area of the emitter.

Intensity can be defined anywhere in space, and is the power flux (energy per second and m²) that passes through the unit surface area. In general the intensity of a wave is related to the amplitude as $I \propto \xi_0^2$, where the amplitude ξ_0 may designate a displacement, a pressure fluctuation, a density amplitude, or the magnitude of an electric or a magnetic field. As already mentioned, the intensity of a wave can change for two reasons. First it may decrease due to the geometrical expansion in three dimensions (3D) as light spreads from the sun, or in two dimensions (2D), Fig. 7.6, as wave fronts spread on the surface of a pond when a kingfisher bird has splashed down for a meal. Second, the intensity is decreased due to the removal of energy from the wave field by scattering, and absorption. The geometrical dilution of the wave intensity must be quantified first.

3D expansion of a wave field applies to a point source of power P that emits radiation in all directions. The wave fronts are spherical shells of area $A = 4\pi r^2$. Conservation of energy requires

$$P = I_1 4\pi r_1^2 = I_2 4\pi r_2^2 .$$ (7.6)

Equation (7.6) can be used to find the intensity I_1 at the radius r_1 if the intensity I_2 is known at the radius r_2.

$$I_1 = I_2 \cdot (r_2^2/r_1^2)$$ (7.7)

Relation (7.7) is called the inverse square law, and holds for any wave in 3D. A simple example illustrates the magnitude of the intensity drop off with distance.

The light bulb emits the power $P = 40$ W of radiation energy, uniformly in all directions. What is the intensity I of this light at a distance of $r = 3.0$ m? Imagine a sphere of $r_1 = 3.0$ m with the surface area of $A = 4\pi (3\,\text{m})^2$. Since the radiation pow-

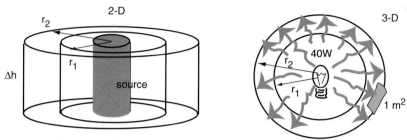

Fig. 7.6. Wave fronts spreading in two dimensional and threedimensions

er is spreading uniformly in all directions, the intensity at the radius $r_1 = 3.0$ m is
$I = P/A = P/4\pi r_1^2 = 40$ W $/\{4\pi (3$ m$)^2\} = 0.35$ W/m^2.

If the wave spreads from a source of height Δh in a cylindrical volume of height
Δh, the wave front surface area is $A = 2\pi r \Delta h$, then $P = I_1 2\pi r_1 \Delta h = I_2 2\pi r_2 \Delta h$, and
the intensity varies with distance as

$$I_1/I_2 = r_2/r_1, \text{ or } I_1 = I_2(r_2/r_1). \tag{7.8}$$

This cylindrical spreading relation (7.8) applies for instance to surface waves of
water, to light trapped inside plate glass, or to sound trapped inside the Sofar
channel (a layer in the ocean where the sound velocity has a minimum). The trapping effect will become clearer in Sect. 7.4 where refraction effects are discussed.

Sound intensities are generally measured with the decibel scale, which is defined as

$$\beta = 10 \log \{I/I_{ref}\} \text{ dB, pronounced } decibel. \tag{7.9}$$

I is the intensity and I_{ref} is a reference intensity which is specified from time to
time. In acoustics one uses as reference the threshold intensity for human hearing,
which has the incredibly small value

$$I_{ref} = 10^{-12} \text{ W/m}^2. \tag{7.10}$$

An increase of intensity by a factor 10 means an increase of the sound level by
10 dB. A thousand fold increase in intensity equals 30 dB. From the threshold of
hearing at about 1dB to the threshold of pain at 120 dB the intensity varies by a factor 10^{12}. Very few technical detectors have a sensitivity range of 12 orders of magnitude. A whisper may have 20 dB, a noisy classroom has typically 65 dB. The
threshold of pain is 120 dB, and a jet at takeoff may generate noise of 150 dB.

7.3.4 Reduction of Intensity by Absorption ($\lambda \gg D$)

Absorption reduces the intensity of sound or light waves and transfers (part of) the wave energy into the medium, so that the medium will heat up. Absorption of light can take place on a surface, where it is characterized by the absorption coefficient α, which equals the emission coefficient ε. It can also take place inside a dilute body, like in air or in water, or in a milky piece of glass, where it is measured by the bulk absorption coefficient κ.

Consider the absorption of light. Imagine a slab of matter of thickness x that is placed into the path of a wave at a position r_0 where the intensity is $I_0 = P/\pi\, r_0^2$

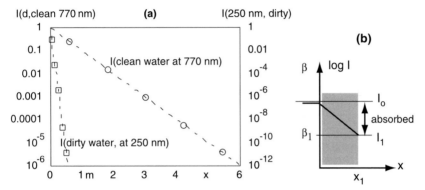

Fig. 7.7. (a) Attenuation of wave intensity. (b) Intensity I, and decibel scales

The material has an absorption coefficient, sometimes also called the extinction coefficient, κ. Due to the (partial) absorption the intensity drops exponentially inside the material according to

$$I(d) = I_0\, e^{-\kappa x} \tag{7.11}$$

where $x = r - r_0$. This exponential drop of intensity can be plotted as a straight line in an intensity I versus distance x plot, shown in Fig. 7.7. Dirty water attenuates the

Table 7.6. Absorption coefficients κ in m^{-1} for light in water. Data for clean water adopted from Denny [1993]. Data for sea water adopted from the Handbook of Optics [95]

absorption coefficient κ for light	UV	violett	blue	yellow	dark red
at the wavelength λ in nm	250	400	470	550	670
clean water	0.23 m^{-1}	0.01	0.005	0.04	0.3
dirty (waste) water, secondary effluent	50				
tertiary effluent	250				
sea water		0.018	0.016	0.064	0.43 m

intensity much more than clean water, because κ_{dirty} is much larger than κ_{clean}. The total intensity ΔI lost from the wave and absorbed between the front surface where $I = I_0$ and a plane at the depth $x = d$ is

$$\Delta I = I_0 - I(d) = I_0 \left(1 - e^{-\kappa d}\right) . \tag{7.12}$$

This intensity will continually heat up the slab. The inverse of the extinction coefficient called the range $R = 1/\kappa$, is a measure for the distance over which the light intensity is reduced by a factor $1/e = 37\%$. For instance a value $\kappa = 0.035 \text{ cm}^{-1} = 3.5 \text{ m}^{-1}$ implies a range $R = 1/0.035 \text{ cm}^{-1} = 28.6 \text{ cm}$. The absorption coefficient depends on the wavelength, see Table 7.6, and Fig. 7.8.

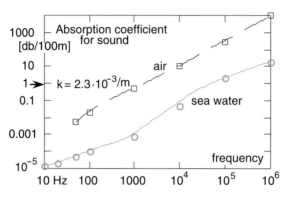

Fig. 7.8. Absorption coefficient κ for sound in air and sea water. Data adopted from Rossig [1990]. Note that an attenuation of $\Delta \beta = 10$ dB per 100 m is equivalent to $\kappa = 2.3 \cdot 10^{-2} \text{ m}^{-1}$

The absorbed intensity ΔI may also be expressed in sound level difference $\Delta \beta$ using the definition of β

$$\beta = 10 \log \left(I/I_{ref}\right) . \tag{7.13}$$

One can describe the absorbed fraction of intensity as a difference of sound levels at the point a and the point b

$$\Delta \beta = \beta_a - \beta_b = 10 \left\{ \log \left(I_a/I_{ref}\right) - \log \left(I_b/I_{ref}\right) \right\} = 10 \log \left(I_a/I_b\right) . \tag{7.14}$$

The relation is useful because sound absorption in water is often quoted in decibel difference per distance $\Delta \beta / \Delta x$, where $\Delta x = 100$ m, (see Fig. 7.9). This measure of absorption can be converted into an absorption coefficient κ by the relation

$$\kappa = (1/\Delta x) \ln(10^{\Delta \beta / 10}) . \tag{7.15}$$

The reduction of intensity by extinction is of importance for all animals living in the water. Light does not penetrate far into the ocean. It is utterly dark down there. In contrast sound penetrates easily through water.

With this additional knowledge about the interaction of waves with matter we want to recalculate the intensity produced by a light bulb that is submerged under water, Fig. 7.9. Consider again the spherical vessel of radius $r_1 = 3.0$ m, filled with water and illuminated by a light bulb at the center, which radiates a power of 40 W at the wavelength $\lambda = 770$ nm. The bulb is housed in a glass sphere of $r_0 = 0.1$ m. We now must take into account both, the extinction $I = I_0 \exp(-\kappa \, \Delta r)$, where $\Delta r = 2.9$ m, and $\kappa = 0.0238$ m^{-1}, and the intensity attenuation by the inverse square law $I = I_0 \, r_0^2 / r^2$. The combined effect is $I_1 = I_0 (r_0^2/r_1^2) \exp(-\kappa \, \Delta r)$. First find the intensity I_0 at the surface of the bulb $I_0 = 40$ Watt$/4 \, \pi \cdot 0.1$ m$^2 = 318$ W/m^2. Then apply the extinction formalism to find: $I_1 = 318 \, (0.1/3) \, 2 e^{-0.0238 \cdot 2.9} = 318 \cdot 0.0011 \cdot 0.933 = 0.330$ W/m^2. This intensity is smaller by the factor $e^{-0.0238 \cdot 2.9} = 0.93$ than for the sphere without water.

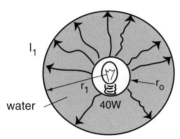

Fig. 7.9. Water filled sphere, $r_1 = 3$ m

7.3.5 Scattering

Scattering makes objects visible. Waves are scattered off objects, which are small compared to the wavelength.

Sunlight passing through the atmosphere is scattered by the atoms and molecules, Fig. 7.10. Blue light is scattered at angles around 90°, therefore the sky above us looks blue. Red light passes through the atmosphere, therefore a sunset looks red. Leaves of trees look green, because they absorb the red light and scatter the green light. A rabbit looks grey-brown because it scatters that part of the daylight, which mixes into a brown color. But one cannot see a rabbit in a dark night because there is not enough random light of all wavelengths that could be scattered off the animal to form a colored image on the retina of the eye. If one shines a red flashlight onto the rabbit it would appear to be red instead of brown, because the red flashlight does not have the blue and yellow components that would add up to brown.

Similarly, sound is scattered when traveling through branches of trees or dense foliage in a forest. These obstacles act as scattering centers or as *secondary emitters*. Therefore the transmitted intensity is reduced, although the total energy car-

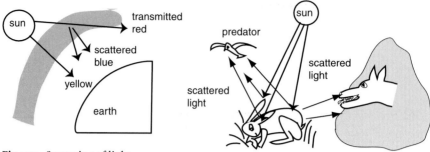

Fig. 7.10. Scattering of light

ried by all parts of the unscattered and the scattered sections of the wave is not diminished.

Other forms of light scattering, not discussed here are 1) Compton scattering, for x-rays, where light is scattered off the nuclei, 2) Thomsson scattering for UV, where light is scattered off electrons, and 3) Mie scattering for infrared light, where radiation is scattered off large molecules.

The attenuation of sound in air depends both on moisture content and temperature. For instance, just before the rain starts one can hear a train whistle over a long distance.

7.4 How Waves Change Their Direction

Waves spread in all directions from their source. Waves on the ocean are a familiar picture. One can easily visualize such wave fronts, or "phase fronts", namely the locations where at a given instant the displacement is a maximum. Imagine now a line drawn at right angle to the local direction of one of these fronts, as in Fig. 7.11. The line is the direction in which the wave is traveling. Such wave front normals are called *rays*.

In optics and acoustics, the propagation of the waves can be completely described by considering the rays. Waves from a point source in a uniform medium spread out like spherical shells. The rays point in a radial direction. Critical for the change of direction of rays is the local phase velocity v.

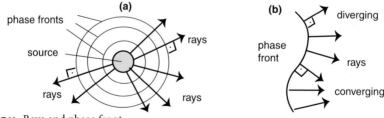

Fig. 7.11. Rays and phase front

7.4.1 The Phase Velocity

The phase velocity v of a wave depends on the medium through which the wave travels. Examples are given in Table 7.7. The velocity v changes if the medium parameters change. For instance, if air gets warm the speed of sound goes up. The speed of light in vacuum is $c_0 = 3.0 \cdot 10^8$ m/s. Typically the speed of light in glass is $c_{gl} = 2 \cdot 10^8$ m/s, and the speed of sound in air is $v_{air} = 340$ m/s.

Table 7.7. Examples of waves, and parameters

name	oscillating quantity	type	phase velocity, and parameters
sound waves, a_i speed of sound of medium i	pressure	longitudinal	$a_{gas} = (\gamma \cdot p/\rho)^{1/2} = (\gamma R_g T/M)^{1/2}$, γ = adiabatic exponent, ρ = density, $R_g = 8.31$ J/mol K, T temperature, M = molecular weight in kg, see also Sect. 7.4.6. $a_{air}(20°C) \approx 340$ m/s. $a_{liquid} = (B/\rho)^{1/2}$, B = bulk modulus, Y = Young's modulus, $a_{water} \approx 1,500$ m/s $a_{solid} = (Y/\rho)^{1/2}$, $a_{steel} \approx 5,000$ m/s,
surface waves on deep water surface waves on shallow water waves on a string	 sections of the string	transverse transverse transverse	$v = (g\lambda/2\pi)^{1/2}$, λ = wavelength, $g = 9.81$ m/s² $v = (gd)^{1/2}$, d = depth of water $v = (\tau/\mu)^{1/2}$, τ = tension in N, μ = mass/unit length in kg/m
electromagnetic waves: γ-rays, x-rays, ultra violet (UV), light, infrared (IR), heat, radar, TV, AM, FM	electro- magnetic field	transverse	$c = c_0/n$; in vacuum $c_0 = 3 \cdot 10^8$ m/s n = index of refraction. $n_{air} \approx 1.00$, $n_{glass} \approx 1.5$ $n_{water} = 1.33$, see Sect. 7.4.4
shock waves $v_{shock} > a_s$ Mach number $M = v_{shock}/a_{gas} > 1$	pressure and density	aperiodic	$v_{shock} = ((\gamma+1)(p_b/2p_a))^{1/2}$, γ = adiabatic exponent p_b = pressure behind shock, ρ_a = density ahead of shock.

Wave propagation is characterized by the phase velocity. Often one does not need to know the absolute value but rather the ratio of the phase velocity divided by a reference velocity. For light waves one uses the index of refraction n, which is defined as the ratio of the speed of light in vacuum to the speed of light v in some medium.

$$n = c_0/v \tag{7.16}$$

Some values of n are given in Tables 7.7 and 8.2. The speeds of light in a medium depends on the wavelength. High frequency waves may have a lower velocity than low frequency waves, a phenomenon that is called *dispersion*. Hence the index of refraction is a function of the wavelength. For light this phenomenon gives rise to the color of the rainbow.

Light is a member of the family of electro-magnetic waves, which include γ-rays, x-rays, infrared, heat, radar, TV, FM, and AM radio waves. The sound speed in a gas depends on the temperature. The sound velocity a in water depends on the temperature and salinity. Pulsed pressure waves of very high amplitude are called shock waves. Shock waves travel as at speeds v that are larger than the acoustic velocity a. For such waves one defines the Mach number M = v/a, which is an inverse refractive index for sound.

Table 7.8. Nerve conduction velocities

fiber type	function	d [μm]	v [m/s]
efferent	carry signals away from brain		
Aα	skeletal muscle	15	100
Aβ	skeletal muscle & muscle spindle	8	50
Aγ	muscle spindle	5	20
B	sympatic, preganglionic	3	7
C	sympatic, postganglionic (unmyelinated)	1	1
afferent	carry signals towards brain		
Ia	muscle spindle	13–20	80–120
Ib	tendon organ	13–20	80–120
II	secondary muscle spindle	6–12	35–75
III	deep pressure sensors in muscles	1–5	5–30
IV	pain(unmyelinated), temperature	0.2–1.5	0.5–2

Nerve pulses travel as depolarization waves. Their speed depends on the nerve diameter d and on the nature of the nerve's insulation layer, see Table 7.8. The larger d, the faster is the signal velocity v. The data are plotted in Fig. 7.12. The data suggest the empirical relation $v_{nerve} = 3d^{1.2}$ m/s, where d is given in μm.

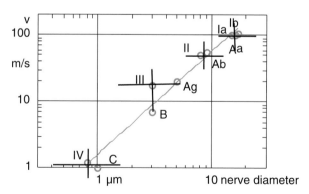

Fig. 7.12. Nerve signal velocity v as function of the fiber diameter d. The symbols are explained in Table 7.8

7.4.2 The Phase Velocity in a Compressible Medium

The phase velocity v of waves is in general defined through the wave equation (7.5), namely v is the factor in front of the differential $\partial^2 \xi / d z^2$. When sound travels in the direction z through a gas or another compressible medium (of density ρ and pressure p_0) slight variations of density $\Delta\rho(z,t)$ and pressure $\Delta p(z,t)$ are induced and volume elements vibrate at time dependent quiver velocity $u(z,t)$. These variations must satisfy the equations of conservation of mass (i) and conservation of momentum (ii).

$$\text{(i)} \quad \frac{\partial \Delta\rho}{\partial t} = -\rho_0 \frac{\partial u}{\partial z} \ , \text{and (ii)} \quad \rho_0 \frac{\partial u}{\partial t} + \rho_0 u \frac{\partial u}{\partial z} = -\frac{\partial p}{\partial z}$$

A wave equation for sound waves in a compressible medium can be derived by combining these two equations. The velocity fluctuations u in sound waves are quite small quantities. Then the spatial gradient $\partial u/\partial z$ can also not be very large, so that one can safely neglect the product $\rho_0 u\, \partial u/\partial z$ on the left-hand side of (ii). The pressure gradient on the right hand side of (ii) may be expanded as $\partial p/\partial z = (\partial p/\partial \rho)_S (\partial \rho/\partial z)$. The index S on the pressure-density differential indicates that one considers here the fast, adiabatic, response of the medium where there is not enough time to exchange any heat. This set of equations can be reduced to a single differential equation. For this purpose one differentiates equation (i) with respect to t and equation (ii) with respect to z. This operation leads to

$$\text{(i)} \quad \frac{\partial^2 \Delta\rho}{\partial t^2} = -\rho_0 \frac{\partial^2 u}{\partial z \partial t} \ , \text{and (ii)} \quad \rho_0 \frac{\partial^2 u}{\partial t \partial z} \approx -\frac{\partial p}{\partial z} = -\left(\frac{\partial p}{\partial \Delta\rho}\right)_S \frac{\partial^2 \Delta\rho}{\partial z^2} \quad .$$

The term $\rho_0[\partial^2 u/(\partial t \partial z)]$ can be eliminated from both equations so that a single differential equation is obtained

$$\frac{\partial^2 \Delta\rho}{\partial t^2} = \left(\frac{\partial p}{\partial \Delta\rho}\right)_S \frac{\partial^2 \Delta\rho}{\partial z^2} \quad .$$

This is the typical form of a wave equation (7.5). The factor in front of the differential $\partial^2 \Delta\rho / \partial z^2$ is the square of the sound velocity in a compressible medium.

$$v^2 = \left(\frac{\partial p}{\partial \Delta\rho}\right)_S \tag{7.17}$$

With the help of (7.17) one can find the sound velocity for any biological material where the adiabat of density as function of pressure $(\partial p/\partial \Delta\rho)_S$ has been measured.

7.4.3 Refraction

Whenever a wave passes through regions where the phase velocity (or the index of refraction) changes, the rays change their direction by refraction, or by reflection. Refraction is responsible for the action of lenses. Reflection – the mirror action – bounces the wave back, and produces a virtual "mirror" image.

Figure 7.13 shows these rays. It is customary to describe the direction of rays coming in and going out from one point with angles ϕ_1 and ϕ_2 measured against the incident normal. The angles are related by the reflection law

$$\phi_1 = \phi_2 \tag{7.18}$$

and by Snell's law of refraction

$$\sin \phi_1 / \sin \phi_2 = n_2 / n_1 = v_1 / v_2 . \tag{7.19}$$

When ϕ_1 is known, one can calculate ϕ_2. The smaller angle is always on the side where the phase velocity is smaller, or where n is larger. Reflection and refraction often take place at the same time, and a part of the wave is reflected while the transmitted wave carries less energy. Any reflection generates an image. Reflection off a plane surface produces a virtual image, which lies behind the reflecting surface.

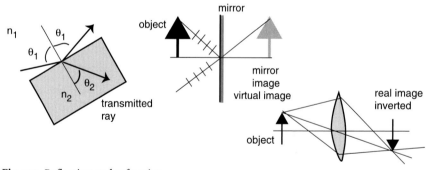

Fig. 7.13. Reflection and refraction

7.4.4 Total Internal Reflection

Consider a light beam traveling from a medium with larger n (slow phase velocity) towards a surface separating it from a medium with lower n (faster phase velocity). For a certain critical angle ϕ_{crit} measured in the slower medium the angle ϕ_1 in the medium with the higher phase velocity approaches $\phi_1 = 90°$, hence $\sin \phi_1 = 1.00$. Then Snell's law yields $\sin \phi_1 / \sin \phi_2 = 1 / \sin \phi_{crit} = n_2 / n_1 = v_1 / v_2$, or

$$\sin \phi_{crit} = n_1 / n_2 = v_2 / v_1 . \tag{7.20}$$

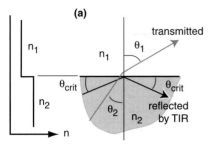 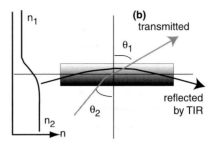

Fig. 7.14. Details of total internal reflection

It is impossible to pass waves from the slow medium into the fast medium with an incident angles $ø > ø_{crit}$. Such waves are trapped inside the slow medium. The effect is called total internal reflection.

Total internal reflection is perfect reflection, Fig. 7.14. Table 7.9 shows the critical angles for total internal reflection between different media. Total internal traps light in light pipes, discussed in Sects. 7.4.5 and 8.5.1. Total internal reflection also guides sound in certain layers in the air or the ocean, see Sect. 7.4.6.

Table 7.9. Critical angles of total internal reflection

interface	$\sin ø_{crit} = n_1/n_2 = v_2/v_1$	$ø_{crit} = \sin^{-1}(n_1/n_2)$
air – water	1/1.33=0.75	41.8°
air – glass	1/1.50=0.666	48°
water – eye lenses tissue	1.33/1.413=0.930	70°

7.4.5 Light Pipes and Wave Guides

If there is a channel of low phase velocity, where the index of refraction has a maximum, some waves are trapped: waves that impinge from inside of the medium at an angle $ø_2 > ø_{crit}$ suffer total internal reflection, an effect that is the basis of light pipes. Insects utilize light pipes in the individual eyelet of their facet eyes. All fiber optics conductors of modern high speed signal communication systems are light pipes. The waves are trapped even if there are no sharp steps in the index of refraction but rather a steady change, as indicated on the right of Fig. 7.15.

The cross section area of a light pipe $A = \pi r^2$ is constant. Since all waves are reflected back into the core by total internal reflection, the intensity remains approximately constant. The intensity decreases very slightly because no material is 100% transparent. Any material absorbs or scatters some of the energy of waves passing through it. Only vacuum is truly transparent for light, but of course vacuum does not transmit sound at all. In contrast the intensity I of a source of power P drops off in free space with the square of the distance $I = P/4\pi d^2$ W/m² because the surface area of the wave grows as $A = 4\pi d^2$, see Sect. 7.3.3.

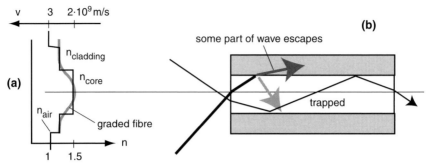

Fig. 7.15. Reflection and refraction in a light pipe

7.4.6 Sound Pipes: The Sofar Channel and Ground Effect

The confinement of waves in regions of lower phase velocity can be observed in optics and acoustics. Total internal reflection of sound occurs in the Sofar channel at a depth of about 1000 m in the ocean, where the speed of sound has a local minimum, Fig. 7.16. The sound speed varies because of changes of the local temperature and salinity.

The slow speed layer acts like a *sound pipe* in which the sound is trapped spreading cylindrically in a plane and not spherically like sound in free space. Therefore, the intensity drops only slowly with distance, so that voice communication over large distances becomes possible. Aquatic mammals know that.

The Sofar channel acts as a cylindrical sound channel of the height h. The surface area of this channel grows with distance as $A = 2\pi d h$. Therefore, the intensity of a sound signal P emitted in the Sofar channel drops off linearly with distance d, whereas in free space the sound intensity drops of with the square of the distance.

$$I = P/(2\pi h d) . \tag{7.21}$$

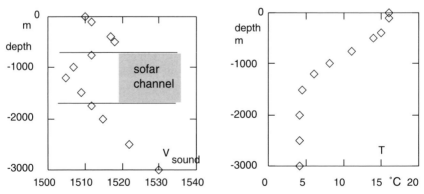

Fig. 7.16. (a) Variation of sound velocity in the ocean with depth. (b) Variation of temperature in the ocean with depth at latitudes of about 45°

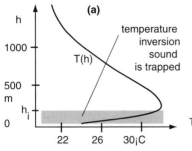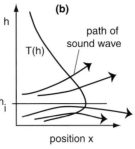

Fig. 7.17. Ground effect. (**a**) Thermocline of atmospheric inversion. (**b**) Sound wave trajectories

A similar sound trapping effect can happen in the air since the speed of sound in air a_{gas}, and hence its index of refraction for sound waves, defined by (7.19), is a function of the temperature T, namely $a_{gas} = \sqrt{(\gamma RT/m)}$.

Under certain atmospheric conditions, when colder air is trapped close to the ground, the sound speed is lower close to the ground than higher up, Fig. 7.17. The sound will be trapped in the low speed layer, similarly as in the Sofar channel in the ocean. This phenomenon is called the ground effect. Such temperature inversions can occur on the grasslands of Africa. When this happens animals can communicate by sound over much larger distances than under normal circumstances.

7.4.7 The Lateral Spread of Wave Fronts

When waves propagate near obstacles that inhibit their lateral spreading, two different phenomena may occur: shadows and diffraction. These phenomena can be understood applying Huygens principle: a wave front at location II, in Fig. 7.18a, can be understood as the superposition of elementary wave fronts (called Huygens wavelets) that have emerged from the preceding location I. Four cases must be distinguished.

a) The wave front width d is large compared to the wavelength λ (Fig. 7.18a). In this case a shadow with a narrow penumbra (half-shadow regime) is created.

b) The slit width d is larger than the wavelength λ, but not by more than about an order of magnitude. A diffraction pattern is generated with a center maximum of width Δ and several much smaller maxima on either side, Fig. 7.18b.

The first order diffraction minimum is found at the half angle ϕ_{min} given by

$$\sin \phi_{min} = \lambda/d . \tag{7.22}$$

For a circular aperture (like the iris of an eye) of radius $r = d/2$ the center spot has a half angle given by

$$\sin \phi_{min} = 1.22 \, \lambda/d . \tag{7.23}$$

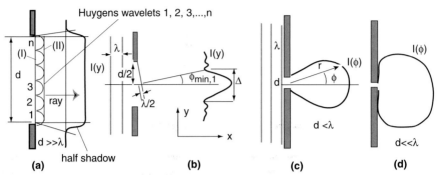

Fig. 7.18. (a) Shadow $d \gg \lambda$. Incident wave front (|), common tangent of Huygens wavelets (|||). **(b)** Diffraction pattern, intensity $I(y)$ for $d \approx \lambda$. **(c)** Narrow slit diffraction $d < \lambda$. The intensity $I(\emptyset)$ is represented by the length of the radius vector r in the $\emptyset$ direction. **(d)** Spreading in all directions $d \ll \lambda$

Instead of forming infinitely sharp images, all lenses generate pictures composed of diffraction spots, no matter how fine and sharp the object points appear to be. Diffraction limits the resolution of every eye.

c) When d is of the order of the wavelength λ and the slit is gradually reduced, the diffraction pattern spreads out and the center order maximum starts to fill the space to the right of the slit. For $d/\lambda = 0.5$ the spreading angle $\emptyset_{min,1}$ equals 30°. One may display the intensity distribution as an r–$\emptyset$ contour line, Fig. 7.18c, where the magnitude of the intensity is represented by the length of the radius vector r in the $\emptyset$ direction.

d) When d is very small compared to the wavelength, the slit acts like a point source, Fig. 7.18d. Waves are spread into the full hemisphere to the right of the slit.

Diffraction is known from laboratory experiments in optics, where light hits small objects or slits. However, one can also readily experience diffraction of sound waves in a simple experiment. Close your left ear with the right hand and rub thumb and index finger of the left hand close to the left ear. You will hear nothing. However, when you stretch out your left arm and rub the fingers again, still having the left ear closed, you hear the high pitched sound of rubbing in your right ear. The sound source is now far enough from the impeding object, so that the sound waves diffract around your head and reach the right ear. Typically the high pitched sound has frequencies of a few kHz, say $f = 3400$ Hz $= 3400$ s^{-1}. Then the wavelength $\lambda = 340$ ms$^{-1}/3400$ s$^{-1} = 0.1$ m is of the order of magnitude of your head diameter, and diffraction becomes noticeable.

There are some other physical effects that change the intensity distribution in waves, namely interference and superposition. These phenomena will be discussed in later sections as they become important for the perception of sound and light.

7.5 Information Background

The biosphere is imbedded into various fields. The earth's gravitational field "ties" every object to the ground. The magnetic field of the earth gives some directional guidance to pigeons, and bacteria. These fields do not contribute to the exchange of information.

The vast sea of information is enabled by the radiation fields of the sun which makes objects visible for organisms with eyes. Only at night do the other spurious light sources like moon, stars, and lightning make any impression on the observer, and only deep at the bottom of the oceans or in very dark nights do bioluminescent organisms play any role in the flow of visual information. Most of this optical information is passive, revealing both moving things and motionless objects.

Another gold mine of information is the sound field. Acoustic waves emanate from moving structures: blowing wind, rushing water, dripping rain, and roaring thunder. Nearly every moving organism of the biosphere and every moving part of a living body radiates its acoustic signature into the surrounding. Vast floods of acoustical data float around, ready to be picked up by sensitive ears.

The third background field harboring information is the electrostatic field. Its sources are of organic nature. Every living cell generates a tiny electrostatic field. In environments, where light and sound fail, the ambient electric fields is the last resort for obtaining information about the surrounding. The information flux enabled by light, sound, and electrostatics will be discussed in the next chapters.

Light and sound signals travel as waves. Waves can transmit data by two principal techniques: first by a change of the intensity, $I = I(t)$ that may happen gradually as the sun sets, or abruptly as in a Morse code. Secondly, information may be encoded with a variation in the frequency as in a whistle tone. Light waves transmit many different *image points* simultaneously at an enormous speed. Eyes are indispensable for observing rapid motion. Sound can transmit many frequencies simultaneously, but with very poor imaging qualities. Ears of higher animals analyze the frequency spectrum of the sound field. Voices and ears are indispensable to facilitate language, education, and culture. We will first discuss the physics of light and the biology of light receptors as light plays an important role for animals in every phylum and at every level of evolution.

Table 7.10. Frequently used variables of Chap. 7

variable	name	units	name of unit
c_o	speed of light	m/s	
f	frequency	s^{-1}	Hertz
I	intensity	W/m2	
k	wave number	m^{-1}	
n	index of refraction		
v	phase velocity	m/s	
β	sound level in decibel		
κ	absorption coefficient	m^{-1}	
λ	wavelength	m	
ξ	wave amplitude		

Problems and Hints for Solutions

P 7.1 Darkness at the Bottom of the Sea

Calculate the intensity fraction I/I_0 of blue ligth ($\lambda = 470$ nm, $\kappa_b = 0.016$ m^{-1}) and dark red light ($\lambda = 670$ nm, $\kappa_r = 0.43$ m^{-1}), penetrating down to $d = 400$ m in clean sea water.

P 7.2 Long Distance Talk of Whales

a) Determine the critical angle of total internal reflection for sound waves in the Sofar channel. What is the beam angle 2δ that an animal should emit in order to match the Sofar channel sound wave guide? b) Suppose a whale in the Sofar channel talks to a friend $d = 2000$ km away. The animal emits a sound signal (of $P = 1.2$ W, at the frequency $f = 10$ Hz) into the conical beam of half angle δ calculated above. What is the intensity of this sound wave at a distance $d = 2000$ km? Express your result as sound level β in dB using $I_{ref} = 10^{-12}$ W/m^2. Assume a height of the Sofar channel of $h = 800$ m. Consider both, the spreading of the wave with distance and the reduction due to absorption. c) How long does it take for the message to travel the distance?

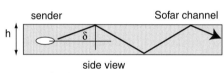

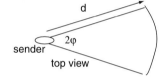

Fig. 7.19. Sofar channel

P 7.3 Resolution of Sonar Images

Assuming that objects with lateral dimensions of 2λ can be seen, calculate the minimum frequency of sound which aquatic animals must use in its sonar beam to resolve individual kelp (sea weed) leaves of $d = 15$ cm width.

P 7.4 Features of Optical and Acoustical Information

Ask one person to leave the room for a minute, show Fig. 7.20 to the class for about 3 seconds and ask the class what they have seen.

> The question how many apples were there is likely not answered immediately; however, if you ask how many rows (4), and how many columns (5) there were, most likely someone from the audience will recognize that two apples were missing from the maximum of 20, and one of the apples did not have any leaves.

Before the outsider returns arrange for 3 members of the audience, which are known to the student outside, to ask a simple question. Then let the outsider return with eyes still covered. Now the questioners ask their questions one at the time. Let the test person show the direction from where the question came, and who asked the question. Note how accurately distance and direction can be estimated, and that the test person easily knows who asked. Sound gives various clues about locations and identity of the source. Then ask the audience to tell the test person the content of the picture previously shown to the class. Note how much time it takes to convey all the details of the picture. This illustrates the proverb: a picture is worth one thousand words.

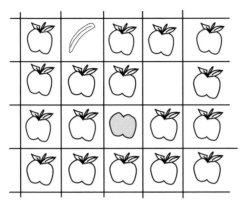

Fig. 7. 20. Apples and a banana arranged in a grid

P 7.5 Survival Without Senses
Compare the survival chances of one higher animal that has eyes and ears to the survival strategy of a primitive animal, which lacks these senses (e.g. shark and shellfish).

Hints and Solutions

S 7.1 Darkness at the Bottom of the Sea
$\text{Intensity}(I/I_0)_{blue} = e^{-\kappa\lambda} = e^{-0.016 \cdot 400} = \exp\{-0.016 \cdot 400\} = 1.66 \cdot 10^{-3}$. Similarly find $(I/I_0)_{red} = e^{-0.43 \cdot 400} = 2.0 \cdot 10^{-75}$. There is no red light at this depth.

S 7.2 Long Distance Talk of Whales
a) From Fig. 7.17 extract the maximum and minimum sound velocities $v_{max} = 1518$ m/s, and $v_{min} = 1506$ m/s. Therefore, $\sin \emptyset_{crit} = 1506/1518 = 0.9920$, and $\emptyset_{crit} = 82.8°$. $\delta = 90 - 82.8 = 7.2°$. Opening angle $2\delta = 14.4°$.

b) Solution strategy: first find the distance r_0 from whereon the sound will be trapped inside the Sofar channel, then determine the intensity $I_0(r_0)$, (without absorption). For $r \leq r_0$ the sound spreads out as a 3D wave. Then find $I(r)$ as 2D wave. Now include absorption. Convert $I(r)$ into $\beta(r)$. Assume that the whale is located at the center of the Sofar channel.

A ray that is just barely confined leaves the source at the angle δ and strikes the top of the Sofar channel at the distance r_0, thus $\tan \delta = (h/2)/r_0$ and $r_0 = h/(2\tan\delta) = 800/(2 \cdot \tan 7.2°) = 3.16 \cdot 10^3$ m. Intensity at r_0 without absorption $I_0 = 1.2\,W/(4\pi r_0^2) = 9.57 \cdot 10^{-9}\,W/m^2$. Cylindrical expansion of wave from r_0 to r without absorption

$I(d) = I_0 \cdot (r_0/r) = (r_0/r) \cdot P/(4\pi r_0^2) = P/(4\pi r_0 r) = 1.51 \cdot 10^{-11}$ W/m². Now include absorption: The absorption coefficient can be read from Fig. 7.9. Note that the figure has two scales. An absorption of 1 dB/100 m is equivalent to $\kappa = 2.3 \cdot 10^{-3}$ m^{-1}. Sound of $f = 10$ Hz has an absorption coefficient $\kappa = 4.6 \cdot 10^{-8}$ m^{-1}. Convert $I(d)$ to the intensity with absorption $I(d_{abs})$, by calculating the exponential drop of intensity, Eq. (7.11), between the source and r: $I(d_{abs}) = I(d)e^{-\kappa d} = 1.51 \cdot 10^{-11}$ W/m² $\cdot$ exp$\{-4.6 \cdot 10^{-8} \cdot 2 \cdot 10^6\} = 1.38 \cdot 10^{-11}$ W/m². Absorption does not do much! The sound level in decibel at $r = 2000$ km is: $\beta = 10 \log_{10}\{I(d_{abs})/10^{-12}\} = 10 \log_{10}\{1.38 \cdot 10^{-11}/10^{-12}\} = 11.39$ dB. c) $\Delta t = \Delta x/v = 2 \cdot 10^6$ m/1506 m $\cdot$ s$^{-1} = 1328$ s $= 22.1$ min.

S 7.3 Resolution of Sonar Images

Assume $v = 1520$ m/s, $d = 2\lambda = 0.15$ m, $\lambda = 0.075$ m, $f = v/\lambda = 1520$ ms$^{-1}/0.075$ m $= 20.3$ kHz.

8. Light, Abundant Information

A picture is worth 1000 words,
but the beauty is only in the eye of the beholder.

Light Makes Good Images

All animals respond to the opportunities and dangers in their environment. They can react only if they are aware what is out there. Vision gives such information during broad daylight. Vision is the stealthy long distance sense of the aggressor, and the defensive sense of the hunted.

The wave phenomena of light – refraction, reflection, and diffraction – explain how images are generated in the eyes of animals, and how much detail the images contain. Wave models also help to understand how different eyes work. Interference and absorption of light waves clarify how body colors come about. However, the images have to be recognized by sensor elements. Light has a double nature. It can be described either as a wave phenomenon, or as a flood of photons. Each photon carries a tiny trace of energy $E = h\nu$ that causes the molecular changes in the photo-sensitive elements of the eyes. The incredible abundance of these minute energy missiles, as they are scattered off every object, makes vision possible.

Chapter 8 begins in Sect. 8.1 with a few facts of light that help to understand the image acquisition and the sensitivity range of eyes by investigating photons, and the intensity-wavelength distribution of sunlight. Section 8.2 describes imaging principles, lens relations, how imaging can separate objects from the background, image quality, and problems with seeing under water. Section 8.3 is a thumbnail sketch of the human eye. Section 8.4 describes several specialized optical receivers of animals: the "heat" eyes of snakes, spider eyes, fish eyes, and trilobite eyes. Insects have gone yet another route to acquire visual information: facet eyes, described in Sect. 8.5. For some animals in the oceans visibility is wanted or undesirable. They employ interference and polarization of light, discussed in Sect. 8.6.

8.1 Facts of Light

Most life depends on the sun for energy. The sun's energy arrives on earth as electromagnetic radiation in a wavelength range from x-rays with wavelengths λ of a few nanometers (nm) to radio waves with wavelength in kilometers. Most of the energy comes as light in a very narrow wavelength range, from violet light (with a wavelength $\lambda = 0.4\ \mu m = 4 \cdot 10^{-7}$ m, and frequency[1] $\nu = c/\lambda = 4 \cdot 10^{14}$ Hz) to dark red

---

[1] In optics the Greek letter ν is commonly utilized to designate radiation frequencies. The letter f is reserved to denote the focal length of a lens.

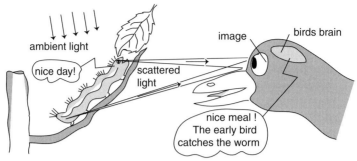

Fig. 8.1. The early worm is eaten by the bird

ligth (with a wavelength $\lambda = 0.7\ \mu m$, and frequency $v = 7.5 \cdot 10^{14}$ Hz). Since light facilitates seeing the surrounding, organs to detect the light have been invented many times during the evolution of life on earth.

The sun's radiation renders objects visible, because during daytime sun light is abundant and everywhere. Sunlight is scattered, and it illuminates every spot. Scattering is a process where light is reflected randomly off surfaces. Every spot on a surface that scatters light looks as if it was emitting light. This scattered light carries information about the scattering object itself. Organisms must have eyes to detect and neurons to analyze this information. Optical information can be recorded simultaneously by an imaging system combined with a high-resolution detector. Then the information must be decoded by the brain, because the beauty is actually in the brain of the beholder.

8.1.1 Photons and Waves

To understand the superb imaging properties of light one must be aware that light has a double nature. It can be depicted as tiny waves of electro-magnetic radiation with wavelength λ wiggling through vacuum or through transparent media. Alternately light can be described as a shower of tiny energy bullets, photons, rushing at the speed of light, $c_0 = 3 \cdot 10^8$ m/s, away from the source.

Before discussing details it is useful to get an intuitive feeling for the double nature of light. Huygens imagined light as tiny wave surfaces traveling in the direction of the surface normals, which are called rays. A few of such wavelets are drawn in Fig. 8.2. Each has a three dimensional structure like the canopy of an umbrella, and the shaft is pointing in the direction of the ray. To emphasize the wave nature, little umbrellas with multiple canopies simulating two wave crests are drawn for the light wavelets, and the letter λ is attached.

Everything that happens to light between a source and a receptor can be understood by depicting light as a wave structure. However, during the birth of the radiation at the source, and during the death at the receptor, light must be treated, as if the wave surfaces were folded up into a bundle of energy, like a pocket umbrel-

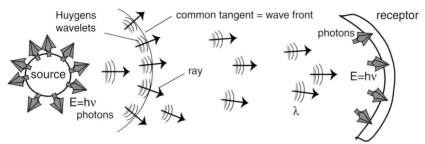

Fig. 8.2. Huygens' pocket umbrellas are either open or closed, wavelets, or photons

la. These structures are called *photons*. Each photon carries the energy $E_p = h\nu$, where $h = 6.6 \cdot 10^{-34}$ J·s is Planck's constant. Physicists describe the transition from wavelet to photon as the collapse of the wave function. The *Huygens pocket umbrella* is a simple model for the dual nature of light: either folded up as photons, or spread out as umbrella canopy wavelets. Light is never both at the same time, the umbrella is either open, or closed.

8.1.2 Why Light Waves Can Produce Sharp Images

The superb imaging qualities of light depend on a number of fortunate circumstances: 1) During the daylight hours there is plenty of sun light – measured as intensity I, with the units W/m², or counted as number of photons hitting an object every second. 2) Most objects of interest to animals, and the eyes themselves are large compared to the wavelength of light λ. The size difference guarantees good image resolution as the images in most cases are large compared to the diffraction spots (described in Fig. 8.18) that are created by the entrance pupils of the eyes. 3) Light is scattered and refracted. It spreads out in all directions from each scatter point, ready to be *seen* by capable eyes. 4) The speed of light depends on the wavelength and on the material. Differences of the speed of light in materials cause refraction. Lenses, and light pipes guide light due to the physical effect of refraction.

In addition to these convenient physical properties light also gives the hunter two important physiological advantages: (i) The speed of light is so high that images are received instantaneously, compared to all sensual reaction. (ii) The *object* is unaware that it is *seen* by some one else, because the observer utilizes the light that is scattered anyway. Vision is limited only when the background light level is low: at night in terrestrial landscapes, and during the day deep down in the ocean. However, vision is useless in opaque media such as murky waters.

8.1.3 Intensity, Wavelength, and Photon Numbers

Every object of temperature T measured in Kelvin degrees, emits "thermal" radiation, described by the Kirchhoff-Planck formula (8.1), which gives the intensity I_λ as function of wavelength λ, temperature T, and the natural constants, namely the speed of light in vacuum $c_0 = 3 \cdot 10^8$ m/s, Planck's constant $h = 6.6 \cdot 10^{-34}$ J s, and Boltzmann's constant $k_B = 1.38 \cdot 10^{-23}$ J/K:

$$I_\lambda \, d\lambda = \frac{8\pi hc}{\lambda^5} \cdot \frac{d\lambda}{e^{hc_0/\lambda k_B T} - 1} . \tag{8.1}$$

The total radiative emission is found by integration over all wavelengths. This integration yields Stefan's law, Eq. (2.21), $I = \int I \, d\lambda = \sigma T^4$, which was already discussed in Chap. 2. Even an object as *cold* as ice ($T \approx 273$ K) produces a radiation spectrum.

The surface of the sun with a temperature of $T_s \approx 5800$ K emits radiation, with a wavelength spectrum where most energy is contained in the range $0.3 \, \mu m = \lambda = 10 \, \mu m$. The radiation detectors of all animals operate in this wavelength range. Human eyes are sensitive in the very narrow range $0.4 \, \mu m \leq \lambda \leq 0.7 \, \mu m$, called *visible* where the sunlight has its highest intensity.[2]

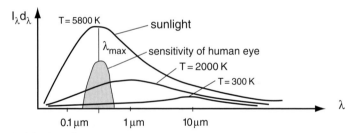

Fig. 8.3. Black body radiation

By the use of (8.1) the spectra of bodies of arbitrary temperatures can be predicted. For instance, the sun with a surface temperature in the photosphere of $T \approx 5800$ K produces the spectrum sketched in Fig. 8.3.

As the radiation spreads out from the sun the intensity falls of with the square of the distance. At the radial distance $R_{earth} \approx 150 \cdot 10^6$ km of the earth the intensity has dropped off to the value known as the solar constant $S = 1.37 \cdot 10^3$ W/m². The peak intensity of the Kirchhoff-Planck spectrum occurs at the wavelength λ_{max}, which can be determined with Wien's displacement law

$$\lambda_{max} \cdot T = 2.88 \text{ mm} \cdot \text{K} . \tag{8.2}$$

---

[2] The shortest and the longest wavelength of visible light differ by less than a factor 2; in contrast the ears of animals are sensitive from about $f \approx 10$ Hz to 10^5 Hz. This is a range of 4 orders of magnitude.

For the sun, with a surface temperature of about 5500 K, the maximum intensity is found at the wavelength λ_{max} = 2.8 mm K/5500 K = 0.5 μm. The nose of a rabbit at a temperature of T = 20° C = 273 K + 20 K = 293 K will generate a *thermal* radiation spectrum with the peak intensity at the wavelength λ_{max}=2.8 mm/T= 2.8 mm · K/293 = 9.55 · 10⁻³ mm = 9.55 μm. This infrared radiation cannot be seen by humans. However, snakes detect infrared with their heat eyes. They can perceive prey, like worms, mice, or rabbits underground, and at night.

Light travels in energy packages, called photons. Each carries the energy E_p

$$E_p = h\nu = h\,c_o/\lambda \,, \tag{8.3}$$

where h = 6.6 · 10⁻³⁴ J s is Planck's constant, c_o = 3 · 10⁸, is the speed of light in vacuum, ν is the frequency of the vibration, and $\lambda = c_o/\nu$ is the wavelength. The energy of a typical photon of yellow light (λ = 0.580 μm) is minute, and is often measured in the small unit eV called electron volt, 1 eV = 1.6 · 10⁻¹⁹ J. The energy of the yellow photon is E_p=hc_o/λ = 6.6 · 10⁻³⁴ J s · 3 · 10⁸ms⁻¹/5.8 · 10⁻⁷m = 3.4 · 10⁻¹⁹ J = 2.1 eV. Since this energy is so tiny the number of photons in the sunlight is enormous. If all the sunlight was concentrated into the yellow section of the spectrum the photon flux, namely the number of photons per second and per square meter, would be

$$\phi_o = \frac{S}{h\nu} = \frac{1.37 \cdot 10^3 \, J s^{-1} m^{-2}}{3.4 \cdot 10^{-19} \, J/phot} = 4.03 \cdot 10^{21} \, phot/s \cdot m^2. \tag{8.4}$$

The small energy of one individual photon is just enough to cause some electronic excitation in exotic molecules, such as the rhodopsin molecules in the retina of eyes. However, in order to create a nerve pulse some charges have to be set into motion. The light sensors must therefore somehow amplify this excitation signal, and convert it into an electrical signal. Here atomic physics comes into play. Our eyes, when adapted to complete darkness can detect individual photons, however sharp images are made from millions of photons, and therefore we cannot perceive images in darkness.

8.1.4 Biological Effects of Different Wavelengths

Light affects living tissue in various ways, depending on the wavelength or photon energy E=h c/λ respectively. Ultraviolet light can break up and damage organic tissue, by ionizing atoms or molecules, and thereby turning neutral atoms or molecules into charge carriers, which upset the metabolism of cells. Visible light can excite the electrons in the atomic shells of atoms and molecules, leading to the very controlled release of electrons in the rodopsin molecules of a retina of an eye, which in turn stimulates a nerve and causes visual perception. Infrared photons may induce molecular vibrations, but they generally cannot excite atomic transitions, and are instead experienced as heat.

Table 8.1. Wavelength and photon energies

color	ultra violet	blue	red	near infrared	far infrared
wavelength λ	≈ 400 nm	≈ 460 nm	≈ 650 nm	$\approx 1\ \mu m$	$\approx 10\ \mu m$
photon energy	> 3 eV	2.7 eV	1.9 eV	1 eV	0.1 eV

8.1.5 How Many Photons Make an Image?

Suppose we watch a seagull flapping her wings in broad daylight at a distance $R = 1.2$ km. The gull is exposed to the full daylight, which contains a photon flux of about $\phi_0 = 4 \cdot 10^{21}$ photons/s m², as calculated in (8.4). Consider a section of wing of the gull $A_w \approx 0.01$ m², which reflects light towards our eyes. The wing intersects the photon flux at the rate Φ_w, which is found from

$$\Phi_w = A_w \phi_0 = 0.01\ \text{m}^2 \cdot 4 \cdot 10^{21}\ \text{photons/m}^2\text{s} = 4 \cdot 10^{19}\ \text{photons/s}. \quad (8.5)$$

Suppose the wing absorbs $\alpha = 90\%$ of the incoming radiation, then it will scatter into all directions only 10% of the incoming photon flux, namely $\Phi_s = \Phi_w \alpha = 4 \cdot 10^{18}$ photons/s. The wing acts as a secondary radiation source, which can be observed by a spectator. The flux density of light scattered off the wing into a hemisphere is $\phi_s = \Phi_s / (2\pi R^2)$ at the distance R. A spectator at the distance $R = 1{,}200$ m is exposed to the scattered photon flux intensity $\phi_s(R) = 4 \cdot 10^{18}$ photons $\cdot$ s^{-1}/ $(4\pi\ 1200^2\ \text{m}^2) = 4.4 \cdot 10^{11}$ photons/s $\cdot$ m².

Now consider the spectator. Suppose the pupils of the observer's eyes in the bright sunlight are contracted to an opening of $d = 1.5$ mm diameter. The pupillary contraction leaves an aperture with the area $A_{eye} = \pi(0.75\ \text{mm})^2 = 1.76 \cdot 10^{-6}\ \text{m}^2$. Each eye will collect $\Phi_{eye} = I_s(R) \cdot A_{eye} = 7.8 \cdot 10^5$ photons $\cdot$ s^{-1}.

The photon flux from the bird's wing will be focused by the lens of the spectator's eye onto the retina, where light sensitive rods and cones are located. The image of the seagull is produced by these photons. The spectator *sees* the gull, because the photons of the scattered light carried the message, revealing the presence of the gull. Images are perceived as fleeting glances, in the "blink of an eye", which

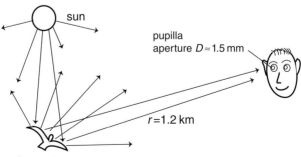

Fig. 8.4. Scattered intensity

lasts of the order of 0.1 s. One tenth of a second allows about $N \approx 5 \cdot 10^4$ photons to enter the eye and catch a glimpse of the bird, still plenty to see the object. The scattered daylight thus reveals the gull. Such image generation works during the day but not in a dark night, and not at the bottom of the ocean, since light is absorbed by the water.

This example illustrates how many photons are scattered by a single object at a large distance. When one thinks of the many objects that are close by as well as far away in a typical environment, one must surely wonder why the eyes are not swamped with photons, and why the brain is not completely overloaded by signals coming simultaneously from every point in the environment.

One knows that the ears are unable to separate many simultaneous messages. Think just how much you understand when you walk into a room where an animated party of thirty excited people talk to each other about their different vacation experiences. Eyes do better than the ears. With a few glances around you can see who is there, who wears what dress, and who talks to whom.

The enormous flood of optical data can be made useful, because eyes generate sharp images and the eyes of all higher animals have millions of photo detectors, which operate in parallel. One can focus onto any part of the scenery. Objects *out of focus* generate just a blur that adds to a background noise. To understand how image selection through focusing works, one must be somewhat familiar with the laws of refraction and reflection, and how refraction at curved surfaces (lenses) helps to produce sharp images.

8.2 Imaging Principles

Eyes serve the purpose of collecting optical information from an enormous flood of scattered light coming from every point in the environment of an individual. This task is not trivial. As calculated in Sect. 8.1.5 a small surface section like the wing of a gull may receive about 10^{19} photons. Although only a tiny fraction may be scattered into the direction of the observer's eye, the reflected light might still contain millions of photons. However, every other object scatters light also. The total amount of photons that hit the eyes is a very large number. The eyes must cope with this torrent of data. Eyes can discriminate the light from different objects by focusing. Focusing involves directing wave fronts toward sharp focus points, by the processes of refraction and reflection. During evolution the information gathering ability of eyes increased in four stages.

The first stage, Fig. 8.5a, has the on-off response of a single photo-receptor carried by single cell animals with a light sensitive spot. The second stage, Fig. 8.5b, achieved by snails, and mollusks, features parallel processing with many light sensitive receptors that give the animal a first rudimentary field of view. In the third stage, Fig. 8.5c, animals like *nautilus* acquire pinhole camera eyes which project a scenery onto a multi receptor image plane. Parallel procession of signals allows updating image information continuously. Pinhole eyes require bright daylight,

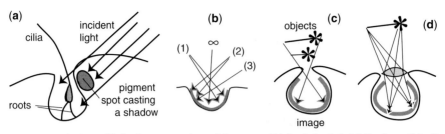

Fig. 8.5. Evolution of light detectors adopted from von Ditfurth [1980]. (a) Euglena, (b) pit snail, (c) pinhole eye, (d) vertebrate eye

since the small aperture, which is needed to obtain sharp projections automatically, limits the image intensity. Pinhole eyes do not yet have depth perception.

In the fourth stage, Fig. 8.5d, lens eyes are invented. They possess a vastly improved resolution compared to pinholes. Lens eyes produce sharp images with depth perception where image points stand out distinctly against background radiation. Vertebrates and octopus have reached this stage. The size of the photoreceptor elements in lens eyes is carefully matched to the theoretical diffraction limit of image resolution that comes from the wave nature of light.

Some other animals solve the problem of parallel processing of image points in a wide field of view by a radically different approach: multiple light collecting devices. Examples are the eight eyes of spiders, the several hundred ommatidia of the now extinct trilobites, and the numerous light pipe channels of flies, and other insects.

8.2.1 Primitive Radiation Detectors

The most basic radiation detector is illustrated in Fig. 8.5a, and belongs to the single-celled *euglena's* photo-sensitive spot located at the root of the muscle that drives the cilia. Radiation is illustrated in this figure by rays drawn as arrows. The associated wave fronts are not shown. In broad daylight the pigment spot casts a shadow. If the muscle works more when in shadow, the organism will predominantly travel in a certain direction relative to the light. There is no data acquisition, just an on / off response. Also any light will do, there is no discrimination of sources.

The next step in evolution is to start distinguishing different objects, and relative locations in a field of view. The pit snail receptor, illustrated in Fig. 8.5b, has its eye depression filled with light sensitive cells that can record different light signals simultaneously. Bright objects nearby and close to the symmetry axis send radiation to all areas of the pit. Objects off to one side from the symmetry axis, points (1), (3), will stimulate detector cells on the opposite side. Objects at large distances will mainly irradiate the center of the detector area. The pit eye of the pit snail is a vast improvement over the on-off photo cell of *euglena*, yet this eye is not capable of imaging the surrounding.

8.2.2 Pinhole Cameras

The pinhole eyes of *nautilus,* Fig. 8.5c, are the first true imaging devices[3]. The imaging principle is illustrated by Fig. 8.6. Each point of the object A, say a point p at the foot of the arrow, located at the distance o_1, generates an image spot in the image plane at the back of the pinhole camera. Each image spot is made by a very small section of the wave front centered at the object point p, which passed through the pinhole opening. If the diameter of the pinhole is d and the image distance is i, the spot size Δ can be found from similar triangles, namely

$$\Delta = d\frac{o_1 + i}{o_1} \approx d \text{ , for } o_1 \gg i.\tag{8.6}$$

The intensity of the light I contributing to each image spot depends on the surface area section of the waves passing through the pinhole. The pinhole area is $\pi (d/2)^2$.

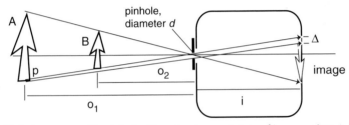

Fig. 8.6. Pinhole camera with pinhole, diameter d, uncertainty of image points Δ. Two similar objects, A and B, at different distances generate the same image

Pinhole cameras generate true images of bright objects. However, in order to produce sharp images, the pinhole diameter d must be as small as possible. A small pinhole aperture, however, cuts down on the light intensity of the image. Hence dimly lit objects will be difficult to see with pinhole eyes. Pinhole cameras paint a flat image of the entire scenery. The image provides no information about the object distance. The two similar objects, A and B, located at different object distances, o_1 and o_2, generate identical images. The image spot Δ contains information from the foot points of the different objects A and B. This overlap and image confusion is unavoidable in pinhole eyes.

The number N_p of resolvable image points of a pinhole camera is easy to calculate. Let d be the diameter of the pinhole, and A the area of the largest image that can be detected. Then for an object at a large distance the image height Δ of the

---

[3] A pinhole camera can be easily made from a shoe-box: cut an opening into a side panel and cover it up with translucent paper. then punch a hole with a thick needle into the opposite side. When pointing the pinhole side towards a bright scenery an inverted image appears on the translucent screen.

object point p is approximately equal to the diameter d of the aperture, and the image point area is $a_i = \Delta^2 \approx d^2$. With the knowledge of the image area A_i the number N_p of distinguishable image points can be found.

$$N_p = A/a_i = A/d^2 \qquad (8.7)$$

A typical value of the image size could be $A \approx 100$ mm^2. With a pinhole of $d^2 = 1$ mm^2 the pinhole eye would resolve $N_p = 100$ image points. The number of image points can be increased by more than six orders of magnitude, when the eye employs a lens instead of a pinhole. How this increase of the image resolution comes about will be described in the following section.

8.2.3 Imaging by Lenses

The general construction of vertebrate eyes resembles the design of a camera. Each eye has a lens and a light sensitive retina at the back. The lens collimates the light waves into sharp image points. We first look at a camera type model eye with an empty space between the lens and retina, and later investigate how the gelatin body in the image space affects vision.

Table 8.2. Refractive index n of various materials. δ is a small number that depends on the density of air

material	vacuum	air	water	eye body	eye lens	crown glass	silicate flint glass	guanine
n	1.000	$1+\delta$	1.33	1.36	1.43	1.45–1.55	1.6–1.7	1.8

The collecting power of a lens is based on refraction. Refraction arises due to differences in the speed at which light waves propagate. The phase velocity of light is $c_0 = 3 \cdot 10^8$ m/s in vacuum, and has smaller values in transparent media. Instead of dealing with large numbers, one references the phase velocity in materials to the speed of light in vacuum c_0 by defining the index of refraction $n_1 = c_0/c_1$, which was already described in (7.15). Different values of n are shown in Table 8.2. Many

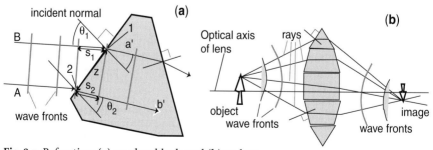

Fig. 8.7. Refraction: **(a)** on glass block, and **(b)** on lens

transparent solids have $n \approx 1.5$, however some fish have small crystals of nitrogenous compounds of guanine in their scales $(C_5H_5N_5O)$, which has $n = 1.8$, [see Denton 1971]. Generally the refractive index decreases with wavelength.

The frequency of light is fixed by the emitting source. The speed of light c_0 determines the wave length

$$\lambda_1 = \frac{c_1}{v} = \frac{c_0}{n_1 \cdot v} . \qquad (8.8)$$

If the index of refraction varies in a medium, the wavelength of the light traveling through the medium must vary as well. The multiple canopies of the Huygens elementary wave umbrellas change their separation λ as they travel through different materials.

Consider a simple situation where a plane light wave in air hits a dense medium with refractive index $n_2 = c_1/c_2 > n_1$, Fig. 8.7a. Then $\lambda_2 < \lambda_1$. The distances s_1 and s_2 may represent the wavelength in air, and dense medium respectively. Since $s_1 > s_2$ the wave fronts in the dense medium are turned towards the incident normal. Light is *refracted* as it enters the dense medium. The incident angle $\emptyset_1$ and the penetration angle $\emptyset_2$ are related by Snell's law.

$$\frac{\sin \emptyset_1}{\sin \emptyset_2} = \frac{n_2}{n_1} = \frac{c_1}{c_2} . \qquad (8.9)$$

The rays are illustrated in Fig. 8.7a. When light hits a dense medium head on the incident angle is $\emptyset_1 = 0$. Then by Eq. (8.9) the penetration angle $\emptyset_2$ is also 0: the light waves continue straight on. For any other incident angle the light does not travel straight on, but is bent at the surface separating the two media, the larger the incident angle $\emptyset_1$ the larger $\emptyset_2$. This is the optical secret of a lens. A lens can be modeled as a dense medium with sections that present surfaces with ever increasing incident angles farther away from the optical axis. As a consequence, rays diverging from the object point in Fig. 8.7b are made to converge in an image point behind the lens, and then the rays diverge as they travel on. The focusing of rays in a single point *and* their subsequent divergence are essential for successful image analysis.

8.2.4 Eyes in Air and Under Water

Seeing is based on the formation of an image inside the eye of the spectator. The truth (the *image*) is in the eye of the beholder. But where exactly is the image?

Imaging with thin lenses in air, $n_{air} \approx 1$, is easy to explain. However, most eyes are not built like cameras, where the lens is surrounded by air both at the front and the back. Instead the medium between the lens and the image plane in vertebrate eyes is filled with a transparent aqueous gel (called vitreous humor), which has an index of refraction n_l, different from 1. As we will see below, the gel reduces the re-

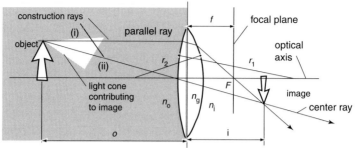

Fig. 8.8. Construction of image with magnifying lens, for different index of refraction in the object and the image space

fractory power of the lens. The sight of marine animals is further impeded by the fact that even the front sides of the lenses are exposed to water with the index of refraction $n_o = n_{water} = 1.33$. Any diver knows that seeing under water without goggles is difficult, because the eyes have less focusing power.

If one wants to know where the image is located one must determine how each surface bends the rays, see Fig. 8.8. Refraction is related to the speed of light in various media. Since the refractive index $n_{med} = c_o / c_{med}$ is a function of the wavelength, the following analysis applies to monochromatic light only.

All lenses have an axis of symmetry, called the optical axis. A positive lens focuses a parallel beam, where adjacent rays are parallel to the optical axis, and where the wave fronts are flat, into the focal point F. The focal point is located on the optical axis in the focal plane of the image space, Fig. 8.8. The distance between the lens and the focal point is called the focal length f. A similar focal point F_o exists in the object space. A negative lens diverges parallel rays, so that they appear to be coming from F_o. The position of an image point can be geometrically constructed by using three rules. (i) A ray from an object point, which travels parallel to the optical axis (called *paraxial ray*), is refracted by the lens so that it passes through the focal point F. (ii) A ray from an object point that intersects a thin lens on the optical axis, called a *center ray*, continues straight on without being refracted. (iii) The image point is located where the paraxial ray crosses its path with the center ray.

The lens separates the image from the object space. The curvature of the lens surfaces determines where the image will be located. A magnifying lens as shown in Fig. 8.8 generates a real image in the image space. De-magnifying lenses generate virtual images, which are located in the object space.

The surface of the lens in Fig. 8.8 looks convex as seen from the object. The center of radius of curvature of the first surface is located to the right of the lens and the radius r_1 is counted positive[4]. The center of the radius of curvature r_2 of the second surface is to the left of the lens. Therefore, r_2 is counted negative.

---

[4] If the surface looks concave from the object space its center of curvature lies to the left of the lens therefore r_1 is negative. The surface de-magnifies.

Assume the lens has the index of refraction n_g, and the index of refraction in the object and image space have the values n_o and n_i respectively. Then image distance i, and object distance o, are related as

$$\frac{n_o}{o} + \frac{n_i}{i} = \frac{1}{f} \tag{8.10}$$

where

$$\frac{1}{f} = \frac{n_g - n_o}{r_1} - \frac{n_g - n_i}{r_2}. \tag{8.11}$$

The radius is counted negative if the center of curvature is to the left of the surface: in Fig. 8.8 the radius r_1 is counted positive, and r_2, is negative. For thin lenses in air $n_o = n_i = 1$ these two relations reduce to

$$\frac{1}{o} + \frac{1}{i} = \frac{1}{f} \text{ , or } i = 1/(\frac{1}{f} - \frac{1}{i}) \tag{8.12}$$

and

$$\frac{1}{f} = (n_g - 1) \cdot (\frac{1}{r_1} - \frac{1}{r_2}). \tag{8.13}$$

There are two methods to produce a sharp image from an object that is located at some distance o.

(i) Use the lens of a given focal length f and place the image plane at the appropriate image distance i, which can be determined with the help of (8.10). This method is used in most cameras and in the eyes of spiders. Focusing here means to move the lens thereby changing the distance i until the image is in focus.

(ii) Fix the image distance i, and select a lens of exactly the correct focal length f to satisfy Eq. (8.11). Vertebrate eyes focus by this principle. In order to focus onto objects at different object distances, for instance o_1, and o_2, a lens with variable focal length, a zoom lens, is required. Equation (8.11) shows that the focal length, f, depends on the radii r_1 and r_2, and on the differences in the refractive index $n_g - n_o$ and $n_i - n_g$. Small radii and large diffraction index differences give the lens a short focal length, and a large refractive power $1/f$. A short focal length allows the eyeball to be small.

The lens of the human eye is made from an elastic material that changes shape when pulled in radial direction. When looking at nearby objects, the ring muscle around the lens contracts, the lens shrinks, the radii r_1 and r_2 get smaller so that the focal length is also reduced. When looking at distant objects the ring muscle relaxes. The lens flattens, so that its focal length increases.

Human eyes have a body with refractive index close to that of water: $n_i = 1.336$, and an elastic lenses with $n_g \approx 1.43$. [e.g. see Strong 1958]. For eyes of aquatic animals one must set $n_o = 1.33$. Since n_g varies with wavelength a single surface cannot focus rays from different wavelength into a single point. Therefore large aperture lenses are composed out of different lens elements, a lesson that was apparently not lost on trilobites who had 2-lens element eyes, see Sect. 8.4.5.

A lens with $r_1 \approx -r_2 = r$ that images an object in air ($n_o = 1$) into an image space filled with water ($n_i = 1.33$) has the focal length $f = r/\{(n_g - n_o) + (n_g - n_i)\} = r/\{(1.5 - 1) + (1.5 - 1.33)\} = 1.49\,r$. To obtain this result the second term of (8.11) has been written as $-(n_g - n_i)/r_2 = (n_g - n_i)/-r_2 = (n_g - n_i)/r$, so that the numerators of both terms in (8.11) become r. If this lens is imaging an object in water, the focal length nearly doubles, namely $f = r\{(1.5 - 1.33) + (1.5 - 1.33)\} = 2.94\,r$. Therefore, the ring muscle of the human eye may not be able to deform the radius of the lens sufficiently to see sharp images under water. The refractive power of a lens is much lower under water. For that reason fish eyes have the smallest radius possible: $r \approx d/2$. More details of fish eyes are given in Sect. 8.4.3.

8.2.5 The F-Number

Lenses collect light and they produce images. The speed of a lens, namely the flux of luminous energy from an object point source passing through the lens, is determined by the aperture angle S of the collected cone of light. This angle can be described by the relative aperture, the ratio of the entrance pupil, d to the focal length f, see Fig. 8.9. The inverse of this ratio is called the F-number

$$F\# = f/d = 1/S .\tag{8.14}$$

A good camera lens may have $F\# = 2.8$. F-numbers like 2.8, 4, 5.6, 8, 11, ... are often engraved onto camera bodies. The larger S (or the smaller the F-number) the more light is concentrated in to image points.

The highest light collecting power is associated with the largest aperture angle $S = d/f$, or the smallest F-number. In order to understand how animals can optimize the light collection in their eyes, Eqs. (8.14), and (8.11) are combined

$$S = \frac{d}{f} = d \cdot \left(\frac{n_g - n_o}{r_1} - \frac{n_g - n_i}{r_2} \right).\tag{8.15}$$

The material constants, namely the index of refraction in the image and the object space n_i and n_o, and the index of the lens material, n_g, are fixed. This leaves the radii of the lens surface r_1, and r_2 as the free parameters for optimizing S. The largest speed is obtained for the smallest possible radii of curvature. The smallest values are $r_1 = |r_2| = d/2$, where the lens is formed into a sphere.

Table 8.3. *F#* of cameras, and aperture angle *S*

F#	2.8	5.6	8	16
S	0.35	0.18	0.125	0.025

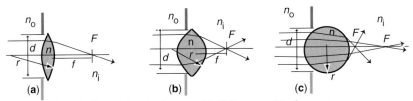

Fig. 8.9. Focal length illustration for lenses with different radii of curvature

$$S_{max} = \frac{d}{d/2}\left((n_g - n_o) + (n_g - n_i)\right) = 2\left(2n_g - (n_i + n_o)\right) \tag{8.16}$$

Unfortunately, a homogeneous spherical lens does not produce sharp images, because rays located farther away from the optical axis do not converge at the same distance as rays that travel close to the axis. However, nature has found an ingenious way to focus light with spherical lenses, in which the index of refraction changes gradually. These are the graded index lenses of fish eyes, discussed in Sect. 8.4.3.

8.2.6 The Diffraction Limit:
How Sharp Are the Images of Lens Eyes?

Geometrical optics can be used to calculate where an image will be formed on the retina. When imaging two points with different object distances, geometrical optics helps to find how far their images will be separated. Geometrical optics however cannot predict the size of the smallest image *point* that is produced on the retina. Diffraction sets a limit to the minimum size of images that can be created. Due to the wave nature of light the iris will always generate a small light spot of a finite diameter Δ, namely the diffraction pattern of the aperture d of the iris, which was derived in Sect. 7.4.7. Diffraction renders a point image into a diffraction ring pattern, shown in the insert of Fig. 8.10. The half width $\Delta/2$ of the center spot of the pattern subtends the angle $\emptyset_{min}$, which was given in (7.21), namely $\sin \emptyset_{min} = \lambda/d$. The angle $\emptyset_{min}$ can also be shown in the image space of the lens, namely $\tan \emptyset_{min} = (\Delta/2)/i$. Since this angle is quite small, one has $\sin \emptyset_{min} \approx \tan \emptyset_{min} \approx \emptyset_{min}$, hence $\emptyset_{min} \approx \lambda/d \approx \Delta/2i$. Further, most objects of interest are located at distances o that are large compared to the focal length, so that $i \approx f$. Therefore, every infinitely sharp object point generates an image of the finite width

$$\Delta = 2i\lambda/d \approx 2f\lambda/d. \tag{8.17}$$

The rods and cones in a human eye have a diameter of about $\Delta r \approx 2.5 \, \mu m$. This diameter is matched to the diffraction spot Δ of the eye with aperture $d \approx 8$ mm, and image distance $i \approx 20$ mm for the wavelength $\lambda = 5 \cdot 10^{-7}$ m. The spot size increases when the iris contracts. In broad daylight the iris diameter may shrink to $d = 1$ mm, in which case the angular resolution of the eye becomes $\sin \emptyset_{min} = 5 \cdot 10^{-7}/10^{-3} = 5 \cdot 10^{-4}$, or $\emptyset_{min} = 0.028°$.

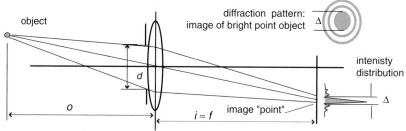

Fig. 8.10. Diagram illustrating the generation of a diffraction pattern

Birds of prey have evolved eyes with an even higher angular resolution in order to spot their meals from great heights. For instance, eagles have larger pupils than humans but smaller eyeballs, and therefore have a diffraction spot diameter of about 1 μm. The eagle's cones and rods are also about 1 μm in diameter. Again these eyes are perfectly matched to the diffraction limit. The diffraction spot is a figure of merit for the resolution of the eye. The speed S or the F-number respectively, characterizes the light gathering power. Both together determine the image resolution.

8.2.7 Image Resolution of Lens Eyes

All images are made up from N discrete image points. The more discrete points an image contains, the finer the details are that can be distinguished. In a pinhole eye, Fig. 8.11a, the diameter of the image points $\Delta_p \approx d$ is approximately equal to the diameter d of the pinhole. Each image point has the area $A_p \approx \Delta_p^2$. If A is the area of the retina of the pinhole eye, then the number of distinguishable image points is found from (8.7), namely $N_p \approx A/d^2$. The lens eye, Fig. 8.11b has a vastly larger image resolution, because the diameter of each image pixel is equal to the width $\Delta_{diff} = 2f\lambda/d$ of the diffraction spot, given in (8.17), which depends on the wavelength. Short wavelengths yield small image points.

Assume again that the image plane area of the retina is A. Then the number of image points for the lens eye $N_{lens} = A/\Delta_{diff}^2$ is

$$N_{lens} = \frac{A \cdot d^2}{4 \cdot f^2 \cdot \lambda^2} = \frac{1}{4} \cdot \frac{A}{d^2} \cdot \frac{d^2}{f^2} \cdot \frac{d^2}{\lambda^2} \tag{8.18}$$

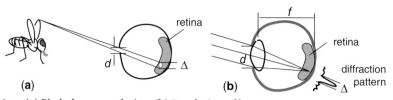

Fig. 8.11. (a) Pin hole eye resolution. (b) Resolution of lens eye

The first factor, A/d^2 on the right hand side of (8.18) is the resolution of a pinhole eye, known from (8.7). The second term, $S^2 = (d/f)^2$, characterizes the light gathering power of the lens: large apertures are good for high resolution. The last factor of (8.17) is the diffraction limited resolution contribution, $(d/\lambda)^2$.

Typically a human eye operates with an aperture $d \approx 4$ mm. For yellow light with $\lambda = 0.58$ μm the diffraction factor of the image resolution is $(d/\lambda)^2 = 4.8 \cdot 10^7$, which represents an enormous increase in resolution compared to the pinhole eye.

8.2.8 Why Imaging Cuts Down on Background Noise

The image construction Fig. 8.12 shows another very important advantage of focusing. Two similar objects, namely the foot points 1 and 2 of the objects A and B, which are located at different distances, generate images I_1 and I_2 at different image distances. The pinhole camera, Fig. 8.6, places these images right on top of each other.

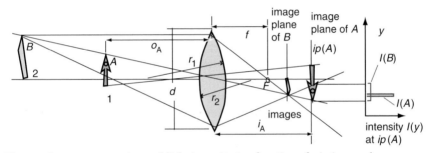

Fig. 8.12. Image construction and light intensity I as function of y in image plane

With the lens eye the images of similar points A and B can be clearly distinguished. A detector, located in the image plane ip, sees the image of A as a sharp point, whereas the image of point B is smeared out over a wider area. The image intensity I of the points 1 and 2, is plotted as function of lateral distance y on the right side of in Fig. 8.12. The signal due to I_A stands out as a delta function against the smeared out background of the image I_B. Our eyes detect intensity differences. This reinforces the ability to distinguish small localized intensity variations against large background intensity, and to recognize dim objects against a bright background. Think for instance of the image of a small bird flying across the bright sky.

Image production with lenses helps animals to focus onto objects of interest without being confused by all the light coming from the same directions, but from other distances. Animals therefore can see important dim objects even if the optical background noise is large in terms of photon numbers.

8.3 The Human Eye

The human visual system is a marvel of engineering design and optimization. It consists of the eyes, two miniature digital cameras with about 10^8 individual photo-receptors each. Each camera has a zoom lens, and a variable F-stop (iris), and an external shutter (eyelid) that opens and closes rapidly, and continuously washes the lens. The size of the photoreceptor is carefully matched to the size of the diffraction spot created by the iris, in order to obtain optimal image resolution. The eyes are mounted on mobile platforms, which direct the center of most distinct vision continuously over the field of view, to give the impression of a sharp full panorama of about $40°$ although the region of sharp vision only covers about $2°$. Furthermore, the parallax between the views of both eyes provides clues about the depth of field and the distance of objects in the surrounding.

8.3.1 Geometry and Physiology

The lens of the human eye is quite thick in the middle. Dimensions of the eye are given in Fig. 8.13. In cameras, the object is focused onto the film plane by moving the lens. In contrast, the human eye has a fixed distance between eye lens and retina. The eye is focused by adjusting the focal length of the eye lens by changing the radii of curvature of the lens [e.g. see Koretz and Handelman 1988]. The human eye has a zoom lens.

Zoom focussing in the human eye is done with the ciliary muscle, which encircle the lens and are connected to it by spoke-like (zonule) fibers. The fibers are normally under tension to flatten the lens, and are off loaded when the ciliary muscle contracts. For sharp vision at close range the muscle contracts decreasing the radius of the muscle rim and allowing the lens to contract into a more convex shape. When the ciliary muscle relaxes it becomes longer, increasing the radius and pulling the spokes at the rim of the lens to make it flatter.

How much must the lens radius change in order to accommodate the eye from infinity to the distance of most distinct vision $o = 25$ cm? Actually the optical system of the eye consists of all the curved surfaces between the object and the image. A significant part of the refractive power comes from the curved cornea of the eye. To demonstrate the principle of focusing we consider a simplified eye, consisting of a lens ($n_g = 1.43$), an aqueous body ($n_i = 1.33$), and the retina, located at $i = 20$ mm, on which a sharp image is to be produced.

The object space has the refractive index $n_o = 1$. (The relevant equations are given in previous sections.) First determine the focal length f_{12} when focusing at an object located at $o = \infty$. From (8.10) one finds for distances measured in mm $n_o/\infty + n_i/20 = 1/f_{12}$. This leads to $f_{12} = 20/1.33 = 15$ mm. Now one uses (8.11) to find the radius of curvature assuming a symmetric lens with $r_1 = -r_2 = r$. The relevant equation is $1/15 = 1/f_{12} = (1.43 - 1)/r + (1.33 - 1.43)/(-r) = (0.43 + 0.1)/r$, so that $r = 15 \cdot 0.53 = 7.95$ mm. What is the new radius r_1 when the eye focuses onto an object at $o = 250$ mm? Now one has $n_o/250 + n_i/20 = 1/f_{12}$ leading to $1/f_{12} = 0.0705$. This is en-

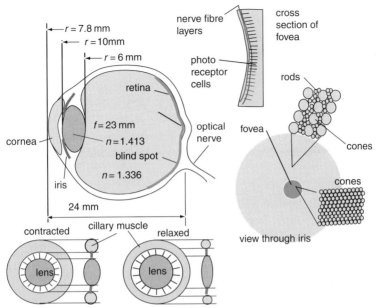

Fig. 8.13. Dimensions of a human eye (Gullstrand's eye), redrawn after Strong [1958]. Cross section of fovea, rods and cones sketched after Miller [2000]

tered into (8.11) to get $1/f_{12} = 0.0705 = (1.43-1)/r_1 + (1.33-1.43)/(-r_1)$, leading to $r_1 = 0.53/0.0705 = 7.52$ mm. The radius of curvature is changed by $\Delta r = 7.95 - 7.52 = 0.43$ mm. This represents a relative change of $\Delta r/r = 0.43/7.7 \approx 5\%$. One only has to flex the ring muscle by a small amount in order to accomplish this change of focus.

8.3.2 Receptors of the Eye, Sensitivities, and Field of View

The retina of the human eye is covered by color sensitive cones and black and white sensitive rods. The rods are much more sensitive than the cones. The difference is typically a factor of 100. The light actually has to pass through layers of tissue before it can reach the cones and rods. Their peak sensitivity is in the blue green range of the spectrum. The highest density of rods is in the fovea region, namely about $400 \cdot 400 = 160{,}000$ per mm^2. Cones are tuned to three principal colors, red, green, and blue, thus providing color vision. There are only about $6 \cdot 10^6$ cones, but $1.2 \cdot 10^8$ rods. The signals from these light sensors are carried by only 10^6 optical nerves. Hence there must be some preprocessing inside the eye by which the signals from about 100 sensors are carried by a single nerve.

The area of most distinct vision has a diameter of about 1.5 mm. It contains yellow pigment and is therefore called the yellow spot. At its center is the *fovea centralis*. The fovea diameter together with the focal length of the eye of about 20 mm yields the field angle Ø of most distinct vision: tan = 1.5 mm/20 mm = 0.07 radian,

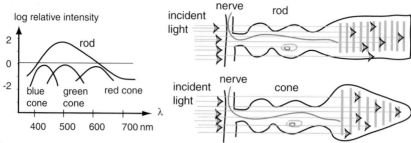

Fig. 8.14. (a) Spectral sensitivities of rods, and cones. (b) Rods and cones

corresponding to an angle of 4.5°. The perceived field of view is much larger (typically 45°) since the eyeball constantly sweeps the field of view and one also experiences the sensory input from the receptors outside the fovea.

The fields of view of different animals are matched to their lifestyles. Grazing animals like rabbits have close to 360° coverage whereas hunters like eagles or foxes have a narrow forward looking field of view. The eyes catch a glimpse of the surrounding within about 0.1 second. If an object moves very quickly across the field of view it becomes a blur. Blurred vision would also result if the head moved quickly across a scenery. This could be a problem for foraging animals like rabbits and pigeons who's fields of vision are pointing towards the sides. If the animal walks quickly forward the scenery moves rapidly across the field of view, obfuscating the vision. To combat this motion-blur, pigeons rhythmically move their head back and forth with every step. If the head moves back as quickly as the whole body travels forwards, the eyes actually stand still and get a sharp snapshot of the surrounding for one half of the motion cycle.

The sensitivity of the eye receptors covers an enormous dynamic range. With a strongly absorbing filter that cuts down the intensity by about a factor 10^4 one can look straight into the sun, which has an intensity of about $\phi_0 \approx 4 \cdot 10^{21}$ photons/m²s, as derived in Eq. (8.4). The filter therefore transmits $\phi_{filter} = 10^{-4} \cdot 10^{21} = 4 \cdot 10^{17}$ photons/m²s. The aperture of the iris when looking into bright light is typically $A = 1$ mm² $= 10^{-6}$ m². Therefore the eye can handle the photon flux $\Phi = A\phi_{filter} = 10^{-6}$ m² $4 \cdot 10^{17}$ photons/m²s $= 4 \cdot 10^{11}$ photons/s. In very dim light the eye can perceive individual photons. Therefore, the range of sensitivities spans about 11 orders of magnitude.

The receptors of the eyes and the connected data analyzing system of nerves and brain are sensitive to *changes* of intensity only. This specific sensitivity is essential for an organism that must notice sudden changes in its surrounding caused by the appearance of enemies or opportunities. Such a visual system is useless for assessing a stationary scenery, which emits a steady stream of photons. However, we do see stationary scenes quite well by employing minute subconscious oscillatory scanning motions of the eyeballs. If these motions are arrested by a chemical agent, the vision blurs, and the scenery disappears. These minute oscillations are superimposed onto the voluntary or subconscious major motions of the eyes associated with scanning the field of view.

8.3.3 Resolution of the Human Eye

If the rods and cones have a diameter d which is smaller than or equal to the diameter of the center diffraction spot Δ, the sensors match the optical resolution of the eye. The human eye is built to match its diffraction limit. At 400 rods per mm on the retina, each rod has about $d_{rod} \approx 2.5 \cdot 10^{-6}$ m, and the image distance of the eye $i \approx 2$ mm. The angular half width $\emptyset_{rod}$ of a rod is found from $\sin \emptyset_{rod} = \Delta / 2i = d_{rod}/2i \approx 2.5 \cdot 10^{-6}$ m$/4 \cdot 10^{-2}$ m $= 6.25 \cdot 10^{-6}$. This angle is identical to the diffraction limit of the eye, as determined from the maximum pupilla diameter $d = 8$ mm and the average wavelength of light $\lambda \approx 5 \cdot 10^{-7}$ m, namely $\sin \emptyset_{min} = \lambda / d = \lambda \approx 5 \cdot 10^{-7}$ m$/8 \cdot 10^{-3}$ m $= 6.25 \cdot 10^{-6}$. (If $d_{rod}/2i$ was smaller than λ / d the rods would be smaller than the smallest light spot on the retina.) In fact the detector size is perfectly matched to the best possible optical resolution. Nature knew its *wave optics* quite well when she designed the eye.

From the resolution angle $\sin \emptyset_{min} = 6.25 \cdot 10^{-6}$ one can calculate the smallest object height H that can be resolved by the eye: $\emptyset_{object} \approx H/o > \emptyset_{min}$. For instance, at the distance of most distinct vision $o \approx 0.25$ m one has $H_{min} \approx o \cdot \emptyset_{min} = 250$ mm $\cdot 6.25 \cdot 10^{-6} = 1.5 \cdot 10^{-2}$ mm. This height is about ½ the width of a human hair, a resolution that is quite adequate to see the objects of interest to hunting, and gathering humans.

8.3.4 Aging of Eye Components

As people get older the near point of distinct vision moves away from the eye. To focus an eye onto a nearby object the ciliary muscle must be relaxed, and the lens has to shrivel into a thicker ball, so that the radii of curvature get smaller and the focal length gets shorter. Old eyes lose their elasticity, and thus the eye cannot focus on nearby objects any more. Glasses with positive magnification are then needed to help to restore vision, as for many people at age 55 the arm seems to be too short to enable reading without glasses. Also the pupil loses its ability to dilate. Old people have small pupils, and appear to have a piercing gaze, but they cannot help it. These are the optical signs of old age.

For distant objects the magnification of the human eye is often still too small. Than one might want to use a telescope.

Table 8.4. Eye dimensions as function of age

age [years]	1	20	50	80
near point [cm]	≈5	15	30	200
maximum iris diameter [mm]	10	6	3	1

Zugereist in diese Gegend	*Traveling through this neck of woods*
Noch viel mehr als sehr vermögend	*Full of money and of goods*
In der Hand das Perspektiv	*Carrying the telescope*
Kam ein Mister namens Pief.	*Was a gentleman called Joop.*
"Warum soll ich nicht beim Gehen"	*"While I'm walking I can see"*
Sprach er "in die Ferne sehen?	*Says he "distant scenery.*
Schön ist es ach anderswo,	*This delights me I must say*
Und hier bin ich sowieso."	*Since I'm here anyway."*
Hierbei aber stolpert er	*Saying this he trips and falls*
in den Teich und sieht nichts mehr.	*In the pond... his vision dulls.*

Wilhem Busch 1884   (translated by the author)

Fig. 8.15. The absent minded professor [Wilhem Busch 1884]

8.4 Animal Eyes

The vertebrate eye is an example of the evolution of an organ which has, in many respects, reached the limits set by physical principles. The human eye's sensitivity spans an enormous range, from detecting individual photons to recognizing objects in the presence of photon fluxes of about $\Phi \approx 10^{14}$ photons/s. However, the human eye perceives only a narrow wavelength range between blue and deep red, compared to the spectrum ranges seen by some other animals. For instance, falcons perceive the UV radiation reflected off urine traces of mice [Wilke 2002]. Bees can see UV but not red. Snakes have special pinhole eyes for infrared. The spatial resolution of eyes reaches the diffraction limit. The eyes update the image information continuously. Some eyes have a zoom lens so that the overall dimensions of the eyeballs do not change. Many eyes can also be swiveled around in their sockets to scan the scenery. Many animal eyes surpass the human design in one way or another. Such features have evolved in response to the opportunities or demands of the niche in the biosphere, in which various animals became masters of perception. Some of these designs illustrating the four stages of evolution, Fig. 8.5, are discussed here.

8.4.1 Pinhole Camera Eyes for Heat Radiation

Snakes find some of their preferred prey in burrows underground where sunlight is not present. How does a snake see a rabbit in the dark? All objects emit electromagnetic radiation as *heat* according to their temperature T. This black body radiation has an intensity maximum at the wavelength λ_{max}, which can be calculated from (8.2). At a body temperature of 300 K the radiation spectrum peaks at about $\lambda_{max} \approx 10$ μm. Our eyes are blind at that wavelength, but our skin is somewhat sensitive to it, recognizing the radiation as heat. Snakes however can *see* this radiation with their tongues [e.g. see Gamow and Harris 1973], or with specially designed pit-organs.

The heat-sensing organ of the snake heat eye consists of two funnel shaped (pinhole) collectors of the radiation and a chamber (pinhole camera), which is divided by a very thin membrane into an outer and an inner space, Fig. 8.16. This membrane can be thought of as a *thermal retina*, namely an organ that records the thermal image produced by the pinhole. The space behind the membrane provides thermal insulation from the rest of the body. Since the membrane is thin, it has little thermal mass, and since it is separated from other tissues by the inner chamber, the membrane can heat up locally with a rapid response time.

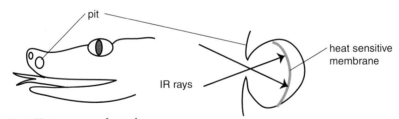

Fig. 8.16. Heat sensors of a snake

The trigemina nerve is located in this thin membrane, so that the brain can resolve a *thermal image* generated by the pinhole. The angular resolution of the thermal sense is about ±5°, and the snake can detect thermal images of objects that are only about ±0.003°C = 3 mK different from the background. By the use of two pits, one on either side of the head, the snake is able to detect the location of prey by triangulation.

The thermal response to an object can be estimated as follows. Suppose a little mouse, modeled as a ball of radius $R_m=30$ mm, is located at a distance of 100 mm in front of one of the pits of the snake, Fig. 8.17. Let us assume that the mouse has a body temperature of 290 K while the surrounding is at $T=280$ K. Treat the mouse as a gray sphere with an emissivity of $\varepsilon=0.7$. The surrounding is assumed to have an emissivity of $\varepsilon=0.6$.

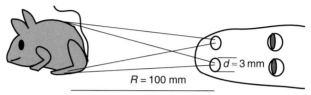

Fig. 8.17. Pit snake and mouse

The mouse emits a certain radiation intensity I_{mouse} that can be calculated with Stefan's law (2.21), namely $I_{mouse}=0.7\cdot5.67\cdot10^{-8}\cdot290^4$ W $=280$ W/m². In contrast the surrounding scenery at $T=280$ K, and $\varepsilon=0.6$ emits the surface intensity $I_s=209$ W/m². The inverse square law reduces this emitted intensity. At the distance of $R=0.1$ m the radiation from the mouse has dropped off to $I(R)=I_{mouse}(0.03/0.1)^2=25.3$ W/m². The pit, with a diameter of 3 mm and the area $A_p=\pi1.5^2$ mm² $=7\cdot10^{-6}$ m² intercepts the power $P_p=A_pI(R)=7\cdot10^{-6}$ m² $\cdot$ 25.3 W/m² $=1.78\cdot10^{-4}$ W. The infrared radiation power impinges onto a segment of the retina membrane with an area equal to the projection of the pinhole A_p. Suppose the snake catches a glimpse of the mouse of $\Delta t=0.5$ s duration, so that the membrane section receives the radiation energy $\Delta Q=P_p\cdot\Delta t=8.9\cdot10^{-5}$ J. The heat pulse contains quite a large number N_{IR} of infrared photons, $\Delta Q=N_{IR}E_{IR}$, which can be found as follows. The average wavelength of these photons can be obtained from Wien's law (8.2), which links the wavelength λ_{max} of maximum emission of a thermal radiator and its temperature T. For the body temperature of the mouse $T=290$ K the typical wavelength falls into the infrared region of the spectrum $\lambda_{max}=0.0028$ mK/290 K $=9.7\cdot10^{-5}$ m. Thus the infrared photon energy is $E_{IR}=hc/\lambda_{max}=6.6\cdot10^{-34}\cdot3\cdot10^8/9.7\cdot10^{-5}=2\cdot10^{-20}$ J/photon, and the number of photons taken in by the snake in the 0.5 second snapshot is $N_{IR}=\Delta Q/E_{IR}=8.9\cdot10^{-5}$ J/2 $\cdot10^{-20}$ J/photon $=4.4\cdot10^{15}$ photons.

By how much would the membrane heat up? Suppose the membrane has a thickness of 0.1 mm. The membrane section then has the mass $m=\rho\cdot0.1$ mm $\cdot$ 7 mm² $=7\cdot10^{-4}$ g, where $\rho=10^{-3}$ g/mm³ is the density of water. The caloric equation from Table 2.2 gives an answer. $\Delta Q=Cm\Delta T$, where $C=4.18$ J/K°g. Solving for ΔT one finds $\Delta T=\Delta Q/Cm=8.9\cdot10^{-5}$ J/(4.18 $\cdot7\cdot10^{-4}$ J/°) $=0.030°=30$ mK. The surrounding, with an intensity $I_s=(209/280)I_{mouse}=0.87\ I_{mouse}$ would raise the temperature by 22 mK. The difference of both temperatures $\Delta T=30$ mK -22 mK $=8$ mK is larger than the thermal resolution of ±3 mK of the snake heat eye retina. Thus the snake would be able to *see* the infra red image of the mouse.

8.4.2 Spider Eyes

Spiders focus their eyes by changing the distance between lens and retina, see Fig. 8.18. A certain female spider hunts under water and in the air. A quantitative example illustrates the optics of her eyes. Suppose the image distance is $i_1 = 5$ mm when she focuses onto a target under water, which is for instance at a distance $o = 100$ mm. Assume that the spider eye lens has the same index of refraction as the human eye, namely $n = 1.43$. The eye of the spider is filled with a watery transparent fluid for which we assume $n_i = 1.33$.

The calculation is guided by the following questions: a) What is the focal length of the eye? b) What is the radius of curvature of the eye lens? c) What is the image distance if the spider focuses at a similar object with $o = 100$ mm in air? d) How could the change of image distance be accomplished?

a) First consider the spider under water. The index of refraction in the object space is $n_o = 1.33$. Using Eq. (8.10) we write $1/f = (1.33/100) + (1.33/5) = 0.0133 + 0.266 = 0.279$, thus $f = 1/2.79 = 3.58$ mm. b) Assume that the eye lens is symmetric, so that the radii of curvature r_1 and r_2 are equal in magnitude. However, the surfaces curve into opposite direction, hence $r_2 = -r_1$. Then by the use of Eq. (8.11) one finds $(n - n_o)/r_1 - (n - n_i)/r_2 = 1/f = (1.43 - 1.33)/r_1 - (1.43 - 1.33)/(-r_1) = 0.2/r_1$, so that the radius is found as $r_1 = 0.2 \cdot f = 0.716$ mm. (c) In air the index of refraction in the object space is $n_o = 1$. Then the lens law is written as $1/100 + 1/i = 1/f = 0.279$. This can be solved for $1/i = 0.279 - 0.01$, or $i_{air} = 3.72$ mm. The image is closer to the lens when the object is viewed in air by the amount $\Delta i = i_{water} - i_{air} = 5$ mm $-$ 3.72 mm $= 1.28$ mm. d) The change in image distance can be accomplished by squashing the eyeball slightly with the help of a ring muscle surrounding the eye body, see Fig. 8.18.

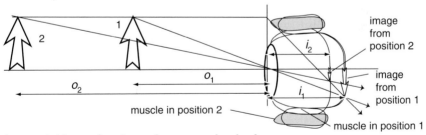

Fig. 8.18. Spider eye focusing under water and on land

8.4.3 Fish Eyes

Have you ever wondered why fish eye lenses are round like glass beads? They do not look flat like a magnifying glasses at all. Mathiessen [1886] noted more than a century ago, that many fish lenses of different sizes have about the same focal length ratio $f/R = 2.55$, where R is their radius. This number and the spherical shape do not come as a surprise when one knows the particular constraints on vi-

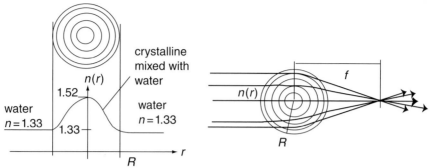

Fig. 8.19. Fish eye lens, a gradient index lens

sual perception under water. There are two problems. First, the light is dim down there. Second, lenses immersed in water have less focusing power $1/f$, because the focusing power depends on the difference of the refractive index of the lens and the outside $\Delta n = n_g - n_o$.

To collect light of low intensity the aperture diameter d of the eyes should be as large as possible. However, the light collecting power, or speed S of a lens is larger if its radii of curvature r_1, and r_2 are small, see (8.15). The best compromise for the smallest r and the largest d is a spherical lens, with $r_1 = r_2 = R = d/2$. In actuality transparent spheres can act as good focusing lenses for monochromatic light if their index of refraction is not constant but varies as function of the radius [Luneburg 1944]. Such gradient index lenses are called modified Luneberg lenses.

One can imagine such gradient index lenses as onion-shell structures wherein the core has the highest index of refraction n_c, see Fig. 8.19. Successive shells have a gradually decreasing refractive index, which approaches the value of the outside material n_o, at the radius R. The function $n(r)$ can be given as a complicated analytical expression, which was evaluated numerically by Gordon [2000]. His data are reproduced in Fig. 8.20a, showing the ratio $n(r)/n_o$ for a spherical lens immersed into water as function of the reduced radius r/R.

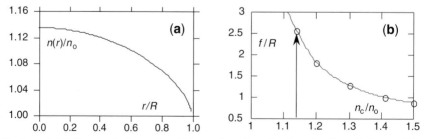

Fig. 8.20. (a) Refractive index ratio $n(r)/n_o$ as function of reduced radius r/R for modified Luneberg lens n_o, index on the outside. (b) Focal length of fish eye lens as function of refractive index normalized to the index n_c at the center of the lens [after Gordon 2000]

Gradient index lenses can be constructed from a biological material, crystallin, a protein with $n = 1.52$. By changing its water content the refractive index of this material can be varied smoothly from 1.52 to 1.34. The center of such a modified Luneberg lens has the maximum value $n_c = 1.52$, so that the ratio of this maximum value to the index of the water on the outside is $n_c/n_o = n_c/n_w = 1.52/1.33 = 1.143$. This ratio determines the focal length of the lens. The focal length f of gradient index lenses can also be expressed as a complicated analytical relation of the lens radius R, and the refractive index at the center of the lens n_c. Gordon's data for the ratio f/R as function of n_c/n_o are displayed in Fig. 8.20b. For the maximum ratio $n_{crystallin}/n_{water} = 1.52/1.33 = 1.143$ that can be generated with the materials water, and crystallin the normalized focal length is $f/R = 2.55$, which is exactly the empirical value found by Mathiessen [1886]. Spherical eye lenses built with this material have a fixed aperture ratio $2R/f = 0.78 = 1:1.28$.

Spherical gradient index lenses are not color corrected. Since n is a function of wavelength, light rays of different colors are focused at different distances. This does not cause a significant problem because the red light is quickly absorbed by water. Only blue can penetrate to larger depth, thus in the deep waters all fish look blue, and eyes of deep swimming fish do not need color correction.

8.4.4 Big Eyes of the Deep

Some animals find their prey at great depths in the ocean, yet light intensity rapidly decreases with depth. Animals that want to see at a great depths must have big eyes. Indeed the critters of the deep do have huge eyeballs. The biggest eyeballs ever, with diameter $2R = D = 26$ cm, are attributed to *temnodonto sourus* [Motani 2000]. Today's giant squids are a close second with $D = 25$ cm. Table 8.5 shows the eye diameters of some existing and extinct animals.

One can use the diameter of eyeballs D, to obtain an estimate of the depth limit at which the animal can still see an object given clear water. Assume a cloudy day where the photon flux is a factor 40 below the maximum value given by (8.4), namely $\phi_o = 10^{20}$ photons/s·m². The pupil diameter is by necessity smaller than the eyeball diameter D, but it is easy to account for this difference later. The area of the eye ball $A = \pi (D/2)^2$ could collect $\Phi_p = A \phi_o$ photons/s.

Due to absorption, the photon number decreases rapidly with depth. Red and ultraviolet are rapidly attenuated while blue green light survives, which contains

Table 8.5. Eyes of some aquatic animals compared to humans and max hunting depths. Data adopted from Motani [2000]

animal	human	African elephant	blue whale	ophthalmo saurus	giant squid	temnodonto saurus
body mass M in kg	80	5000	100 000	950		
diameter of eye D in mm	25	50	150	230	250	260
d_{100} in m	610		688	705	708	710

only the fraction $f_{bg} \phi_o$ of the original number of photons. At a depth d the photon flux is reduce to $\phi(d) = f_{bg} \phi_o e^{-\kappa d}$. By multiplication with the pupil area on finds the number the photon that would be captured by the eyeball of the animal $\Phi(d) = A\phi_{bg} = A f_{bg} \phi_o e^{-\kappa d}$. Suppose an aquatic predator hunts by looking for the shadow of a prey passing overhead. Assume that the animal needs a minimum flux of $\Phi_{100}(d_{100}) = 100$ photons per second to recognize a potential catch. This minimum light level determines the maximum hunting depth d_{100}

$$100 = \Phi(d_{100}) = A f_{bg} \phi_o e^{-\kappa d_{100}}. \tag{8.20}$$

This depth can be found from (8.20). Divide the equation by the factor $A f_{bg} \phi_o$ in order to separate the e-function. Then take the natural logarithm of the whole equation, and remember $\ln(e) = 1$. By this procedure the right hand side becomes $-\kappa d_{100}$. Thus the 100 photon depth is obtained

$$d_{100} = \frac{1}{\kappa} \ln(A f_{bg} \phi_o / 100). \tag{8.21}$$

By setting the aperture of the eye equal to the cross section area of the eyeball A, one clearly gets an upper limit for this hunting depth. However, the logarithm is not a sensitive function. If A is reduced by a factor 10 the hunting depth with blue green light would go from 2220 m to 2076 m. Table 8.6 shows values for the absorption coefficients κ as function of wavelength for clean sea water and the calculated maximum hunting depth d_{100} assuming $f_{bg} = 0.05$, and $A = 0.05$ m^2.

The absorption coefficient depends not only on the wavelength but also on the turgidity of the water. Blue light has the smallest value. Plankton and other stuff in the sea water scatters and absorbs mainly in the UV and blue section of the spectrum. The effective κ depends on the composition of the floating micro matter integrated over the whole depth.

The depth d_{100} is a maximum-depth limit, because seawater with rich marine life, where predators would operate, has a high degree of turgidity. Plankton, other dissolved organic matter, and inorganic matter from land drainage may increase the value of κ substantially, so that less light penetrates to greater depths. To somewhat account for turgidity in the calculating the maximum hunting depth in Table 8.5 a larger value for the absorption coefficient $\kappa = 0.06$ was used.

Table 8.6. Absorption coefficient for clean sea water, and 100 photon depth at different wavelengths. Data from Handbook of Optics, 2nd ed., Vol. 1, p. 43.28.

Absorption coefficient	near UV	blue	green	yellow	dark red
wavelength λ in nm	400	470	540	580	670
clean sea water κ in m^{-1}	0.018	0.016	0.056	0.108	0.43
max depth d_{100} in m	1973	2220	634	328	82

8.4.5 Non Spherical, Large Aperture Lens Eyes of Trilobites

Another now extinct marine species with large eyes were the trilobites. These animals had schizochroal (multi-lens) eyes, numbering up to a few hundred of eyes per animal, which were mounted on two dome-shaped sclera.

Large aperture lenses do not focus into sharp points. The lens laws for spherical surfaces derived in Sects. 8.2.4 and 8.2.5 hold only for rays that pass through the optical system close to the axis of symmetry (paraxial rays). The optical surfaces of a vertebrate eye may be approximated as spherical shells. They produce sharp images in broad daylight, when the iris apertures are small.

Complications arise when the aperture diameters are increased to admit more light. Rays that strike the lens farther away from the optical axis (off-axis-rays) are bent too much by the spherical surfaces, producing an image that is closer to the lens than the image produced by near-axis rays, Fig. 8.21c. Spherical lenses with large apertures cannot produce sharp images. However, there are two remedies, gradient index lenses, discussed in Sect. 8.4.3, and compound lenses.

Compound lenses are used in all expensive cameras. They consist of two or more components made from materials with different refractive indices. Compound lenses consisting of only two components cannot have a spherically shaped interface[5]. Trilobites had such 2-element lenses in their schizochroal eyes. Appar-

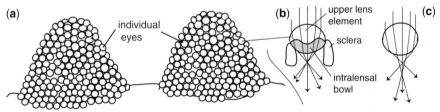

Fig. 8.21. (a) Sketch of Schizochroal left and right eyes of Trilobite. (b) Trilobite lens *Crazonaspis stgruvei Henry* after Clarkson. (c) Spherical lens with large aperture

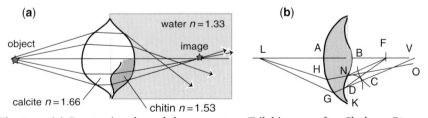

Fig. 8.22. (a) Ray tracing through large aperture Trilobite eye, after Clarkson F# ≈ 1.1. (b) Sketch of Huygens' construction of non spherical – large aperture lens

---

[5] Expensive camera lenses like the Leitz Elmar had up to 6 components with spherical surfaces, which focus off-axis-rays of different colors into sharp image points. Todays camera lenses have 3 or 4 elements with non spherical surfaces.

ently the lenses were built like an-acromats, having two components with different indices of refraction. More details can be found in the beautiful book by Clarkson [1968].

A section of one trilobite lens is presented in Fig. 8.22a. The ray tracing diagram in the upper half of Fig. 8.22a shows how rays from an object point do *not* converge into a single image point due to spherical aberrations. In the lower part of Fig. 8.22a one can see how a second lens element, made with a different index of refraction, corrects for the spherical aberration. The optical components of the trilobite eye supposedly were calcite $n_{ca} = 1.66$, and chitin $n_{ch} = 1.53$. The shape of these lens elements is very similar to a non-spherical compound lens proposed by Huygens, Fig. 8.22b, and an independently derived design by Descartes.

8.5 Facet Eyes

The eyes of vertebrates have evolved to a high degree of perfection, reaching the physical limits in resolution and sensitivity. Since light is almost omnipresent, and since much information can be gleaned from it there exists a strong incentive to develop eyes. As in many situations, an exclusive solution to a given problem does not exist. Indeed eyes have been developed in different phyla and lens eyes are not the only possible solution to view the surroundings. Insects went their separate way. Instead of collecting light with lenses, insects channel radiation through light pipes: facet eyes. Each photo-receptor of a facet eye has its own light pipe. Facet eyes do produce a very fine picture of the surrounding with an angular resolution approaching the diffraction limit of lenses. Facet eyes also have the ability to dim the light, not by individual iris diaphragms, but by actively induced losses in the individual light pipe eyes employing frustrated internal reflection.

8.5.1 The Principle of Light Pipes

Light waves can be conducted in *light pipes*, somewhat like electrical currents flow in wires, or water in a river bed. The principle of trapping the light is reflection at the walls. Light can be reflected off metal mirrors, or off interfaces between two transparent media due to total internal reflection TIR, see Sect. 7.4.4.

In principle, some light is lost at every reflection off metal mirrors. Internal reflection at sufficiently large angles, Fig. 8.23a has zero losses. It is total internal reflection (TIR). TIR occurs when the core, a medium of large index of refraction, say $n_{core} = 1.5$ is surrounded by a transparent cladding of lower refractive index, say $n_{clad} = 1.4$, see Fig. 8.23b. Therefore, the core has a smaller phase velocity c_{core} than the cladding c_{clad}.

$$c_{clad}/c_{core} = n_{core}/n_{clad} > 1 \qquad (8.21)$$

The diameter of the core and the thickness of the cladding must be larger than the wavelength of the light. Figure 8.23 shows the path of rays in a light pipe.

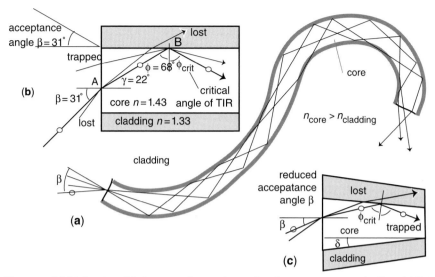

Fig. 8.23. (a) Light pipe. (b) Construction to determine the acceptance angle $\emptyset_{crit}$. (c) Reduction of acceptance angle in conical pipe section with cone angle δ

Cylindrical light pipes have an acceptance angle of typically $\beta \approx 20°$. The acceptance angle β can be reduced by attaching a conical, somewhat trumpet-shaped section at the entrance of a pipe, see Fig. 8.23c. The acceptance angle β of a light pipe can be calculated as follows. Light rays are refracted, when they intersect a surface between two transparent media with different refractive indices, like n_{co} and n_{cl}. The entrance angle $\emptyset_1$ measured against the surface normal is larger on the side of the smaller refractive index. The entrance angle has the limit $\emptyset_1 = 90°$. Total internal reflection occurs when $\emptyset_1$ is exactly $90°$. The angle $\emptyset_2$ belonging to this incident angle is called the critical angle $\emptyset_{crit}$, of total internal reflection, known from Eq. (6.16) as $\sin \emptyset_{crit} = n_{clad}/n_{core}$, or

$$\emptyset_{crit} = \arcsin (n_{clad}/n_{core}) . \tag{8.22}$$

Light, which strikes the surface from inside the core at $\emptyset_{crit}$ or at any larger angle is trapped by total internal reflection. It cannot leave the core medium. Rays that impinge at any smaller incident angle are not confined.

With the knowledge of $\emptyset_{crit}$ one can determine the acceptance angle β for light pipes. Consider a light pipe made from organic materials, which points at a scene in air $n_0 = 1$. Let the core, with an index of refraction like the human eye lens $n_{core} = n_{eye} = 1.43$, be surrounded by a cladding of water $n_{clad} = n_w = 1.33$. First, find the critical angle $\emptyset_{crit}$ for total internal reflection at point B, Fig. 8.23b. It is $\emptyset_{crit} = \arcsin (n_{clad}/n_{core}) = \arcsin (1.33/1.43) = 68°$. Now follow the path of the rays backwards. The ray reflected at point B with the angle $\emptyset = 68°$ can be traced back to point A, where it emerged at the angle $\gamma = 90° - 68° = 22°$, see Fig. 8.23b. This ray must have entered from the outside at the trapping angle β, which is found from

Snells law (7.18). One has $\sin\beta = (1.43/1)\sin 22°$, so that $\beta = \arc\sin(1.43 \cdot \sin 22°) = 31°$. It is easy to see that rays striking the surface of the light pipe near A with angles smaller than β will all be trapped, whereas rays that arrive with larger angles will be lost. In general one can give the trapping angle β as

$$\sin\beta = \frac{n_{core}}{n_0} \cdot \sin\left\{90 - \arcsin\left(\frac{n_{clad}}{n_{core}}\right)\right\} = \frac{1.43}{1}\sin 22°. \qquad (8.23)$$

One can see from (8.23) that the acceptance angle is smaller if the indices of refraction of the core of the light pipe and the cladding are closer together than 1.43 and 1.33. The acceptance angle changes if the core of the light pipe is a conical section with the half angle δ. Then one finds

$$\sin\beta(\delta) = \frac{n_{core}}{n_0} \cdot \sin\left\{90 - \left[\delta + \arcsin\left(\frac{n_{clad}}{n_{core}}\right)\right]\right\}. \qquad (8.24)$$

If a light pipe consists of a conical section with the half angle δ attached to a straight section, its overall acceptance angle is reduced from the value given by (8.23) to the smaller value $\beta(\delta)$, calculated from (8.24).

8.5.2 Insect Eyes

Insects have developed compound eyes, made up of many narrow cones attached to a little lens. These ommatidia are mounted on a curved platform generating an outer radius R. In each ommatidium a lens of diameter d, concentrates the light into the light pipe at the bottom. If the core of the light pipe has the diameter b, the intensity is increased by a factor d^2/b^2. In a perfect facet eye the width $\Delta = 2f\lambda/(n_{core}d)$ of the center of the diffraction pattern created by the lens aperture d should be smaller than the core diameter b of the light pipe.

Radius R and lens diameter d determine the viewing half angle β_1 that each light ommatidium occupies in the hemispherical geometry of the compound eye. With typical values $R = 2$ mm, and $d = 20$ µm, each eyelet is looking into the field of view angle $\beta_1 = d/R = 0.02$ mm/2 mm $= 0.01$ rad $\approx 0.57°$.

8.5.3 Intensity Attenuation by Frustrated Internal Reflection

Compound eyes do not have individual stops. However, animals with facet eyes can vary the incident intensity by a completely different process. The intensity transmitted through an individual light pipe can be varied by changing the refractive index n_{clad}, by varying the thickness Δ_{clad} of the cladding, or by moving absorbing pigments into the evanescent wave that accomplishes the internal reflection.

The underlying physical effect is known as frustrated internal reflection. At every internal reflection the waves actually penetrate somewhat (about one wave-

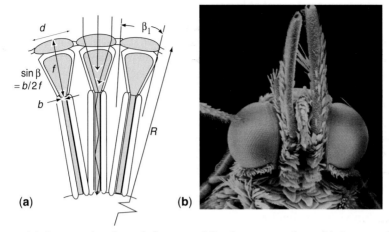

Fig. 8.24. (a) Cross section through the ommatidia of a compound eye. (b) Compound eye of insect, courtesy of Humphries [2002]

length) into the cladding material. This is because of the wave nature of the light: just think that the canopy of the Huygens wavelet-umbrella with a typical lateral width of λ penetrates somewhat into the cladding material, see Fig. 8.25a. The light travels a distance $\partial y \approx \lambda$ out of the high-n core material and translates some distance ∂z, in the forward direction, like a dolphin porposing out of the water. The forward displacement is called the Goos-Hänchen shift [Goos and Hänchen 1947].

If some part of the evanescent wave is prevented from going back, the internal reflection is *frustrated*. This effect occurs when some light is absorbed or scattered in the cladding, Fig. 8.25b. If the thickness of the cladding Δ_{clad} is small compared to the wavelength λ, and high n material is located on the outside of the cladding, some light will actually *tunnel* through the cladding. Some insects can activate pigments in the tissue outside their ommatidia which absorb radiation. The pigments migrate towards the ommatidia in bright light, so that the internal reflection is frustrated. Through this measure these insects can control the intensity reaching the photoreceptors in their composite eyes. Frustrated internal reflection is now in technical use primarily for switching light into, and out of fiber optics networks, but also in some other technical applications. However, insects have used this effect long before humans came onto the scene.

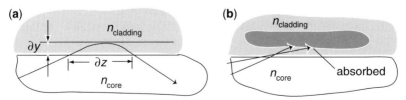

Fig. 8.25. (a) Evanescent wave. (b) Frustrated internal reflection

8.6 Unwanted and Wanted Visibility

Sight gives an enormous advantage to the hunter to see its prey. How does the prey overcome this disadvantage and tip the arms race in its favor? One way is to get out of sight, by hiding in bushes, or in sea-weeds, or behind a self produced smoke screen.

The second-best solution is to look like something else. Such *optical deception* may be accomplished by body coloration that blends into the background, or by markings that help the hunted pretend to be a dangerous foe rather than a tasty morsel. Looking like something else can be accomplished by assuming body colors derived from partial absorption, from interference, or from reflection.

Blending into the background can also be accomplished by looking like absolutely nothing, like a ghost roaming through a medieval castle. This is particularly effective in the dim light that aquatic animals encounter deep in the ocean. No ghost worth his magic would show up in the broad daylight of 10^{20} photons/ m^2s. However, an invisible prey may be rendered visible by eyes that recognize polarized light.

Much of this optical trickery can be accomplished by using interference colors, and reflection effects. Interference is a wave phenomenon brought about by superposition of waves. Interference mirrors give animals perfect reflectors. These are used on fish scales to become invisible, and are also employed in cat eyes to see better in the dark. Seeing your meal, and not being seen by your predator, is a continuous arms race.

8.6.1 How Animals Make Perfect Reflectors Using Interference

Interference can increase or reduce the intensity of a wave field. One application of this effect is perfect reflection of light by interference mirrors. Animals don't look into the mirror to check their makeup. Instead some animals use mirrors to diminish background light in eyes so that they can see very faint signals in dim light. Other animals employ mirrors to make themselves invisible.

Light can be reflected by thin film interference. All interference effects with light are only observed under special circumstances: one must somehow make two identical light waves. This can be accomplished by partial reflection on thin dielectric materials. Interference phenomena arise because waves can be added or subtracted. Consider two waves of equal amplitude and identical wavelength, Fig. 8.26a. There are two distinctive cases *constructive* interference, and *destructive* interference. In constructive interference, wave A always adds to wave B to mutually reinforce the wave motion. In destructive interference one of the waves (C = −B) always moves opposite to the other (A), so that both together cancel each other out.

Look at the details of a dielectric mirror, Fig. 8.26b. A light beam o strikes a thin film with an index of refraction n. Part of the light is transmitted (t), and part is reflected (r, beams 1 and 2). The transmitted beam is partially reflected on the

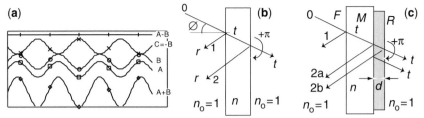

Fig. 8.26. (a) Interference of two wave. (b) Interference at a thin film. (c) Interference at a double layer

rear surface and emerges on the front side as beam 2. Imagine that the dielectric film is quite thin and the incident angle Ø is small so that the reflected wavelets 1 and 2 are close together and interact. The reflected beam on the rear side suffers a change in phase by π radians (which corresponds to a path difference of $\lambda/2$).

Constructive interference arises when crests of wave A fall onto crests of wave B. The resulting wave is enhanced, the combined amplitude, $A + B = 2A$, is double as large. Destructive interference arises if the crests of one wave always coincide with the troughs of the other wave, $A + (-B) = 0$. Then both waves cancel each other and there is no wave motion. If the beams 1 and 2 are completely in step, like the waves A and B in Fig. 8.26a, all the light is reflected, and nothing is transmitted. Such *in step* waves must have a path difference of $m\lambda$ or a phase difference of $m2\pi$, where m is 1, 2, or a larger integer number.

Here we consider only the case $m = 1$ for the thinnest dielectric mirror. This phase difference is made up from the effect of the reflection at the rear surface, accounting for ½ wavelength (or a phase difference of the angle π), and from twice traversing the thin film (thickness d). Note that one has to use here the wavelength in the medium $\lambda_{med} = \lambda_{vac}/n = \lambda/n$. Since the passage through the film must account for another ½ wavelength the thickness d of the for perfect reflection is

$$2d = \lambda_{med}/2 \quad \text{or in general} \quad d = \lambda/4n. \tag{8.26}$$

Conservation of energy requires that the incident intensity equals the transmitted intensity plus the reflected intensity. If all the light is reflected, nothing can be transmitted. The thin film then acts as a perfect reflector. If the thin film is made up of a double layer with an internal surface M the reflection is not perfect, Fig. 8.26c. Some of the light (beam 2b) is already reflected at M, and will not be perfectly in step with beam 1. Then not all the light will be reflected and some must be transmitted. The internal surface M destroys the perfect mirror.

Animals use dielectric mirrors for improving the contrast in their eyes, and for modifying their external appearance.

8.6.2 Improving the Contrast with Dielectric Mirrors

The contrast of images is reduced if stray light can reach the retina. Stray light is light that gets into the eye and is diffusely scattered. Stray light may either be removed by absorption, or it can be eliminated by reflecting it straight back out of the eye. It seems that the second strategy is more efficient. Special mirrors are needed inside the eye to remove the unwanted light.

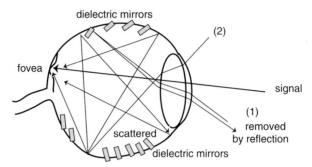

Fig. 8.27. Positioning of dielectric micro reflectors in the eye to reduce stray light

Nocturnal hunters have such mirrors inside their eyes. Technical high quality mirrors have metallic reflection layer, but animals cannot refine metals. Instead animals uses wave optic tricks. Cats and other nocturnal animals have layers of small retro-reflector mirrors in the *tapetum* located behind the retina of their eyes that remove any light before it can be scattered inside the eyeball [Denton 1971]. These reflectors work on the principle of thin film interference as described in Sect. 8.6.1. Apart from removing stray light, the sensitivity is increased since the light must cross the retina twice. The color of the reflected light depends on the color transmission of the retina and the interference characteristic of the mirror. Cat eyes reflect red light, buff colored cocker spaniels generally show yellow tapetal reflection, while black Labrador's eyes glow green in reflected light [Pickett 2001].

8.6.3 Hiding in the Water

Many animals use interference reflectors. Fish that live near the surface of the water have these mirrors in their scales to look like nothing. Fish make efficient use of the light in the water by either redirecting or absorbing the radiation coming from certain directions, and even by emitting their own light, Fig. 8.28. The purpose is to let the animal blend nearly perfectly into its environment. This feat is accomplished using absorption, reflection, and bioluminescence.

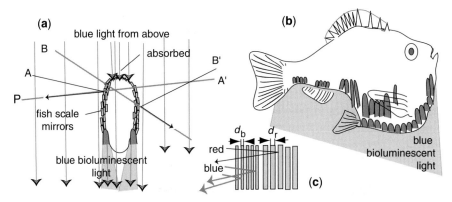

Fig. 8.28. The hatch fish absorbs light from above to eliminate back reflection. It reflects light on the sides, and emits bioluminescent blue light downwards to eliminate its shadow. (a) Cross section, (b) side view, (c) dielectric mirrors imbedded in the scales reflect by destructive interference of the forward beam

The scales of the hatchfish contain many very small individual mirrors, which are vertically mounted. Underneath the mirrors is a black pigment, which absorbs all the light passing between the mirrors from above, Fig. 8.28a. Looking down onto the hatchfish a predator sees nothing, because no ambient light is reflected. Each one of these mirrors is a tiny section of thin film [Denton 1971]. A predator looking from P in the direction of the hatchfish believes that he sees the rays coming from A' and B'. However, in reality its eyes collect reflected the rays A and B which are practically identical in intensity to rays A' and B' coming from behind its prey. Without the mirrors in its scale the hatchfish would appear to its predator like a dark contour.

A predator looking up from below would expect to see a shadow when another animal is overhead. However the bluish bioluminescent light emitted downwards by the hatchfish looks practically identical to the ambient daylight that makes it down to the depth where the hatchfish cruises, Fig. 8.28b.

The individual micro-mirrors units, Fig. 8.28c, consist of a double layer of guanine, index of refraction $n_{gu}=1.8$, and cytoplasm $n_{cp}=1.33$. The thickness d of such a double layer is chosen such that its optical width $n \cdot d = \lambda/4$ is exactly one quarter wavelength. By the principles of interference (8.26) this condition yields maximum reflectivity for a thin film embedded in lower n-material.

If each unit mirror was made of homogeneous material it would have 100% reflectivity. Since the unit consists of a double layer, some light reflected at the internal surface will not achieve perfect reflection. However, when a few of these double layer units are placed on top of each other they have a very high reflectivity (75%) for light of the wavelength λ. The hatch fish achieves high reflectivity for different wavelengths by having in its scales double layers of different width d_{blue} and d_{red}, for blue and red light, Fig. 8.27c.

8.6.4 Ghosts of the Deep with Anti Reflection Coatings

Many small and slow critters of the ocean like medusas, jellyfish, and shrimp have clear transparent bodies. They drift in the ocean currents swimming very slowly with minute muscle power. For instance Polyorcas, a grape-sized jellyfish, possesses a single muscle that closes its body cavity against an elastic force, allowing the animal to eject water in pulses to obtain rocket propulsions [Megill 2002]. The body elasticity opens the umbrella when the muscle relaxes to refill the internal cavity for the next stroke.

Invisibility is enhanced when light reflection off surfaces is inhibited. At the interface, the light must pass uninhibited say from water (density ρ_w, index of refraction n_w) to the gelatinous body of a jellyfish (density ρ_{jel}, index of refraction, n_{jel}). The indices of refraction are distinctively different: $n_{jel} > n_w$. The trick of a ghost-like creature is to minimize reflection off its body. Opticians know what to do in this case: the front surface of a camera lens has an anti-reflection coating, namely a layer less that ½ wavelength thick with an index of refraction that changes smoothly from n_{air} to n_{glass}. Transparent aquatic animals do not have anti reflection coatings. However, they achieve a gradual transitions form n_w, to n_{jel} by having a surface with tiny bumps [Johnson 2000] less that ½ wavelength high. Therefore the average tissue density changes gradually from ρ_w to ρ_{jel}, and the index of refraction similarly changes smoothly from n_w, to n_{jel}. This structure prevents reflection of light, thus the animal becomes invisible like a ghost. Anti-reflection coating is not an invention of, Zeiss or Nikon. With such non-reflecting surfaces the glassy bodies of these aquatic animals is nearly invisible. However, the spiral of competition between body structures and senses never rests. There is a way of perceiving apparitions.

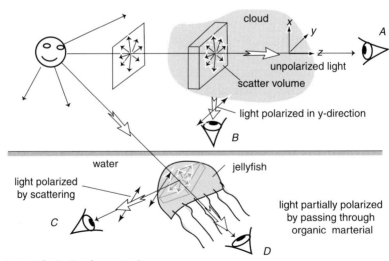

Fig. 8.29. Polarization by scattering

8.6.5 Unmasking the Ghosts with Polarized Light

Can eyes really be fooled for long? Objects may look perfectly transparent and not reflect any light. However, radiation is polarized by scattering, or by the passage through organic structures that contain long chain molecules, like Polaroid filters. Similarly, the light scattered off transparent organisms, or passing through them will be polarized.

In Fig. 8.29 the direct sunlight traveling in z-direction towards A is unpolarized. Light scattered at 90° and viewed from B is polarized in the y-direction. Polarized light looks darker when viewed through a Polaroid filter, because the light waves, which vibrate at right angles to the polarization direction set by the polarizer, have been filtered out. Photographers know that and use a polarizing filter when they want to capture clouds in the sky, because the light from clouds is polarized by scattering. Some hunters of transparent prey in the ocean can see polarized light (from position C), and thereby recognize their favorite ghosts-food.

8.6.6 More About Color

Although the eye is only sensitive to a small section of the electromagnetic spectrum, we can distinguish many different colors. Animals make use of color for visual displays, camouflage, and for warnings.

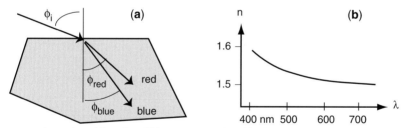

Fig. 8.30. Dispersion. (**a**) Red and blue rays in a glass block. (**b**) Refractive index $n(\lambda)$

The primary reason for the appearance of colors is the fact that *white* light, or sunlight contains all the colors of the rainbow. As soon as some colors are taken away, by whatever means, the remaining light is no longer white. Colors are created in a variety of ways, due to several physical effects:

1) Partial absorption in reflection; e.g. red brick reflects only red light.
2) Partial absorption in transmission; e.g stained glass windows.
3) Scattering of certain wavelengths; e.g. blue sky.
4) Interference on thin films; e.g. soap films, gasoline on water.
5) Diffraction; butterfly wings have elements similar to diffraction gratings.
6) Dispersion; the speed of light changes with wavelength $n = n(\lambda)$.

Dispersion occurs in prismatic transparent structures: a prism sends light of different colors into different directions. Dispersion occurs because the speed of light, and hence the refractive index, in all materials changes with wavelength. The function $n(\lambda)$ is called the dispersion curve. As a consequence, rays of different wavelengths are refracted into different directions, see Fig. 8.30a. A typical dispersion curve for glass is shown in Fig. 8.30b. Due to dispersion a single lens cannot bring red and blue light into the same focus. Camera lenses are color corrected by the combination of several lens elements made from different glasses. Expensive modern camera lenses have up to 6 component lenses for color correction. These are manufactured from materials with different dispersion curves. Apparently a similar technique was already employed by Trilobites, Sect. 8.4.5. In addition, our perception of color is strongly influenced by *afterimages*: if one looks into a bright red light and then turns the eyes immediately towards a white wall, one experiences a green after-image.

8.7 The Active Production of Light, and Limits of Seeing

Diverse animals have learned to produce light actively. Plankton, some squid species in the ocean, and fireflies in the air are examples. These biological light sources are very dim. They serve as camouflage, for communication, as position lights like the tail-lights of cars, and to occasionally illuminate the scenery. In addition in this section of Chap. 8 some thoughts are presented on binocular seeing, and why there is room for other long range senses.

8.7.1 Bioluminescence

There area a number of animals that can produce *cold* light. Generally the intensity of radiation generated by bioluminescence is very small compared to the intensity of sunlight. Therefore, bioluminescence is only used in dark places: at the bottom of the ocean (e.g. plankton), or during dark nights (e.g. glow worms). Bioluminescence in plankton can be observed on the sea-shore on many dark nights. Deep down in the oceans live a variety of animals with radiating organs. For instance the hatch fish. It's light is directed downwards and distributed in various directions by fully reflecting, or half-transmitting mirror elements, Sect. 8.6.3. The color of this light is matched to the color of daylight penetrating to the depth where this fish lives, so that the animal eliminates its own shadow: it becomes invisible to predators cruising below.

The Deep sea angler (*Ceratias holboelli*) carries an *esca* bulb, a small fleshy organ at the end of a spine-like extension of the upper lip, measuring in length about $15-40\%$ of the full body length ($L \approx 10-20$ cm). The *esca* contains luminous bacteria that emit light at the wavelength $\lambda \approx 490$ nm with a power of typically 1 µW.

This light can be seen by other would-be-hunters up to distances of typically 100 m. The deep sea angler just waits motionlessly for its victims to come close enough to be successful in a speedy attack. The light is used as a decoy to mimic a small luminous prey. However, the would-be attacker becomes a victim itself, when *Ceratias holboelli* opens its mouth.

Bioluminescence is a rich field of study that involves quantum effects and molecular chemistry, and thus goes beyond the scope of this book.

No higher animals have managed to generate electromagnetic radiation strong enough to produce search beams. Of the long range wave phenomena, only sound has been found suitable by animals for active scanning of the surroundings.

8.7.2 Signal to Noise Reduction Through Binocular Seeing

The signal to noise resolution of seeing in dim light can be significantly improved when two eyes look at the same object. When the image information from both eyes is subtracted, anything not seen by both eyes can be disregarded as noise [see Ditfurth 1972]. The overlapping field of view has object points which can be seen by each eye. Light signals from such points that are not observed in each eye may be interpreted by the brain as noise and discarded. This signal discrimination increases the sensitivity of seeing by about a factor 10.

However, two-eyed seeing generates the problem of parallax, which can only be overcome with added brain capacity. Once the problem of parallax was mastered there was an immediate added advantage: 3D perception. With three-dimensional perception the organism becomes part of a three dimensional world. This unexpected opportunity may well have been the starting point to self-awareness, and image-ination. Imagination can lead to tools, technology, culture, astro-physics, arts, philosophy, and finally to the world wide web.

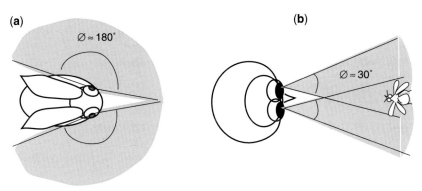

Fig. 8.31. (a) All round view of a grazing animal. **(b)** Forward view of the night hunter

8.7.3 Limitation of Seeing and How the Brain Sets You into the Picture

Light provides very important clues about the environment through the information that is encoded in the wave fields. However, Physics imposes limits on the use of the optical sense in at least two respects: (i) The minimum signal strength, (ii) The size of the smallest perceptible image.

The minimum signal strength is best illustrated by an example: In a dark railway tunnel a green flash lamp bulb emits light of 1 W, at $\lambda = 5.0 \cdot 10^{-7}$ m. Without reflector this light may be seen from a distance of 1000 m. How many photons are received by the eye in a *blink* $\Delta t = 0.1$ s? Assume the iris has a diameter $d = 2$ mm and an aperture area $A_{iris} = \pi (d/2)^2 = 3.14 \cdot 10^{-6}$ m².

First, determine the intensity at the distance $r = 1000$ m, namely $I_{1000} = 1$ W/$[4\pi (1000 \text{ m})^2] \approx 8 \cdot 10^{-8}$ W/m². The power P received by the eye is then found $P = I_{1000} A_{iris} = 8 \cdot 10^{-8}$ W/m² $\cdot 3.14 \cdot 10^{-6}$ m² $= 2.5 \cdot 10^{-13}$ W. A blink of the eye lasting 0.1 s, captures the energy $E = P \cdot \Delta t = 2.5 \cdot 10^{-13}$ W $\cdot 0.1$ s $= 2.5 \cdot 10^{-14}$ J. This can be converted into the number of photons N, that are captured $E = N E_{phot}$, where $E_{phot} = h c_0/\lambda \approx 4 \cdot 10^{-19}$ J is the photon energy. Solve for N to find $N = E \cdot \lambda/(h \cdot c) = 2.5 \cdot 10^{-14}$ J $\cdot 5.0 \cdot 10^{-7}$ m$/(6.6 \cdot 10^{-34}$ Js $\cdot 3 \cdot 10^8$ m s$^{-1}) = 6.3 \cdot 10^4$ photons. This light could be easily seen. However, it would be impossible to see a small square target T of area $A_T = 1$ mm² that is illuminated by this light and held at a distance of $R = 1$ m from the eye. This result can be derived as follows. Assume that $\varepsilon = 50\%$ of the intensity $I_{1000} = 8 \cdot 10^{-8}$ W/m² at the target is reflected uniformly in all directions, giving the target the power $P_T = \varepsilon A_T I_{1000} = 0.5 \cdot 10^{-6}$ m² $\cdot 8 \cdot 10^{-8}$ W/m² $= 4 \cdot 10^{-14}$ W.

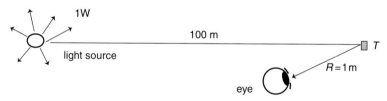

Fig. 8.32. Target T of area 1 mm² illuminated by a 1 W light source

At the distance $R = 1$ m off the target the reflected light has the intensity $I_R = P_T/[4\pi (1 \text{ m})^2] = 3.15 \cdot 10^{-15}$ W/m². With a pupil of the eye of diameter $2r =$ mm, and iris aperture $A_{iris} = 3.14 \cdot 10^{-6}$ m² the radiation power $P_p = A_p I_R = 3.14 \cdot 10^{-6}$ m² $\cdot 3.15 \cdot 10^{-15} = 10^{-20}$ W is captured. This power can be converted into a photon flux Φ photons/s, namely $P_p = \Phi E_{phot}$. Here we use again $E_p = 4 \cdot 10^{-19}$ J, and find $\Phi = P_p/E_p = 0.025$ photons/s. This means that one photon is capture by the eye every $1/\Phi = 40$ s. Obviously this is not enough to form an image of the 1 square millimeter large target.

8.7.4 High Resolution of Optical Signals
Is Not Always Good Enough

The smallest structures, which can be resolved by a wave field, are of the order of the wavelength $\lambda = c/v$. Since the wavelength of light falls into the range 0.4 – 0.7 µm *optical* images can resolve objects down to 1 µm sizes. It appears that humans had no need for such high resolution, since the smallest images seen by the naked eye are typically 100 µm. However, when we resolve distant objects we make full use of the angular resolution

$$\sin \emptyset_{min} \approx \emptyset_{min} \text{ radian} = \lambda / d . \tag{8.27}$$

For an eye with an iris aperture of $d = 3$ mm in yellow light, $\lambda = 580$ nm, this amounts to an angular resolution of $\sin \emptyset_{min} = 5.8 \cdot 10^{-7}/0.003 = 1.9 \cdot 10^{-4}$ radian equivalent $1.9 \cdot 10^{-4}$ radian $\cdot (180/\pi)°/\text{radian} = 0.01°$. One might think that such a resolution would be good enough for all possible circumstances.

Still there are conditions where light does not yield enough information about the environment: (i) if there is too much absorption, (ii) if there are no reference points. These are conditions where passive sound signals, sonar, electrostatic, or magneto static information is still available. Some animals can produce sound images with acoustic frequencies as high as $f = 10^5$ Hz. The smallest structures resolved in these *acoustical images* have dimension of the acoustic wavelength $\lambda = v/f = 340 \text{ ms}^{-1}/10^5 \text{ s}^{-1} = 3.4$ mm, which is about a factor 3000 larger than obtained by optical images. While not impressive in broad daylight, such an acoustic resolution is very good in the eternal darkness of the deep ocean where sperm whales hunt for giant squid, or in a moonless night where bats chase moths and beetles.

Table 8.7. Frequently used variables of Chap. 8

variable	name	units	name of unit
Δ	width of diffraction spot	m	
c	speed of light	m/s	
d	aperture diameter	m	
E_p	photon energy	J	Joule
F	focal point		
f	focal length	m	meter
I	intensity	W/m²	
n	index of refraction		
$\emptyset$	angle		
S	speed of lens		
Φ	photon flux	photon/s	
ϕ	photon flux density	photons/s m²	
v	frequency	s⁻¹	

Problems and Hints for Solutions

P 8.1 Focusing of the Human Eye

A person reads a book, Fig. 8.33a, holding it at a distance of $o = 25$ cm. a) Determine the focal length of the lens in this position, and the radius of curvature of the lens. Assume: (i) the lens is in air, (ii) its index of refraction is $n = 1.43$, (iii) the radii of curvature are equal $r_1 = r_2 = r$, and (iv) the eye accommodates to always produce a sharp image on the retina.

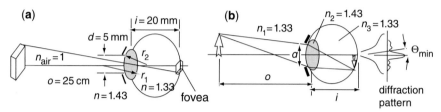

Fig. 8.33. (a) Eye in air. (b) Eye under water

b) Suddenly the person looks through the window and spots an eagle high in the sky. As the eyes focus onto the distant object, do the muscles holding the lenses have to pull or relax? c) Calculate the new radius of curvature of the lens after it has accommodated to bring the eagle into focus.

P 8.2 Resolution of the Human Eye

An archer can hit a white square 0.1 m high at a distance of 150 m. a) What is the angle subtended by the target? How high is the image on the retina? b) What is the area of the image of the square on the retina? c) There are $400 \cdot 400 = 160\,000$ rods and cones per square mm in the fovea area of the eye. How many rods and cones does the image cover? d) If the iris has a diameter of $d = 4.0$ mm what is the diameter of the center spot of the diffraction pattern on the retina, see Fig. 8.33b. The first diffraction minimum appears at the angle $\sin \emptyset_{min} = \lambda/d$. Draw the center diffraction spot into your sketch of the image of the square. e) Assume the white square is illuminated by bright sunlight, average wavelength $\lambda \approx 5.5 \cdot 10^{-7}$ m with an intensity $I = 1.2$ kW/m$^2 = \Phi$ photons $\cdot$ s$^{-1} \cdot$ m^{-2} $(h c_o/\lambda)$, where $h = 6.6 \cdot 10^{-34}$ J s is Planck's constant. If 60% of this light is reflected, how many photons are reflected per second? f) How many photons/s get into the eye of the archer and make up the image? g) How many photons would get into the eye during one blink of the eye which lasts say $\Delta t = 0.25$ s?

P 8.3 Spider Focusing

Spider focus by changing the image distance. Suppose a spider watches a pray at a distance $o = 10$ cm under water. Assume that the spider eye produces a sharp image at the image distance $i = 0.5$ cm. a) Determine the focal length of the spider eye. b) Find the radius of curvature of the eye, assuming a symmetrical lens with $r_1 = -r_2$. c) Now let the spider climb out of the water and focus at another object, which is 10 cm away in air. Find the new image distance. d) In a similar way calculate the image distances for objects at distances 2.0 cm, 5.0 cm, 20 cm inside and outside of the water, and draw two curves with the image distance as function of the object distance under water and in air. e) Comment on the biological problem of affixing such eyes (which focus by changing the image-object distance) on the head of the animal. How does the spider actually look at objects that are very close by?

P 8.4 Owls Eyes
Owls hunt at night using eyes and ears. A big horned owl has about the size of a small turkey. Assume that the owl's eye is a scaled version of the human eye, with eyelid, iris, lens, and retina. Estimate the size of the eyeball. a) Determine the focal length, and the diameter of the iris when fully open (assume that the iris has a circular shape). b) Find for blue light, $\lambda = 480$ nm, the diffraction limited angular resolution (the angle at which the first order diffraction minimum occurs), and determine at which distance L, the owl would recognize two glow worms, which are located 12 mm apart as two objects rather than a single blurred glowing point. c) Determine the amount of light intensity, which this eye would collect when looking from a distance of 50 m at a point source (say a glow worm) emitting 6 µW of light power in a perfectly dark night (no moon, dark overhead clouds to obscure the stars). d) How many photons would this eye collect in one "blink" (1 blink $\Delta t = 1/15$ s). At what distance would the photon flux be so small that the owl could no longer see the glow worm?

P 8.5 Eye to Eye
The giant octopus lives at large depth in the ocean. It has the biggest eyes of all animals. Typically the eyeball is $2R = 25$ cm across. Assume that the octopus eye is a scaled-up version of the human eye, with eyelid, iris, lens, and retina. a) Calculate for the octopus the focal length, and the diameter of the iris when fully open (assume that the iris has a circular shape). b) Determine the diffraction limited angular resolution for blue light (the angle at which the first order diffraction minimum occurs). c) Calculate the amount of light intensity which this eye would collect when looking from a distance of 1000 m at a point source (say a bioluminescent object) emitting 1 µW of bluish green light. d) How many photons would this eye collect in one blink (1 blink, $\Delta t = 0.3$ s). Could an animal with an iris diameter of $2d = 1.0$ mm see this object?

P8.6 Fish Eyes, and Scales
Fish eyes have spherical lenses, and the index of refraction increases gradually from $n(r_0) = 1.33$ at the edge to $n(r = 0) = 1.52$, at the center see Sect. 8.4.3. Consider a fish eye of $r_0 = 2.5$ mm. a) What is the focal length. b) What is the speed $S = d/f$ of this lens? c) Why do fish have spherical lenses? d) Suppose the lens did not have a graded index of refraction but instead the index was constant at $n = 1.52$. Where would the rays R_1 and R_2 traveling parallel to the optical axis be focussed? R_1 is located $y = 1$ mm off the optical axis and R_2 is at $y = 2$ mm. e) What optical components generate the colors of fish scales? f) Why do fish scales change the colors when rotated in ambient light?

P 8.7 Fly Eyes

Flies have compound eyes with many individual photo detectors. Assume the facet eyes of a certain fly consist of tiny light pipes, each of $d=50$ µm diameter, length $L=200$ µm, and $n_e=1.52$, mounted behind a conical structure of height $B=50$ µm and top diameter $D=100$ µm, and $n_c=1.52$. This cylindrical compound light guide is capped by a lens of $f=50$ µm. The tissue between these optical structures has $n_t=1.33$. Assume that the outer radius of the compound eye is $r=1.5$ mm, and the structure occupies ½ of a sphere. a) How many facet eyes are there on each compound eye if the cones touch each other? b) With reference to Fig. 8.23 calculate the critical angle of total internal reflection in the light pipe. c) What is the cone angle δ? d) What is the acceptance angle β for light pipe and cone? e) Name five other animals that have compound eyes.

P 8.8 Now You See Me Now You Don't

Optical illusions can be found in many situations. Animals use them for camouflage. a) Find and sketch 3 examples of optical illusions. Explain how the eye has been deceived. b) Describe an example of an optical trick used by an animal to hide from its predators. c) Describe an example of an optical trick used by an animal to scare off its predators. d) Describe an example of an animal that appears colorful due to either interference or diffraction, and explain how the colors come about.

P 8.9 Changing Colors

Explain why the feathers of humming birds change color when one looks at them from different directions (think of interference).

Sample Solutions

S 8.1 Focusing of the Human Eye

a) Focal length calculated in centimeter: $1/f=1/o+1/i=1/25+½=1/1.852$, thus $f=1.852$ cm. The focal length f is related to the radius of curvature r as $f=r/[n_g-n_o)+(n_g-n_i)]=r/[1.43-1+1.43-1.33]=r/0.53$. Then $r=0.53f=0.9892$ cm. b) The eye muscle must pull the lens into a flatter shape. c) New focal length found from $1/8+½=1/f$, thus $f=2$ cm. The materials in object and image space still have the indices of refraction as in a). Therefore, one has the same relation between radius of curvature r and focal length f as calculated in a) $r=0.53f=0.53\cdot2=1.06$ cm.

S 8.9 Changing Colors

The iridescent green and blue colors of the wings of the bird arise from interference. Thin transparent films of thickness d and index of refraction n are imbedded in the feathers. They act as perfect mirrors for the wavelength $\lambda_1=2\cdot2\cdot n\cdot d=4\cdot n\cdot d$, for light that strikes the feather with an incident angle $\phi=0°$. The angle is described in Fig. 8.26b. Looking straight on to a wing one will see the color associated with the wavelength $\lambda_1=2\cdot2\cdot n\cdot d$. In this product the factor $2\cdot n\cdot d$ is the extra path distance Δx that a wave reflected at the rear surface of the thin film has to travel compared to a wave reflected at the front surface, so that both together generate constructive interference. If the light impinges at an angle $\emptyset > 0°$ the path difference is smaller, namely $\Delta x = 2 \cdot n \cdot d \cdot \cos \emptyset$, which is smaller than $2 \cdot n \cdot d$, since $\cos \emptyset_2$ is smaller than 1 for any positive angle between 0 and 90°. As a wing is rotated the angle grows from $\emptyset_1 = 0$ to some value $\emptyset_2$. Now the mirror acts as a perfect interference mirror for the wavelength $\lambda_2 = 4 \cdot n \cdot d \cdot \cos \emptyset_2$. This wavelength must be smaller that λ_1. The feathers acquire a more bluish tinge when the wing is rotated.

9. Sound

The needs and wants of many animals
Are dressed in sound as chirping, honking, growls.
Voice carries messages of love and anger,
To scare a rival or attract a mate.
Voice is the glue that holds together
The social fabric of all higher life.

Sound, the Supreme Medium of Interaction

Sound provides a physical link between animals in the biosphere, and it connects the biosphere to the soundscape of the earth itself. The soundscape is made up from the noises of running water, breaking waves, thunder, wind, etc, and the sounds of animals as they walk, run, fly, breathe, chew, cough, or talk to each other. This physical link requires only a small amount of energy.

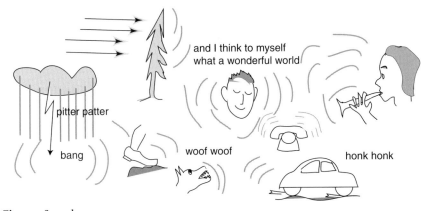

Fig. 9.1. Soundscape

There are hidden messages in the noise that may inform listeners about their surroundings. These messages are free, like the internet for anyone smart enough to hear and decode them. Animals that can hear may improve their chance of catching a meal or not being eaten, increase their ability to find a mate, and/or their ability to interact as groups through language.

The first section of this chapter is a qualitative description of the parameters of sound. Section 9.2 deals with intensity and impedance, quantities that describe the energy transfer associated with sound, and why it is difficult to hear on land. This sets the scene for a discussion of ears in Sect. 9.3. Communication by sound requires knowledge of the production of sound. The main principles of sound generation are discussed in Sect. 9.4. How voices work, is described in Sect. 9.5. Information about distance, direction and identity of the source can be extracted from sound, Sect. 9.6. The use of sound for active image scanning, sonar, has been invented by aquatic mammals and by bats. Some features of this new technique are described in Sect. 9.7. Finally, in Sect. 9.8 the imaging and information-gathering properties of light and sound are compared.

9.1 Signals of Sound, Noise, and Language

Every object that moves rapidly in air or in a fluid generates some sound. Sound of a single frequency is called a pure tone. Pure tones can be produced by tuning forks, or by electronic simulators. Musical instruments, and other sound sources generate more than one pure tone simultaneously. Tunes are something pure and simple, noise is a collection of many frequencies of short intervals. Noise is all around us, because things that move nearly always *make* noise.

> *Musik wird oft nicht schön empfunden*   Quite often music is despised
> *weil sie stets mit Geräusch verbunden.*   As it appears as noise, disguised.
>
>   Wilhelm Busch   (translated by the author)

To characterize a pure tone one must specify its frequency, and amplitude, and how the amplitude varies with time. To describe noise or music one must know the amplitudes and component frequencies that make up this sound, namely the composition of the superimposed frequencies.

Language is a form of noise that we have been accustomed to decipher from early childhood. However, *language* is not a human invention, it even reaches across species. A dog owner, listening to his dog, can easily distinguish many different signals: the japping sound his dog makes when she wants to come into the house differs from the angry growl when a stranger enters the yard, and from the whining howl when she is lonely, or from the happy yelp when a family member comes home. There is a primitive from of language in the dog's voice that is not hard to understand for humans and thus transfers across the boundaries of species. Many different animals *talk* to each other, some over close ranges, and some, like whales, over hundred of miles when swimming in the Sofar channel. Speech is energy-cheap. Most learning, and much of the social behavior has evolved on acoustic interaction between animals.

9.1.1 Phenomena Associated with Sound

The physical processes of production, transmission and reception of sound are well *understood* by animals. In the application of sound, animals have discovered many tricks of acoustics, which likely have advanced their evolution.

- Every sound *source* has a generator/resonator with a controlled mode pattern for the production of tones of single frequencies, mixtures of tones of diverse qualities, or noise pulses of short duration i.e. the vocal tracts.
- Sound can be *directed* by wave-guides, and by lenses, i.e. dolphins.
- Sound is *collected* with directional collection systems, and must be *amplified* to match the vastly different acoustic impedances of air and water.
- Two ears can distinguish small phase differences for *directional orientation*.
- The cochlea of the inner ear is a sophisticated analog *Fourier analyzer* of extreme sensitivity, which can *detect frequencies of up to* 100 kHz.
- *Sonar* systems open new living arenas: the night (bats), and the deep sea (dolphins, whales).

These developments were pushed to the limits with regards to intensity thresholds, and frequencies. At the highest frequencies that bats can generate (80 kHz) they achieve a sonar image resolution of $\Delta x \approx \lambda = c/f = 4$ mm. Smaller objects cannot be resolved. Ears can analyze sound intensities in the range 10^{-12} W/m²– 10 W/m². The very faintest detectable signals are obscured by Brownian motion.

Most sound processing organs were developed by animals that live on land. Examples are frogs, insects, birds, and of course the mammals. Fish have pressure sensors called the lateral lines. Yet, not many examples of sound communication are known for fish or non-mammalian aquatic animals. Mammals, who evolved on land, had to overcome a significant handicap: only a tiny fraction of the sound intensity is transmitted when sound passes from air into water (that is from the air into the inner ear where the sound is detected). Only after animals had developed an impedance-matching organ – namely the outer and the middle ear – was there enough energy transmitted to perceive any sound. However, once this hurdle was overcome mammals quickly developed ears to hear, and voices for social contact, language, and some even developed sonar.

9.1.2 Parameters of Sound

What is sound? The sound we hear consists of rapid pressure vibrations of the air at our eardrums. Sound can be generated by a variety of devices: bells, strings, tuning forks, whistles, rattles, resonance tubes, membranes of drums, and speakers.

Animals use mainly vocal cords to generate sound. One may picture a vocal cord as a hybrid between a string and the membrane of a drum. All these sound sources shake the air surrounding them in rapid sequence, and this motion is passed on to adjacent molecules, spreading waves of oscillation at the speed of sound in all directions. Most of these sound-generating devices oscillate in simple harmonic motion, SHM.

The sound of a tone is what we perceive when the air molecules next to our ears move rapidly back and forth with a certain amplitude s_0 (typically less than 1 μm). This motion can be described by many different sound parameters: (1) phase velocity (air, room temperature) $v = \sqrt{(\gamma R g T/m)} \approx 340$ m/s, (2) frequency f, (3) wave-

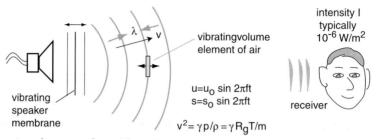

Fig. 9.2. Sound waves and particle motion

length $\lambda = v/f$, (4) displacement amplitude s_0, (5) particle speed $u_0 = f \cdot 2\pi s_0$, (6) pressure amplitude Δp_0, (7) intensity I, measured in W/m², (8) composition of different frequencies (sound quality), and (9) sound signal duration.

Sound travels as longitudinal waves. Sound spreads out from a source in all directions. To illustrate the sound motion, consider a very small volume element of air or water, through which a sound wave travels. This volume element will oscillate at a frequency f, or angular velocity $\omega = 2\pi f$, with a displacement amplitude s_0, and a maximum velocity u_0, (which incidentally is very small compared to the speed of sound).

Sound energy is characterized by the intensity I. If, for instance, the voice of a lecturer generates the intensity $I \approx 10^{-5}$ W/m² at the distance where a student sits, his ear with typical area $A_{ear} = 3 \cdot 10^{-3}$ m² intercepts the acoustic power $P_{ear} = I A_{ear} = 3 \cdot 10^{-8}$ W. However, such small powers can carry much information, especially if the lecturer does not dispense nonsense. In comparison, the full sunlight shining onto the area of an ear, would deliver radiation power that is one hundred million times larger.

Animals benefit from *reading* sound signals because from this information they can infer the existence and/or intentions of other animals. They can also detect threats, or opportunities lurking in their inanimate environment. Out of the nine physical parameters listed above animals can extract many different clues about the sources, namely (i) distance, (ii) direction, (iii) speed, (iv) identity, (v) intentions. In order to appreciate the problem of source identification, we first look at familiar sources, and what sound signals they produce.

9.1.3 Sound Quality

Frequency, amplitude and duration are easily understood for a wave phenomenon such as sound; however, sound quality needs some further explanation. Sound may occur in different forms. *Tones* contain only single, or a few frequencies, *noise* is composed of many different frequencies, and *white noise* contains all frequencies.

A good starting point for learning about sound is music. A sheet of music is a physical record. It is an instruction to produce tones of a certain frequency for a particular length of time. Any sheet of music is a frequency – time graph.

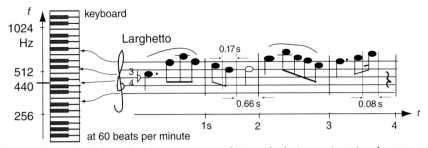

Fig. 9.3. The melody from the slow movement of Mozart's clarinet quintet is a frequency f versus time t trace of separate tones. Piano keyboard tuned to $f = 440$ Hz for A

Each tone represents a frequency, shown on the left of Fig. 9.3. The frequency increases in the y-direction. The time axis is in the x-direction. On either the treble or the base stave, the tones represent "fixed values" of frequency, Fig. 9.4a. The frequency scale is "quantized".

When recording animal voices scientists use a similar representation to display the sound frequency as function of time, Fig. 9.4b. However, instead of indicating the time value of a particular tone by flags with open or full heads, the x-axis is a real time axis: a long tone is drawn as a long line. Tones, which slide from one note to another like sounds of a Hawaiian guitar, can be easily shown as lines that slope and undulate through the f–t plane.

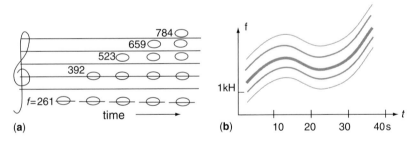

Fig. 9.4. (a) Frequencies of harmonics on the treble stave. (b) Fundamental and harmonics of the song of whales

Devices that generate sound always have a fundamental tone f_0, and often some harmonics with frequencies $f_i = n f_0$, where n is an integer. This rule holds for musical instruments and for voices of animals. To describe the characteristics of a musical instrument, one may graph the intensities I, of the simultaneously generated tones as a function of the frequency, see Fig. 9.5. Noise is a continuous band in such a representation. The tone A_4 has the frequency of 440 Hz. Alternately one can show the varying levels of intensity of the different harmonics by the strength of the lines in an f–t diagram.

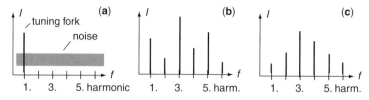

Fig. 9.5. Relative intensities. (**a**) Tuning fork and noise, (**b**) clarinet, (**c**) cornet

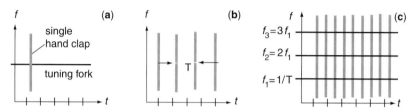

Fig. 9.6. (**a**) Single hand-clap. (**b**) Multiple hand claps. (**c**) Rapidly repeated hand-claps with period T are perceived as tones of frequency $f_1 = 1/T$ with the overtones $f_2 = 2f_1, f_3 = 3f_1$

Hand claps have a very wide frequency spectrum of short duration, see Fig. 9.6. Clicks that are repeated rapidly at time intervals T meld into tones at the frequency $f_1 = 1/T$ with overtones $2f_1, 3f_1$, Fig. 9.6c.

9.1.4 Fourier Analysis

Sound waves, like other waves, can be superimposed by adding one wave to another wave. When many waves with frequencies $f_1, f_2, ..., f_n$, and amplitudes $A_1, A_2, ..., A_n$ are superimposed a single wave is the result with a much more complicated wave form. The reverse process of superposition is the resolution of a complicated wave-form into the individual frequency components from which it is composed. This process can be carried out automatically. It is called *Fourier analysis*. In technical systems Fourier analysis can be done by analog circuits. The cochlea in the inner ears of man, and many animals, is a mechanical Fourier analyzer. Our frequency analyzing system allows us to hear the sounds of music. The relationship between frequency and intensity signals is illustrated by Fig. 9.7.

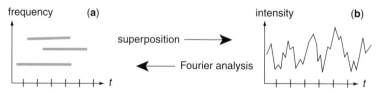

Fig. 9.7. When multiple frequencies (**a**) are superimposed a composite intensity – time waveform is generated. (**b**) Wave addition is called superposition. The reverse operation is the Fourier analysis

9.2 Intensity and Impedance

Sound waves are pressure waves. They transmit power (energy transmitted per unit time). Power density, or power per area is called *intensity I*. Land animals can perceive sound signals only if pressure waves penetrate from the air into the inner ear, a fluid environment. Unfortunately, when sound travels from air into a dense medium like water, or body tissue, most of the signal is reflected, and only a fraction (less than one thousandth) is transmitted. Yet, we can hear sounds, because the ear amplifies the signal by a factor approximately 10^4. In order to fully appreciate the amazing resolution, and sensitivity of ears, one must know several details about the mechanism of sound: how it relates to the motion of particles in the medium, and how the intensity depends on the *impedance*. Impedance is the product of density ρ and sound velocity v of a material.

9.2.1 Intensity and Particle Velocity

Consider a window of area $A = 1$ m² through which a wave front is traveling. The intensity I of the wave field at the window is equal to the energy per second, which travels through the window.

Draw a box of cross section area A and length $\Delta z = v \cdot 1$ s. It has the volume $V = A v \cdot 1$ s, and contains the total mass $M = \rho A v \cdot 1$ s. Consider the short moment when a wave train fills the box, like a freight train passing by the platform of a railway station. After the wave train has traveled through the region, the medium will be at rest again, and all the energy in the wave must have left through the window on the right hand side, Fig. 9.8.

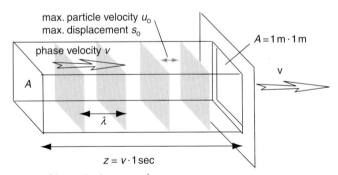

Fig. 9.8. Energy and intensity in a sound wave

The length of the box has been chosen so that a phase front of the wave can travel from the left side of the box to the right side during the time interval $\Delta t = 1$ s. Therefore, all the energy contained in the box passes through the window in one second. While the wave train is inside the box each volume element $dm = \rho dV$ has the total energy $dE = \frac{1}{2} dm \cdot u_0^2$, where the particle oscillation velocity

$u_0 = s_0 2\pi f = s_0 \omega$ is related to the displacement amplitude s_0, the wave frequency f, and the angular velocity ω of the associated simple harmonic motion. The total wave energy in the box is therefore

$$E = \sum dE = \frac{1}{2} M u_0^2 = \frac{1}{2} \rho A v \cdot 1 s \cdot u_0^2 = \frac{1}{2} \rho A v 1 s \cdot (s_0 \omega)^2 . \tag{9.1}$$

The intensity of the sound wave is obtained by dividing the energy by the area A and the time interval 1 second:

$$I = \frac{E}{A \cdot 1 \sec} = \frac{1}{2} \rho v (\omega s_0)^2 = \frac{1}{2} Z \cdot u_0^2 . \tag{9.2}$$

The intensity depends not only on the vibrations of the medium molecules (oscillation amplitude s_0, and angular frequency $\omega = 2\pi f$), but also on the properties of the medium (the density ρ, and velocity of sound v). The product $Z = \rho v$ is called the *impedance*. It is an important acoustic property of the medium.

9.2.2 Pressure, Impedance, and Velocity Fluctuations

The molecules in a medium only move if they have to, namely when a force is applied. The force is provided by pressure differences. The molecules oscillate because the pressure oscillates: when the speaker membrane in Fig. 9.2 moves forward the pressure in the adjacent air molecules is slightly increased to the value $p + \Delta p(t)$. The excess pressure forces the adjacent volume element to move in the $+z$ direction. When the membrane moves in the reverse direction, there will be a slightly lower pressure $p - \Delta p(t)$ which pulls the volume element into the opposite direction. The membrane creates an oscillating pressure field $\Delta p(t,z)$, which spreads out. The varying pressure in turn generates the oscillating displacement field $s(t,z)$ and velocity field $u(t,z)$, Fig. 9.9. Locations of maximum compression are separated in spatial direction z by the wavelength λ. The spacing of the pressure maxima can also be described as a spatial frequency $k = 2\pi/\lambda$.

Where the pressure is large, the displacement will start to build up. However, there is a delay: the displacement follows the pressure by $\frac{1}{4}$ of a period $T = 2\pi/\omega =$

Table 9.1. Impedance of air, water, and body tissue. Data adapted from [Lutz 99]. The mammalian cochlea impedance adapted from Hemilia et al. [1999]

material	density ρ	sound speed v	impedance $Z = \rho \cdot v$
air	1.29 kg/m^3	340 m/s	439 kg/m^{2}s
water	1000 kg/m^3	1496 m/s	$1.49 \cdot 10^6$ kg/m^{2}s
mammalian cochlea			$1.5 \cdot 10^5$ kg/m^{2}s
fat	940 kg/m^3	1476 m/s	$1.39 \cdot 10^6$ kg/m^{2}s
muscle	1058 kg/m^3	1568 m/s	$1.6 \cdot 10^6$ kg/m^2 s
bone	1785 kg/m^3	3360 m/s	$6 \cdot 10^6$ kg/m^2 s

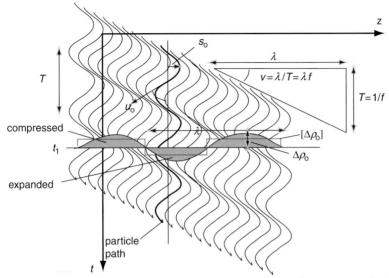

Fig. 9.9. z-t plane displacement field. Volume elements oscillate with maximum velocity u_0 and maximum displacement s_0 in a sound wave. Phase velocity $v = \lambda/T = \lambda f$

$1/f$. Therefore the phases of pressure and displacement differ by $T/4 = \omega\, 2\pi/(4\omega) = \pi/2$. By the use of the relation $\sin\{\phi - \pi/2\} = -\cos\phi$, one can express the time delay, and then describe the fluctuating displacement $s(t)$ of a volume element and its excess pressure $\Delta p(t)$ as function of time t and position z

$$s = s_0 \sin\{\omega t - kz\}, \text{ and } \Delta p = -\Delta p_0 \cos\{\omega t - kz\}. \tag{9.3}$$

This function solves the wave equation (7.5) when the phase velocity is given by $v = \omega/k = \lambda f$. Indeed, Fig. 9.9 shows that a particular phase, say the maximum compression, travels with a speed given by the slope $v = \lambda/T$.

In order to understand the action of sound waves in various media one must know how the local particle velocity in a sound wave u_0 is related to the three other parameters, the local pressure increase $\Delta p(t)$, the phase velocity of the disturbance v, and the local density ρ. These quantities are connected by a simple equation (9.4) that holds quite generally for the small pressure fluctuations in sound waves, as well as for the very large pressure steps across shock waves. (In the case of shocks Δp is the pressure jump across the shock front, v is the shock front velocity, and u_0 is the particle velocity behind the shock.)

$$\Delta p_0 = (\rho\, v)\, u_0 = Z \cdot u_0 \tag{9.4}$$

This relation looks similar to Ohm's law $V = R\, I_{el}$, which connects the current I_{el} to the driving voltage V. The flow of current, is impeded by the electrical resistance R. Similarly, the flow velocity of volume elements in a sound wave u_0 is driven by the excess pressure Δp_0 and it is impeded by the quantity $Z = \rho v$, which is therefore called the impedance. The impedance of air is $\rho v = 1.29 \text{ kg/m}^3 \cdot 340 \text{ m/s} = 439 \text{ kg/m}^2\text{s}$. Values for other materials are shown in Table 9.1.

Equation (9.4) follows from the definition of the phase velocity v of a compressible medium:

$$v^2 = (\partial p / \partial \rho)_S = (\omega / k)^2 = (\lambda f)^2 \tag{9.5}$$

derived in Sect. 7.3.2. The particle motion in a sound wave is so fast that no heat can flow out of the instantaneously compressed regions. Hence, we approximate $(\partial p / \partial \rho)_S \approx \Delta p_0 / \Delta \rho_0$, in order to relate the maximum pressure to sound speed and maximum density $\Delta p_0 \approx \Delta \rho_0 v^2$.

The maximum density $\Delta \rho_0$ in the sound wave is found by considering the motion in detail. Within the distance of one wavelength there is one region of compression, and one region of expansion, each covering a distanced of $\lambda/2$ in the z-direction. The higher compression arises because particles are pushed into the volume element from the rear, and from the front by the maximum displacement distance s_0. Therefore, the original volume $V = A \lambda/2$ of a particular volume element of lateral area A is changed by $-\Delta V = A\, 2 s_0$. The compression ratio is therefore $-\Delta V/V = 2 s_0/(\lambda/2) = 4 s_0/\lambda$.

The density increases at the same rate as the volume decreases $[\Delta\rho_0]/\rho = -\Delta V/V = 4 s_0/\lambda$. However, by this argument one only finds the average density increase $[\Delta\rho_0]$ in any given volume element. Figure 9.9 shows the instantaneous local density at t_1 as a shaded trace. Since the density varies smoothly, the maximum $\Delta\rho_0$ is actually higher than the average value $[\Delta\rho_0]$, just as the maximum value of a sine curve is higher than the average deflection by the factor $\pi/2$. Therefore, one has $\Delta\rho_0 = \rho\, 2\pi s_0/\lambda = \rho s_0\, 2\pi f/v = \rho s_0 \omega/v$. If the particles in a sound wave perform simple harmonic motion their maximum velocity $u_0 = s_0\omega$ is the product of maximum displacement s_0 and angular velocity ω, so that $\Delta p_0 = \rho u_0/v$. This relation can be combined with the definition of sound speed $\Delta p_0 \approx \Delta\rho_0 v^2$ to derive Eq. (9.4).

Further using Eqs. (9.2) and (9.4) one can relate Δp, u_0, and s_0 to the intensity $I = E/(A \cdot 1 \text{s}) = 0.5\,(\rho v)\,(\omega \cdot s_0)^2 = 0.5(\rho v)(u_{max})^2 = 0.5\,\Delta p_0 \cdot u_{max}$, which yields the particle displacement s_0, and particle velocity u_0 as function of intensity and impedance

$$s_0 = \frac{1}{2\pi f} \cdot \sqrt{\frac{2I}{Z}} \text{ , and } u_0 = 2\pi f \cdot s_0 = \sqrt{\frac{2I}{Z}} . \tag{9.6}$$

The local pressure fluctuation Δp can then be expressed as function of the particle velocity u_0

$$\Delta p_0 = \sqrt{2Z \cdot I} = Z \cdot u_0 . \tag{9.7}$$

Note that the particle velocity u_0 does not depend on the frequency. The intensity of the wave can be expressed as a function of the displacement amplitude s_0, the velocity amplitude u_0, or the pressure fluctuation amplitude Δp_0.

Example: A speaker generates a tone of 1200 Hz with a power of 3 W of sound energy in air. Take $v = 340$ m/s, $\rho_{air} = 1.29$ kg/m^3. What is a) the intensity at a distance of 200 m, b) the displacement amplitude of the air molecules, c) the pressure amplitude, (d) the particle velocity? **Answer:** Assume that the speaker emits sound energy uniformly into all directions. a) $I(r = 200$ m$) = 3$ W$/(4\pi \, 200^2)$ m$^2 = 5.9 \cdot 10^{-6}$ W/m^2, b) $s_0 = (\frac{1}{2} \pi f)(2I/v\rho)^{\frac{1}{2}} = (\frac{1}{2} \pi \, 1200)\{(2 \cdot 5.9 \cdot 10^{-6})/(340 \cdot 1.29)\}^{\frac{1}{2}} = 2.18 \cdot 10^{-8}$ m $= 0.022$ µm. c) $\Delta p_0 = \sqrt{(2\rho v I)} = \sqrt{(2 \cdot 1.29 \cdot 340 \cdot 5.9 \cdot 10^{-6})} = 0.0723$ N/m$^2 = 7.14 \cdot 10^{-7}$ atm, (d) $u_0 = s_0 \, 2\pi \cdot 1200 = 0.17$ mm/s.

9.2.3 The Decibel Scale

Sound levels β are generally measured with the decibel scale with units dB which is a relative measure of intensity defined as

$$\beta = 10 \cdot \log_{10}\left(\frac{I}{I_0}\right) dB. \tag{9.8}$$

The local intensity I is compared to some scale intensity I_0. In acoustics one uses the threshold intensity for human hearing, which has the incredible small value

$$I_0 = 10^{-12} \text{ W/m}^2. \tag{9.9}$$

An increase of intensity by a factor 10 means an increase in sound level by 10 dB. A thousand-fold increase in intensity equals 30 dB. From the threshold of hearing at about 0 dB to the threshold of pain at 120 dB the intensity varies by a factor 10^{12}. Very few technical detectors have a sensitivity range of 12 orders of magnitude. A few examples are given in Table 9.2. Note that the motion amplitudes s_0 of the faintest sound that one can hear are less than the diameter of a single atom.

In acoustic studies of aquatic animals two other reference scales are occasionally used, namely the acoustic reference pressures (a) measured in $\Delta p = 1$ µPa and (b) $\Delta p = 1$ µbar $= 0.1$ Pa.

To distinguish between these scales we assign the letters β_p and I_p to the sound level and associated reference intensity, respectively, in the µPa scale. Similarly label β_b and I_b as the sound level and reference intensity in the µbar scale, and call β_0 the sound level with the reference intensity $I_0 = 10^{-12}$ W/m^2. These sound level scales differ by constants $\Delta\beta$ that depend on the impedance of the medium, and the chosen reference intensity.

The conversion constants $\Delta\beta$ are found as follows. Expand the definition of β from Eq. (9.8) so that one can for instance connect the sound levels β_p measured in the µPa scale to the hearing threshold scale β_0 with its reference intensity $I_0 = 10^{-12}$ W/m^2, namely $\beta_p = 10 \log_{10}\{I/I_p\} = 10 \log_{10}\{(I/I_0)(I_0/I_p)\} = 10 \log_{10}\{(I/I_0)(I_0/I_p)\} = 10 \log_{10}(I/I_0) + 10 \log_{10}(I_0/I_p)$, or

$$\beta_p = \beta_0 + \Delta\beta_{op}, \tag{9.10}$$

Table 9.2. Intensities and amplitudes ($f = 1000$ Hz; 1 atm $= 1.103 \cdot 10^5$ N/m^2)

β [dB]	I [W/m^2]	Δp_0 [atm]	s_0
0 (barely audible)	10^{-12}	$2.8 \cdot 10^{-10}$	$1.1 \cdot 10^{-11}$ m
20 (whisper)	10^{-10}	$2.8 \cdot 10^{-9}$	$1.1 \cdot 10^{-10}$ m $= 0.11$ nm $\approx$ diameter of atom
65 (conversation, noisy classroom)	$3.2 \cdot 10^{-6}$	$5.2 \cdot 10^{-7}$	0.02 µm
80	10^{-4}	$2.8 \cdot 10^{-6}$	0.11 µm
120 (threshold of pain)	1 W/m^2	$2.8 \cdot 10^{-4}$ water column of $h \approx 3$ mm	11 µm
150 jet takeoff	1 kW/m^2	$2.8 \cdot 10^{-3}$	0.11 mm

The constant $\Delta\beta_{op} = 10 \log_{10}(I_0/I_p)$ is connected to the reference intensity $I_p = (\Delta p)^2/(2\rho v)$, where $\Delta p = 1$ µPa $= 10^{-6}$ Pa. The impedance ρv must be specified for the medium of interest. For a reference pressure $1 \mu P = 10^{-6}$ N/m^2 measured in water with $Z_w = (\rho v_w) = 1.5 \cdot 10^6$ one has $I_p = (\Delta p)^2/(2\rho v) = (10^{-6})^2/(2 \cdot 1.5 \cdot 10^6) = 3.33 \cdot 10^{-19}$. Hence $\Delta\beta_{op} = 10 \log_{10}(I_0/I_p) = 10 \log_{10}(10^{-12}/3.33 \cdot 10^{-19}) = 64.78$ dB. Similarly the hearing threshold scale can be converted into the µbar scale, where $\Delta\beta_{ob} = 10 \log_{10}(I_0/I_b)$, and $\Delta p = 1$ µbar $= 0.1$ Pa. Then the reference intensity in water is $I_p = (\Delta p)^2/(2\rho v) = (10^{-1})^2/(2 \cdot 1.5 \cdot 10^6) = 3.33 \cdot 10^{-9}$, so that $\Delta\beta_{ob} = 10 \log_{10}(I_0/I_p) = 10 \log_{10}(10^{-12}/3.33 \cdot 10^{-9}) = -35.2$ dB. The conversion of scales in water and air are shown in Table 9.3

Table 9.3. Conversion of sound levels β in water and air measured in different scales

medium	Z	I_0 to µP scale	I_0 to µbar scale	µbar to µPa scale
water	$1.5 \cdot 10^6$	$\beta_p = \beta_0 + 64.8$ dB	$\beta_p = \beta_0 - 35.2$ dB	$\beta_p = \beta_b + 100$
air	439	$\beta_p = \beta_0 + 28.6$ dB	$\beta_p = \beta_0 - 71.4$ dB	$\beta_p = \beta_b + 100$

9.2.4 Beats

When two sound waves with slightly different frequencies f_1 and f_2 are added or "superimposed" the resulting vibration is modulated at the beat frequency $\Delta f = f_1 - f_2$. The intensity (amplitude) modulation is quite noticeable, and it can be used to determine the frequency f_2 if f_1 is known. This effect is utilized by bats, to detect the frequency of the *Doppler shifted* return signal f_2 coming back from a moving target, such as a flying insect. It will be shown later that the speed u of an approaching object can be obtained from a relation $\Delta f/f_1 = 2u/v$ where v is the speed of sound, and f_1 is the carrier frequency of the sonar pulse emitted by the bat. It is likely that aquatic mammals also use this Doppler effect to detect the speed of their prey.

Imagine two tuning forks being struck at the same time. An interesting effect occurs if the frequencies of both waves are just a bit out of tune. Consider two tuning forks which are struck simultaneously with the frequency $f_1 = 440$ Hz, and

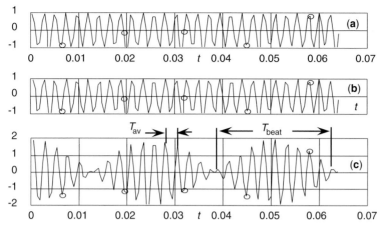

Fig. 9.10. Supereposition of two wave to form beats. (a) $y_1 = \sin(2\pi\ 440\,t)$, (b) $y_2 = \sin(2\pi\ 398\,t)$, (c) Superposition of waves $y = y_1 + y_2$

$f_2 = 443$ Hz (corresponding to the angular frequencies $\omega_1 = 2\pi f_1 = 2{,}765$ radians/s, and $\omega_2 = 2\pi f_2 = 2{,}783$ radians/s). The resultant sound fluctuates in intensity, as if the two tuning forks were continuously turned on and of. The oscillations are individually described by $f_1 = s_0 \sin 2\pi f_1 t$, and $f_2 = s_0 \sin 2\pi f_2 t$. Here it is assumed that both sound signal have the oscillation amplitude s_0. The addition of two sine waves $s = s_1 + s_2 = s_1 \sin 2\pi\ f_1 t + s_2 \sin 2\pi f_2 t = 2\ s_0 \sin\{2\pi\ t(f_1 + f_2)/2\} \cdot \cos\{\ 2\pi t(f_1 - f_2)/2\}$ leads to the product of a sine and a cosine function. The first is called the "carrier wave". It has the frequency $f_{av} = (f_1 + f_2)/2$. The second factor represents the modulation, which causes the amplitude of the carrier wave to oscillate between 0 and $2s_0$. The resultant sound wave periodically changes its amplitude. The tone therefore vibrates in intensity at the beat frequency f_{beat}, and beat period T_b

$$f_{beat} = f_1 - f_2,\ \text{and}\ T_b = 1/(f_1 - f_2)\,. \tag{9.11}$$

The superposition of two waves Figs. 9.10a and b are shown as a graphical illustration. The frequencies are $f_1 = 440$ Hz and $f_2 = 398$ Hz. The oscillations are described by the functions $y_1 = \sin\{440 \cdot 2\pi\ t\}$, and $y_2 = \sin\{398 \cdot 2\pi\ t\}$. The addition of these two waves yields the modulated wave Fig. 9.10c with a beat period $T_{beat} = 1/(440 - 398) = 23.8$ ms. The carrier wave has the frequency $f_{av} = (440 + 398)/2 = 419$ Hz, and the period $T_{av} = 2.39$ ms. Actually the frequency resolution of the human ear is fine enough to distinctly separate these two tones.

9.2.5 Sound Absorption, Scattering and Refraction in Free Space

Ideally a sound wave does not permanently affect the material through which it travels, or conversely the wave does not lose any energy as it travels through space. However, sound can be absorbed, scattered, or refracted by (small) objects wherever sound travels.

Absorption takes energy and momentum away from the sound wave and imparts it to the transmitting medium: the medium gets hotter and it is "pushed" by the sound wave.

- Scattering is reflection of small parts of the wave field in random directions.
- Refraction changes the direction of propagation of a sound wave.

Here we consider a plane wave, impinging onto a surface area A. Due to the (partial) absorption the intensity drops exponentially inside the material according to

$$I(r) = I_0 e^{-\kappa x} \tag{9.12}$$

where I_0 is the intensity at the source, x is the distance to the source, and κ is the absorption coefficient, or extinction coefficient. The total intensity ΔI lost from the sound wave and absorbed in a slab between the front surface at R and the depth $x = d$ is $\Delta I = I_0 - I(r) = I_0(1 - e^{-\kappa d})$. During the "exposure time" Δt the absorbed intensity feeds energy $\Delta E = A\, \Delta I\, \Delta t$ into the slab.

The inverse of the extinction coefficient κ, called the range $R = 1/\kappa$, is a measure for the distance over which the sound intensity is reduced by a factor $1/e = 37\%$. The absorption coefficient and hence the range of sound waves depends on their frequency, see Fig. 7.8.

The absorbed intensity ΔI may also be expressed as sound level difference $\Delta \beta$ measured in dB (decibel). Section 9.2.3. If the incident intensity is β_0, and the transmitted intensity β_1 then the absorbed fraction in decibel is

$$\Delta \beta = \beta_1 - \beta_2 = 10 \log\{I_1/I_{ref}\} - I_0 \log\{I_2/I_0\} = 10 \log(I_1/I_2) \tag{9.13}$$

and

$$\Delta I = I_0 - I(r) = I_0 (1 - e^{-\kappa d}). \tag{9.14}$$

The relation is useful because sound absorption in water is quoted in $\Delta \beta / \Delta x$, where $\Delta x = 100$ m. This measure of absorption can be converted into an absorption coefficient κ by the relation

$$\kappa = (1/\Delta x) \ln(10^{\Delta\beta/10}). \tag{9.15}$$

Without using this formula one can go through the steps one by one. Consider the conversion of $\Delta\beta = 1$ dB per 100 m into the corresponding κ-value. -1 dB $= 10$ $\log(I_1/I_0)$, or $-1/10 = \log(I_1/I_0)$. Raise this relation to the power ten: The left side yields $10^{-1/10} = 0.7943$, and right side $10 \log(I_1/I_0)$. Hence, one has $0.7943 = I_1/I_0$. By definition of κ one has $I_1/I_0 = e^{-\kappa\Delta x}$, where $\Delta x = 100$ m. Now solve for κ. One finds $0.7943 = I_1/I_0 = e^{-\kappa\,\Delta x} = e^{-\kappa\,100}$. Take the natural logarithm on both sides to get $\ln 0.7943 = -0.230$, and $\ln e^{-100\kappa} = -100\,\kappa$. Finally $\kappa = 2.3 \cdot 10^{-3} \mathrm{m}^{-1}$.

Values for the absorption coefficients for sound in air and sea-water in dB/100 m are given in Table 7.6, and shown in Fig. 7.8. Sound is less attenuated in water than in air by more than 2 orders of magnitude. There is so little absorption, particularly at low frequencies, that whales could talk to each other over distances from Hawaii to Alaska before man filled the waters with noises of propellers and exhaust gases from outboard engines.

9.2.6 Impedance Mismatch Between Air and Water

Ears are devices, which absorb energy out of the surrounding sound wave fields and extract information about the location of the source, the frequency, and the intensity. There is a problem since the sound comes through the air but must be analyzed inside the body, which has essentially the acoustic properties of water. It is a problem of impedance missmatch.

The acoustic impedances $Z = \rho v$ of air and water are very different. Whenever there is a large difference between the impedances of two media, through which the sound travels, very little sound energy can be transmitted from one to the other. The problem becomes apparent if one looks in detail at the sound intensity, as function of particle velocity u_o frequency f and pressure amplitude Δp_o as derived with (9.6) and (9.7):

$$I = \frac{1}{2}Z\left(2\pi f s_o\right)^2 = \frac{1}{2}Zu_o^2 = \frac{1}{2}\Delta p_o u_o .$$

(9.16)

Typical values for the impedance $Z = \rho v$ are shown in Table 9.1. Note that the impedance of the cochlea is smaller than that of water by about a factor of 10, because the cochlea is enclosed by a soft window that borders onto the air in the middle ear. For a given sound of frequency f and intensity I the product of impedance and the square of the particle displacement is a constant.

When the sound wave travels through a medium like air with small impedance the displacement s_o will be large. When the same wave travels through water, which has a very large impedance, the displacement amplitude and the maximum velocity will be small.

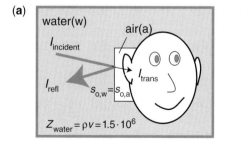
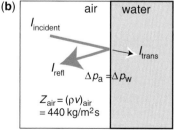

Fig. 9.11. Transmission and reflection of intensity of transmitted wave at an interface air/water. (**a**) Displacement s_o is transmitted. (**b**) Pressure Δp is transmitted

Only very little sound energy is transmitted across boundaries between media of different impedances. Regardless if the sound travels into, or out of the denser medium, the ratio of transmitted to incident sound intensity is equal to the ratios of the smaller to the larger impedance. This result will now be derived in detail.

The transmission of sound from the dense to the dilute medium is governed by the displacement s_0 of the interface. In the reverse case, where sound is transmitted from the dilute to the dense medium, the oscillating pressure exerted by the dilute medium at the interface is the important parameter. Suppose a sound wave traveling through water arrives at an interface with air on the other side. The amplitude of motion in the water is given by (9.6), namely

$$\left(s_0^2\right)_{water} = \left(\frac{2I}{\omega^2 \cdot \rho \cdot v}\right)_{water}.$$

If the free water surface acts like a speaker membrane it will excite a sound wave in the adjacent air with the same frequency and the same displacement amplitude $f s_{0,a} = f s_{0,w}$. Then, by the use of (9.6), the ratios of intensity and impedance must be the same in air and in water.

$$\left(\frac{I}{Z}\right)_a = \left(\frac{I}{Z}\right)_w$$

Now introducing the sound transmission coefficient τ

$$\tau = Z_a/Z_w = (\rho v)_a/(\rho v)_w \tag{9.17}$$

one can write in general

$$I_{a,trans} = \frac{Z_a}{Z_w} I_w = \tau \cdot I_w = \frac{I_w}{3416}. \tag{9.18}$$

Since only the small amount of intensity I_{air} is actually found in the air behind the interface, the rest of the energy must have been reflected. This is indicated in Fig. 9.11a. One can easily calculate the magnitude of the reflected wave. By the conservation of energy the wave must either be transmitted or reflected. $I_w(incident) = I_a(transmitted) + I_w(reflected)$, and with the help of Eq. (9.18) one has:

$$I_w\,(reflected) = I_w\,(incident)\left(1 - \frac{1}{3416}\right) = 0.9997 \cdot I_w\,(incident). \tag{9.19}$$

This calculation shows that 99.97 % of the incident wave intensity is reflected.

It is also instructive to compare the *amplitudes* $s_{0,a}$, and $s_{0,w}$ of two sound waves of the same frequency and intensity in air and water. By writing (9.17) for s_0 for the media water, and air separately, and then taking the ratio of both relations one finds:

$$\frac{\left(s_0^2\right)_a}{\left(s_0^2\right)_w} = \frac{Z_a}{Z_w} = \frac{1}{\tau} = \frac{1.5 \cdot 10^6}{439} = 3416 \quad \text{or} \quad \frac{s_{o.a}}{s_{o.w}} = 58 .$$

Similarly unfavorable is the energy transfer by a sound wave traveling from air into water, Fig. 9.11b. Due to its much larger density the water does not follow the displacement of the air molecules but rather responds to the pressure fluctuations. The pressure must be the same on both sides of the interface. With the help of (9.16) the pressure fluctuations at the interface can be related to the intensity $\Delta p_{o,air} = \sqrt{\{2Z_a I_a\}}$. These pressure fluctuations will cause identical pressure fluctuations in the water, namely $\Delta p_{o,ai} = \Delta p_{o,water} = \sqrt{\{2Z_{wr} I_w\}}$, where I_w is unknown. Therefore:

$$\sqrt{\left(2\rho \cdot v \cdot I\right)_w} = \sqrt{\left(2\rho \cdot v \cdot I\right)_a} . \tag{9.20}$$

This equation is solved to find the intensity transmitted into the water I_w:

$$I_{w,trans} = I_a \frac{Z_a}{Z_w} = \frac{I_a}{3416} = 2.9 \cdot 10^{-4} \cdot I_a . \tag{9.21}$$

The fraction of the intensity, which is not entering the water, namely $(1 - 2.9 \cdot 10^{-4}) I_a$ will be reflected. The interface between water and air acts as a very good mirror for sound waves. This poses a problem for the hearing of land animals, because the inner ear is essentially a watery substance, and the outside is air. Obviously animals had to come up with a very effective impedance matching design.

There is a surprising general result of these transmission calculations. No matter whether the sound is sent from air to water, (9.21) or from water to air, (9.18), only the tiny fraction $(\rho v)_a / (\rho v)_w$ of intensity is transmitted.

Sound, like light and other wave phenomena, can be concentrated by *acoustic lenses*. An example is the acoustic lens or *melon* in a Dophins's head. Snell's law (7.18) governs the ray propagation through acoustic lenses.

9.2.7 Hearing and Voice Transmission in Air

Ears need detectors for a sufficient frequency range, and voices must have generators to produce the sound. All body parts of an animal are condensed matter with impedance similar to water. Due to the difference in impedance the energy content of a sound wave in air is very much smaller than the energy content of a sound wave in water or other condensed matter. Therefore, sound must be amplified for transmission through air and concentrated in the absorption process.

Most animals produce multi-frequency sounds, which are amplified before being beamed away by a spreader/speaker, Fig. 9.12. The sound generated in a body is often spread out and beamed away by vibrating surfaces that are acoustically connected to the generator (oscillator)-amplifier. Diffraction helps with the initial spreading, as long as the wavelengths are of the same order of magnitude as the speaker diameter D. Signals are then sent in all directions by the spherical diverging waves. All frequencies are transmitted simultaneously through the air. At any point in space the sound signal is a single vibration - time signature, containing all the different frequencies, similar to the schematic trace shown in Fig. 9.7b. The intensity of this sound wave decreases as it spreads out from the sender.

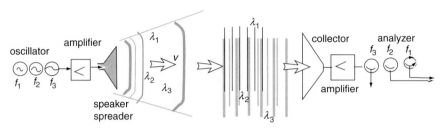

Fig. 9.12. Examples of an emitter and a receiver of sound

The sound wave of intensity I_{inc} incident on an ear of collector area A contains the power $P = A\, I_{inc}$. This power is concentrated by the ear system onto a tiny analyzer area. Elements of the ear are the collector, concentrator, amplifier and the frequency analyzer.

An important function of the collector/receiver is to pick up the tiny vibrations of the incident sound wave. Since the energy density of the sound in air is very low the receiving element should have a large area and a very low mass. Details of ears are discussed in the next section.

9.3 Ears

The environment is full of acoustic signals from a multitude of sources. These sounds contain much information. Ears convert these faint oscillations of the air molecules into nerve signals.

The principal sound reception element of an ear is a structure that vibrates precisely in phase with the incident sound wave. In order to create a detectable signal this structure must meet three conditions. First, it must be free to move relative to the body. Second its impedance must differ from the impedance of the surrounding air, and third it must have a very low-mass, so that it can follow the minute sound vibrations in the air.

Body, brain, and inner ear have the density and impedance of water, whereas the medium, which carries the sound to terrestrial animals is air. Sound amplitudes in air are very much larger than in the water, see Sect. 9.2.6. However, the acoustic energy density in air is much smaller than in water, because the acoustic impedances of air and water differ by about a factor 1000. Therefore, the sound signals must be amplified. Unfortunately in a sound wave all the different frequencies are superimposed, like an orchestra, or like the voices of traders in the stock exchange. In order to take full advantage of this jumble of frequencies the inner ear contains a very precise analyzer that Fourier-decomposes the sound into its various frequencies.

Mozart once listened to a new piece of written by an another composer. Afterwards he sat down and wrote out every note, of all the instruments. He knew what he heard. The ear of a mammal does this all the time. One recognizes different voices, noises, sound sources, and melodies without actually being conscious of the entire score sheet.

9.3.1 Principles of Amplification in Mammal Ears

The principal reception element of a mammal ear is the ear drum, a thin membrane that vibrates in phase with the incident sound wave. The drum is connected to the middle ear, which converts the *soft* vibrations of the air (high amplitude, low intensity) into hard vibrations of the oval window of the cochlea. This impedance matching also involves an amplification of the acoustic signals by about a factor 10^3.

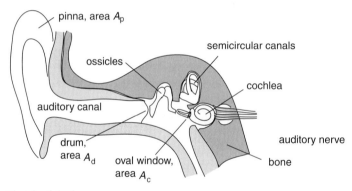

Fig. 9.13. Sketch of the human ear

The amplification proceeds in 3 steps. First the sound is collected by the pinna, and amplified by channeling the signal onto the much smaller drum, area A_D. Second, the force acting onto the ear drum is conducted, and amplified mechanically through the linkage bones of the middle ear, which are connected to the oval window, area A_c. Finally the force acting onto the oval window is magnified by area reduction $A_d \rightarrow A_c$.

The sound then excites waves inside the cochlea, which travel only up to a certain distance, according to their frequency: the lower their frequency f the farther they travel. They are absorbed at a place where they excite a standing wave in resonance. The standing wave shakes a bundle of stereocilias, which are attached to a nerve cell. The nerve cell changes its voltage when the cilia are moved and sends an impulse towards the brain. These principles are illustrated with the example of the human ear, Fig. 9.13.

9.3.2 Pinna and Middle Ear Amplification

The human ear has three essential components: (1) The outer ear with the pinna of area A_p, collects and somewhat concentrates acoustic power $P = I_o A_p$. (2) The middle ear converts the acoustic pressure into a force, which is amplified by the middle ear ossicles, three small bones that form a lever system with a mechanical advantage of about a factor 1.5. The ossicles are attached to the oval window. This membrane is part of the inner ear, a snail shaped organ, called the cochlea. (3) Fluid in the cochlea vibrates when set in motion by vibrations of the oval window. Pressure wave components of various frequencies propagate through the *spiral staircase tract* of the cochlea, and terminate in standing waves at specific locations in the cochlea. The lowest sound the frequency reach the farthest into the cochlea. The standing waves agitate hair bundles that transmit nerve signals. These nerve signals are associated with the perception of specific sound frequencies.

Suppose the outer ear is exposed to sound of the outside intensity I_o. The eardrum is the receiving low-mass element, which vibrates in the rhythm of the incident sound wave. The pinna, area A_p, concentrates the collected incident acoustic power $P_o = A_p I_o$ onto the eardrum of surface area A_d. Consider the acoustic power P_d received by the eardrum. Ideally $P_d = I_d A_d$ is equal to the power $P_o = A_p I_o$ intercepted by the pinna $I_d A_d = P_d = A_p I_o$. Then the intensity at the eardrum is:

$$I_d = \frac{P_d}{A_d} = I_o \cdot \frac{A_p}{A_d} .$$

(9.22)

The acoustic power is related to the pressure fluctuations of amplitude $\Delta p_d = \sqrt{(2 Z_a I_d)}$. The quantity $Z_a = (\rho v)_a$ is the impedance of air in the outer ear. Enter I_d from (9.22) to get

$$\Delta p_d = \sqrt{2 Z_a \frac{A_p}{A_d} I_o} .$$

(9.23)

The middle ear provides amplification and protection. Amplification is accomplished by three small bones acting as levers with the moment arms l_d and l_c re-

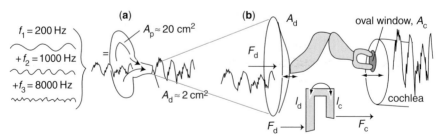

Fig. 9.14. (a) Outer ear amplification. **(b)** Middle ear amplification

spectively that connect the drum mechanically to the cochlea. The drum exerts the force F_d onto the lever system so that a larger force F_c is acting onto the oval window. This force can be easily found from the moment ballance $F_d l_d = F_c l_c$, thus $F_c = F_d l_d / l_c$.

The acoustic pressures acting at drum and the cochlea window $\Delta p_d = F_d / A_d$ and $\Delta p_c = F_c / A_c$ respectively. Since the force is amplified through the ossicles one has

$$\Delta p_c = \frac{F_c}{A_c} = \frac{F_d \cdot l_d}{l_c \cdot A_c} = \Delta p_d \frac{A_d \cdot l_d}{A_c \cdot l_c} \; .$$

Here one can substitute Δp_d using Eq. (9.23) to obtain

$$\Delta p_c = \Delta p_d \cdot \frac{A_d \cdot l_d}{A_c \cdot l_c} = \frac{A_d \cdot l_d}{A_c \cdot l_c} \cdot \sqrt{2 Z_a \frac{A_p}{A_d} \cdot I_0} \; . \tag{9.24}$$

With the knowledge of the acoustic pressure at the cochlea one can calculate the acoustic intensity in general terms: $I_c = \Delta p_c^2 / (2 Z_c)$. The cochlea essentially contains water, however, the oval window provides a somewhat soft containment. Therefore, the effective internal impedance of the cochlea Z_c is smaller than the impedance of water, see Table 9.3. When Δp_c is substituted from (9.24) one has

$$I_c = \frac{Z_a}{2 Z_c} \left(\frac{A_d \cdot l_d}{A_c \cdot l_c} \right)^2 \cdot \frac{A_p}{A_d} \cdot I_0 \geq \frac{Z_a}{2 Z_w} \left(\frac{A_d \cdot l_d}{A_c \cdot l_c} \right)^2 \cdot \frac{A_p}{A_d} \cdot I_0 \; . \tag{9.25}$$

In this equation the middle ear levers contribute the factor $l_d / l_c \approx 1.5$. The area concentration between drum and cochlea bring the factor $A_p / A_c \approx 10$, and the pressure amplitude amplification factor A_d / A_c furnishes the factor 30, leading to an overall amplification, so that the total intensity amplification is of the order of 10^4. This is good enough to compensate for the impedance miss match between air and inner ear so that the faint pressure fluctuations in the air become audible.

The middle ear has another significant function: protection. A small muscle inside the inner ear makes, attenuates, or breaks the mechanical contact between outer and inner ear, thereby controlling the intensity that is actually transmitted to the cochlea. For instance, the cry of an infant would generate an intensity equivalent to about 135 dB at the inner ear. Children would be stone deaf from the sound of their own voices before they were ever able to speak. The muscle disengages the mechanical link to the cochlea, just before the voice starts, so the ear does not hear the own screams.

9.3.3 Inner Ear Frequency Analysis

Sound waves are launched inside the cochlea from the oval window towards the end of the cochlea. The winding stairs of the cochlea channel get gradually narrower away from the oval window. This yields an *acoustic beach*, in which signals of a particular frequency can only run to a particular position. Unable to propagate any further as traveling waves, they turn into standing waves, and bend the stereo cilia (hair cell bundles) that are part of the auditory cells. Thus each location along the cochlea channel is dedicated to a particular frequency within the hearing range of about 9 octaves from $20\,\mathrm{Hz} \leq f \leq 18\,000\,\mathrm{Hz}$. At a channel length of $L \approx 35$ mm, each octave occupies about a 4 mm long section of cochlea channel.

Figure 9.15 shows a model of the cochlea, unrolled, with hair cell bundles and nerves leading to the brain. Each nerve serves one sensory cell with its hair bundle associated to one particular frequency.

Suppose a sound wave carries the frequencies $f_1 = 200$ Hz, $f_2 = 1000$ Hz, and $f_3 = 8000$ Hz. Each hair cell is connected to its own nerve that goes directly into the brain. An electrical impulse on each nerve is perceived as a particular tone. Recently it has been suggested [Frizsch and Manley 2001] that the hair cells vibrating at the frequency of the standing sound wave actively amplify this motion. This effect would increase the sensitivity, and presumably produce sound itself, called otoacoustic emission.

The frequency analysis in the cochlea may be modeled as an inverse grand piano with fixed linkages between keys and strings. If a piano is placed into a very strong sound field containing let say the frequencies 8000 Hz, 1000 Hz, and 200 Hz, the strings that can create these tones will start to vibrate in resonance. Then the keys associated with these three tones appear to be depressed. They could act as switches each connected to an individual wire to a central computer. Since each frequency has its own wire to the CPU (the brain) a low voltage DC sig-

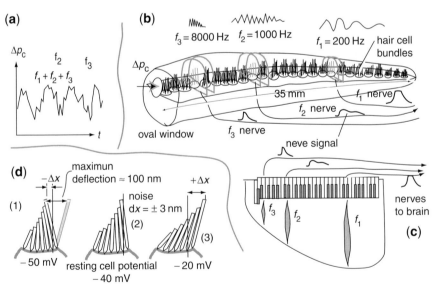

Fig. 9.15. (a) Sound signal with frequencies f_1, f_2, and f_3. (b) Cochlea, unrolled. (c) Cochlea compared to a piano. (d) Stereo cilia bundles (1) with negative displacement, (2) rest position, and (3) with positive displacement

nal could announce the presence of a particular tone. While the pressure signal from the 8 000 Hz frequency oscillates 8 000 times per second the DC signal needs no modulation at all, and yet it would indicate a tone of 8 000 Hz. Such frequency conversion is needed because the nerves can only transmit signals that are modulated at frequencies of less than about 100 Hz.

Electrical signals are only produced when the stereo cilia tips slide relative to each other [Hudspeth and Markin 1994]. When there is a high sound level they come to a halt in some bent position, and will move further only if there is an additional increase of the sound level. In this way the ear can adjust to different base levels of noise: after a while the ear gets adjusted to the noise in a deafening environment. The maximum deflection of the stereo cilia is $\Delta x \approx 100$ nm. When there is no vibration in the fluid near the stereo cilia tips, they occupy an intermediate position and the nerve cell is at the resting potential $V_0 = -40$ mV. When sound is present the tips can have a negative displacement where the nerve cell potential is at $V = -50$ mV or they can have a positive displacement Δx, where the potential might go up to $V = -20$ mV.

The tips are elastic with a spring constant of $k \approx 600$ µN/m. At the maximum displacement the tips would be bent by the force $F = k \cdot \Delta x = 600[\mu N/m] \cdot 100$ nm = $60 \cdot 10^{-12}$ N. This force is about 30 times larger than the force exerted by a myosin head in muscle contraction. The force associated with bending the cilia at the threshold of hearing is $\delta F = k \, \delta x = 3$ nm $\cdot 600$ µN/m = $1.8 \cdot 10^{-12}$ N. Hearing is limited by thermal noise (Brownian motion), which makes the stereo cilia heads dance back and forth with an amplitude of about $\delta x = 2$ nm. Displacements of more than 3nm are perceived as sound. Thus shielding the brain from perceiving Brownian motion as noise.

If there is too much noise for a sufficiently long time the stereo cilia break off. They will never be replaced: the person becomes deaf at this frequency. The position of $f \approx 4000$ Hz auditory cells is about ½ turn away from the oval window (on the opposite side of the snail shaped organ). There appears to be another oscillation mode within the cochlea, which directs acoustic power from all frequencies to the 4000 Hz location. This spot can be overloaded by any noise, and is most easily damaged. Therefore, people first lose the ability to perceive in the 4000 Hz range.

9.3.4 The Ear of an Aquatic Mammal

For the aquatic mammal, like a bottle nose dolphin, or a whale, hearing is easier than for a land animal: the sound signals from the environment already travel through water which has a similar impedance as the inner ear of the animal.

Yet the impedance of the mammalian cochlea $Z_{mam} = (\rho v)_{mam} \approx 1.50 \cdot 10^5$ kg/(m²s) is about ten times smaller than the acoustic impedance $Z_{water} \approx 1.5 \cdot 10^6$ kg/(m²s) of water [Hemilia et al. 1999]. Therefore, the acoustic signal must be amplified. The impedance difference is unexpected since the cochlea is filled with water.

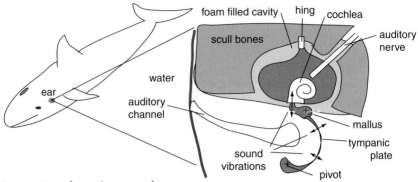

Fig. 9.16. Ear of aquatic mammal

However, the fluid in the cochlea is contained by the oval window, which has air on the outside for terrestrial animals and soft foam for aquatic animals.

The receiving element of the aquatic mammal ear is the tympanic plate, see Fig. 9.16, a thin bony membrane made from high density bone material $\rho \approx 2.7$ g/cm³. For terrestrial mammals the malleus has densities of only $\rho \approx 2.2$ g/cm³. The tympanic plate has a typical area $A_t = 350$ mm². It pivots about the heavy end and is connected to the malleus. Through the rotation about the pivot the amplitude of the vibration increases with distance r from the pivot axis. Further amplification of the amplitude is achieved by the lever system of the malleus.

Acoustical signals can only be detected if the receiving element vibrates relative to the entire ear structure. Since the aquatic mammal is submerged in water the sound shakes the whole animal. Therefore, the inner ear is acoustically separated. It is housed in a bony structure of heavy mass (tympano petromastoids) that sits in a foam filled cavity. An elastic link connects the inner ear to the animal body. This structure acts somewhat like a seismograph. The inertia of the heavy mass keeps the ear from following the acoustic vibrations of the water. Acoustic signals are carried in through the auditory channel, a narrow tube, which acts as a wave-guide. Due to the elastic mount of the inner ear even low frequency oscillations of the whole animal are not transmitted to the ear and the brain.

9.3.5 Lateral Lines of Fish

The lateral line of a fish is a row of small holes running along the length of the body. Hair cells located behind the holes respond to pressure vibrations in the surrounding water. A pressure wave that arrives from the front will activate first the frontal elements of the lateral line. Pressures from the left will be perceived predominantly by the left side, similar to the responses of human ears.

When a fish moves, it generates motion (boundary layer flow, jets, and eddies) in the adjacent water. Such movement causes low-pressure sound wave signals that can be picked up by the lateral lines of nearby fish to adjust their own motion.

Such low frequency sound wave communication would then explain the choreographed motion of schools of little fish who twist and turn and never touch their nearest neighbor: presumably the lateral lines constantly interpret the local pressure field to instantly know where the school mates are.

It is therefore quite likely that the lateral lines have the same functions for the fish as ears have for mammals. Stretched out over the whole length of the body, lateral lines would achieve a similar resolution in the aqueous medium (where sound propagates about 5 times as fast as in air) as the ears of land animals, which are separated only by the width of the body.

9.3.6 The Sensitivity of Ears

The basic sensors in the ear of humans are the stereocilia, fine fibers, which are bent by the action of sound waves in the cochlea. At the threshold of sound perception they are displaced by about $0.003\mu m$. The maximum displacement is about $0.1\ \mu m$.

The intensity range of the human ear, Fig. 9.17, stretches over about 12 orders of magnitude. At the threshold of hearing, $\beta_{noise} = 0$ dB for a frequency of 1000 Hz the air molecules in the sound wave oscillate with the minute displacement of $s_o = 10^{-11}$ m. This oscillation amplitude of about 1/10 of the diameter of an atom, is just above the Brownian noise of the air molecules, so that the ear is not affected by the ambivalent thermal noise.

The upper intensity limit is the threshold of pain at $\beta_{pain} \approx 120$ dB. This intensity (equivalent to 1 W/m²) is quite small compared to the radiation intensity of the sun of $I_{sun} \approx 1.3$ kW/m². The sensitivity of the human ear depends on the frequency. The audible frequency range of the human ear stretches from about 30 Hz to 15 kHz. Whales can perceive much higher frequencies.

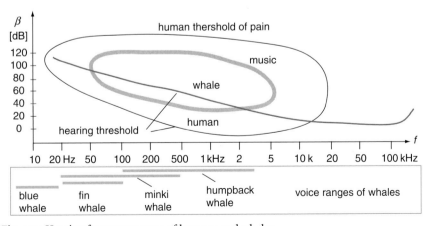

Fig. 9.17. Hearing frequency range of humans and whales

Could one literary hear the grass grow? Note that the intensity does not depend on the frequency, but only on the particle speed. Grass can grow at a speed $u_g \approx 1$ cm/day $= 0.01$ m/10^5 s , or $u_g = 10^{-7}$ m/s. From (8.7) one has:

$$u_0 = \omega \cdot s_0 = \sqrt{\frac{2I}{Z}} , \text{ or } I = \frac{1}{2} Z u_0^2 . \tag{9.26}$$

If the air molecules in a sound wave oscillate with the speed u_g, the sound wave has the intensity $I = (439/2) \cdot (10^{-7})^2 = 2.2 \cdot 10^{-12}$ W/m^2. This intensity corresponds to 3.4 dB. Such low intensities could be just barely perceived. Expressed differently if the tips of the blades of grass where to push against the tips of the stereo cilia for 1/10 of a second they would have grown by 10^{-8} m. This is well above the noise detection threshold of $3 \cdot 10^{-9}$ m of the stereo cilia.

Example: Suppose you hear a faint noise of a quiet conversation in a room with little background noise, say a sound level of 38 dB, and you listen to it for 12 s. How much energy has your ear received?

$\beta = 38 = 10 \log(I[W]/10^{-12}[W/m^2])$, $I/10^{-12}$ W $= 10^{3.8} = 6.31 \cdot 10^3$W/m^2, or $I = 6.31 \cdot 10^{-9}$W/m^2. With a typical pinna area of $A = 0.005$ m^2 each ear will intercept the power $P = IA = 0.005$ m$^2 \cdot 6.31 \cdot 10^{-9}$ W/m$^2 = 3.16 \cdot 10^{-11}$ W. If the conversation lasts for 12 s each ear has intercepted the energy $E = 12$ s $\cdot 3.16 \cdot 10^{-11}$ J/s $= 3.8 \cdot 10^{-10}$ J. Not much! Information is energy-cheap.

9.4 Voices and Sound Production

There are many ways to produce acoustic signals, and it does not take much energy to actively produce sound. With ears ready to listen, and voices able to produce signals, the stage is set for the active use of sound: (i) as language for communication with members of the own species, or as warning for would be predators, and (ii) as sonar beams for scanning the environment in situations where light fails.

Sound is always produced when things move. Air, water, or solid volume elements are in oscillatory motion when a sound wave propagates through these media. This motion starts at an oscillator, and is generally amplified by the voice system. The sound is then focused, or spread out by another part of the vocal tract, acting as the speaker surface, see Fig. 9.12.

Pure tones require a fairly regular motion. There are three principal methods to generate regular motion. (1) Elastic solids, like strings or leaf springs, and vocal cords, may vibrate in simple harmonic motion. (2) Air columns in flutes, or mouth cavities can oscillate like standing waves. (3) Air streams that rush through narrow elastic passages produce interrupted flow due to the Bernoulli effect, coupled with vortex shedding. Examples are wind rushing through venetian blinds near an open window, and breath squeezing through the vocal cords. All these methods are used by animals in the multitude of sound-producing organs.

9.4.1 How Sound Is Shed off a Sender

Any object that moves rapidly in air or in water will make some noise. Such objects may be the vocal cords that flutter in the air expelled by the lungs. They could be the wing tips of a Canada goose swooshing overhead. They could be the ridges on the legs of a beetle, issuing sound when stroked, or they could be branches breaking under the foot of a deer that ambles through the forest.

To understand how sound waves are initiated consider the membrane of a speaker, which produces the tone A with the frequency $f_A = 440$ Hz, corresponding to the period $T_A = 1/f_A = (1/440)$ s $= 2.72$ ms. The membrane vibrates in simple harmonic motion. The air molecules nearby follow this motion, and they in turn push other air molecules that are farther away. Thereby vibrational motion spreads out over all the space into which the sound wave penetrates. In fact this vibration of the air *is* the sound wave. The position of the membrane can be given as

$$s = s_0 \sin\{2\pi \cdot 440\, t\} = s_0 \sin\{\omega_A t\}, \tag{9.27}$$

where $\omega_A = 2\pi f_A = 2764$ radians/s is the angular velocity of the simple harmonic motion. This equation also depicts the motion of a volume element adjacent to the membrane. Instead of describing the vibrating speaker membrane, Eq. (9.27) could equally well characterize the motion of other sender systems such as a wire section of a vibrating string in a violin, or the prong of a tuning fork. Figure 9.2 shows the motion of a speaker membrane. Adjacent air molecules are forced to move in the same rhythm. They shake their neighbor molecules, in the nearby volume element, call it V_1, which then start to move with some delay in phase δ_1. The delay comes about since it takes a certain time to build up the pressure perturbations Δp that drive the sound wave. The molecules in region V_1 in turn shake their neighbours in the next volume element V_2, which get going with an additional phase delay $\delta_2 = 2\delta_1$ compared to the speaker membrane. And so the perturbation spreads all over the place. The nth layer of volume elements moves with the phase delay $\delta_n = n\delta_1$. The volume element V vibrate as

$$s_1 = s_0 \sin\{2\pi\, 660\, t - \delta_1\}. \tag{9.28}$$

The molecules in the next element V_2 with their larger phase delay $\delta_2 = 2\delta_1$ oscillate as $s_2 = s_0 \sin\{\omega t - 2\delta_1\}$. Particles located at larger distances z from the source have still larger phase delays $\delta(z)$. The phase delay increases linearly with distance $\delta(z) = k_z z$. The proportionality constants $k_z = 2\pi/\lambda$ is called the wave number, or spatial frequency of the wave. A large wave number implies a small wavelength. In general

$$\delta_z = k_z z = (2\pi/\lambda)z = (2\pi f/v)z. \tag{9.29}$$

The oscillatory displacement of the particles anywhere along the path of the wave can therefore be given as a general function,

$$s = s_0 \sin\{\omega t - \delta(z)\} = s_0 \sin\{\omega t - k_z z\}. \tag{9.30}$$

We already used this equation in Sect. 9.2.2. Note that the oscillation is in the propagation direction of the wave. But each particle only oscillates around a rest position. It never truly *leaves home*, while the wave train initiated by the speaker membrane travels through the air. The particles can oscillate because air is elastic. Air has a bulk modulus, which acts just like a spring constant of an elastic object. When the oscillation of an elastic object has begun (like the prongs of a tuning fork struck against a table top, or the air in a closed volume element that is suddenly compressed and then released), the motion always overshoots the rest position $s = 0$ and the mass vibrates with an amplitude s_0. The velocity of the particle is then $u = ds/dt = \omega s_0 \cos(\omega t)$ with a maximum value

$$u_{max} = u_0 = \omega s_0. \tag{9.31}$$

One must clearly distinguish this particle velocity u of a volume element, which varies as function of time, from the sound wave velocity v, which is a constant value that depends on the temperature and molecular mass of the medium. If a volume of air is stretched or compressed by the distance s it has elastic potential energy $PE = \frac{1}{2} k s^2$, where k is the "elastic constant" of the air volume. In addition, the oscillating volume has kinetic energy $KE = \frac{1}{2} m_p u^2$, where m_p is the mass of the volume. The total energy of the particle is $E = KE + PE$. It turns out that E is exactly equal to the maximum kinetic energy

$$E = \frac{1}{2} m_p u_0^2 = \frac{1}{2} m_p (\omega s_0)^2. \tag{9.32}$$

Once one recognizes how sound waves agitate volume elements of elastic media and induce local motion it becomes clear that power is transmitted by sound. The transmitted power can be visualized as the kinetic energy passed on from volume element to volume element as the sound wave travels through the elastic medium. In summary, sound will propagate once a sender with some regular motion sets the surrounding air into vibration.

9.4.2 Resonators

The heart of any voice is a type of resonator that produces regular motion. Regular motion can be generated with vibrating elastic solids, vibrating air columns, or periodically interrupted air streams. Muscles pump energy into the sound producing oscillator. For most effective energy use the muscles are activated at the resonance frequency f_r of the oscillator. Occasionally the muscle frequency is governed by an escapement type mechanism.

All elastic solids that may vibrate have a characteristic mass m and a restoring force F that increases when the system is displaced from its rest position by some distance s. For small displacements the force increases linearly

$$F = -k\,s.\tag{9.33}$$

The minus sign indicates that the force acts against the direction of the displacement. Such resonators oscillate in simple harmonic motion (SHM) at a fundamental frequency f and a period T which are related to the spring constant k and the mass m:

$$f_o = \frac{1}{T} = \frac{1}{2\pi}\sqrt{\frac{k}{m}}\,.\tag{9.34}$$

This square root structure is common to all resonance frequency formulas. To achieve resonance the driving force must be applied with the frequency $f_d = N f_o$, where N is an integer. In most cases N is equal to 1. There are exceptions. For instance the wing beat frequency of bees and wasps is in resonance with their thorax oscillations, however the driving muscle compresses the thorax only about every 10th cycle. The motion persists similar as a swing keeps going even if it receives a push only every tenth oscillation.

9.4.3 Oscillations of Elastic Solids

An example of a vibrating elastic solid is a mass m attached to a spring with the spring constant k. The system will oscillate at the frequency given by (9.34), namely $f_o = \frac{1}{2}\pi\,\sqrt{(k/m)}$. The spring constant k can be measured by scompressing the spring with a weight mg and measuring the distance Δx by which the spring is compressed: $k = m g / \Delta x$.

Any solid of cross section area A and length L with a Young's modulus Y can act as a spring. A vocal cord of length L and cross section area A might be modeled as such an oscillating cylindrical member. If the force F is applied the elastic member will stretch by some distance ΔL. Y is defined through the relation $(F/A) = Y(\Delta L/L)$; or $F = (YA/L)\,\Delta L$ by comparison with (9.33) one finds the spring constant

$$k = Y\frac{A}{L}\,.\tag{9.35}$$

A spring, or a stretched ligament oscillates in longitudinal direction. Elastic solids can also be made to vibrate in transverse direction. Leaf springs act in this fashion, Fig. 9.18a. The prongs of a tuning fork or the tongues of a harmonica vibrate like leaf springs. As their vibrations are transmitted to the surrounding air, a tone is generated. Since the air is set into motion the vibrating member loses energy and the oscillation is damped. A damped harmonic motion leads to a broadened resonance curve, see Sect. 6.1.2.

The vocal cords may be modeled as vibrating springs. Note that the frequency in (9.34) gets lower when the mass is increased. The pitch of the voice gets lower if one has a cold: the vocal cords swell up or are covered with mucus, which increases the vibrating mass. Heavy smoking has a similar effect, see Sect. 9.5.2.

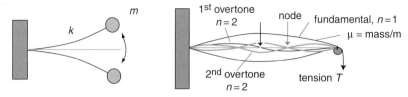

Fig. 9.18. (a) Mass on leaf spring. (b) String oscillating with overtones. The fundamental is also called the first harmonic. The first overtone is the second harmonic

The strings of musical instruments, Fig. 9.18, vibrate in the transverse direction. A string with the mass density μ (mass/unit length) is stretched by a tension force T. If the string is plugged in the middle it will oscillate at the fundamental frequency f_0.

$$f_0 = \frac{n}{2L}\sqrt{\frac{T}{\mu}} \, .$$
(9.36)

It is easy to measure the mass per unit length μ. Cut a section of string, 1 meter long and measure its mass (weight). When the tension on the string is increased, the frequency goes up. This is done, for instance, when one tunes a guitar. The low tones on a piano are made by very heavy strings. Often these strings are compound structures: a heavy brass wire wrapped around a steel wire. The steel wire core gives the tensile strength and the wrapping increases the mass per unit length.

For the fundamental frequency (first harmonic) one sets $n = 1$ in Eq. (9.36), for the second harmonic one sets $n = 2$, for the third harmonic one sets $n = 3$, and so forth.

9.4.4 Oscillations of Air Volumes

Sound is transferred by air. The generators described in Sect. 9.4.3 must be connected to amplifiers and speaker membranes in order to fill the air nearby with sound. A more effective way to move the air molecules is to generate the sound directly in the air. This can be accomplished in various ways.

Air is quite elastic. Not only solids, but any elastic medium can vibrate and produce sound waves. A piston of mass m riding (without sliding friction) on an air cylinder will vibrate in simple harmonic motion, Fig. 9.19. The oscillating mass is m and the effective spring constant are $k_{air} = (\gamma pA/L)$. Hence the oscillation has the frequency

$$f_0 = \frac{1}{2\pi}\sqrt{\frac{\gamma pA}{mL}} = \frac{v}{2\pi}\sqrt{\frac{\rho A}{mL}} \, .$$
(9.37)

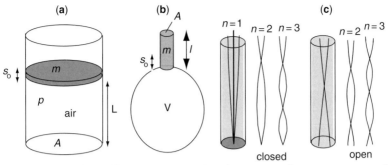

Fig. 9.19. Systems with oscillating air masses. (a) The mass m vibrates on an air cushion. (b) Helmholtz resonator. (c) Open and closed organ pipes with harmonics

The speed of sound $v = \{\gamma\,(p/\rho)\}^{1/2}$ in a gas depends on pressure p, density ρ, and the adiabatic coefficient $\gamma = (f_f + 2)/f_f$, which in turn is related to the degrees of freedom f_f of the gas. Air with its diatomic molecules has $f_f = 5$ so that $\gamma_{air} = 1.4$. Noble gases like helium have $f_f = 3$, hence $\gamma_{He} = 5/3$.

A variation of an air spring is the Helmholtz resonator, which serves well as a model for the frog's voice. In the Helmholtz resonator the mass m riding on the air cushion is not a solid at all, but a plug of air, as shown in Fig. 9.19b. The big vessel of volume V has a neck of length l and cross section A. The air in the neck behaves like a resonator mass $m = \rho A l$, which oscillates with small amplitude s_0 in the vertical direction. The air in the volume acts as the spring with a spring constant that is derived next.

For fast changes of pressure p and volume V the air is compressed and expanded so fast that no heat is exchanged. It is an adiabatic process. Hence the adiabatic pressure-volume relation applies:

$$p_1 \cdot V_1^{\gamma} = p_2 \cdot V_2^{\gamma} = \text{const}. \tag{9.38}$$

Take the logarithm of (9.38) and differentiate: $dp/p + \gamma\,(dV/V) = 0$, thus $dp = -\gamma p(dV/V)$. When the mass m is moved up by the distance ds, the gas in the bottle increases its volume by $dV = A\,ds$. The force exerted onto the mass m is

$$dF = A\,dp = Ap\gamma \frac{dV}{V} = -Ap\gamma \frac{A\,ds}{V} = -\frac{\gamma pA^2}{V}\,ds. \tag{9.39}$$

The force dF is linearly proportional to the displacement ds. The negative sign indicates a restoring force. The gas in the bottle acts like a spring, which pushes the mass $m = \rho_{air}\,Al$ back and forth, producing simple harmonic motion (SHM) with the angular velocity ω. The factor in front of ds is the Helmholtz spring constant $k_H = A^2\gamma p/V = A^2\rho v^2/V$. For SHM one has

$$\omega = 2\pi f_\mathrm{o} = \sqrt{\frac{k_\mathrm{H}}{m}} \,. \tag{9.40}$$

The term γp can be replaced by ρv^2, since the sound speed in air is $v^2 = \gamma p / \rho$. Thus the frequency of the Helmholtz resonator is

$$f_\mathrm{o} = \frac{1}{2\pi} \sqrt{\frac{\lambda p A}{\rho_\mathrm{air} \, l V}} = \frac{v}{2\pi} \sqrt{\frac{A}{l V}} \,. \tag{9.41}$$

Voices of humans and many animals get some of their characteristic features from standing waves in air columns of the vocal tract. Flutes, organ pipes and open bottles serve as illustrations. For instance, a standing wave can be excited in a bottle (or cylinder with one end closed) if a tone is produced at the open end. This can be done by striking a tuning fork. One will hear a loud tone, if the length is right, namely the oscillation must have a node at the bottom and an antinode at the open end. Hence $L = \lambda/4$ or $\lambda = 4L$ respectively. Resonance is obtained if the frequency satisfies the condition $f = v/\lambda = v/4L$.

Actually a whole sequence of different frequencies called overtones can be excited, where the vibrations have nodes at the closed end and antinodes at the open end, namely $L = \frac{1}{4}\lambda_1 = \frac{3}{4}\lambda_3 = \frac{5}{4}\lambda_5$. The resonance conditions are

$$f_\mathrm{n} = \frac{n \cdot v}{4L} \,, \tag{9.42}$$

where the mode number n is an odd integer number (1, 3, 5, ...). These frequencies are called the "harmonics", or "overtones" of closed organ pipes.

Distinctive frequencies can also be generated by blowing into an open pipe. The fundamental frequency f_o of the open organ pipe has $L = \lambda/2$, or $\lambda = 2L$, then $f_\mathrm{o} = v/2L$. The first overtone has the wavelength $\lambda_1 = L$, hence $f_1 = v/L$, and the next overtone has $L = 3\lambda_2/2$, or $\lambda_2 = 2L/3$, so that the frequency is $f_2 = 3v/2L$.

Most sound generators, including the human voice, can produce fundamentals and overtones. The intensity of the overtones depends on the excitation. For instance, if a recorder is blown softly the overtones are hardly noticeable. However, if it is blown hard the overtones become quite pronounced.

9.4.5 Frequencies of Periodically Interrupted Motion

Distinct frequencies can also be generated by interrupting a steady motion, for example by pulling a solid object at speed v over a set of ridges spaced at equal distance Δx. Think of a pencil pulled over a heater grill, or a finger nail dragged over the prongs of a comb, Fig. 9.20a. The frequency is

$$f = v/\Delta x \,. \tag{9.43}$$

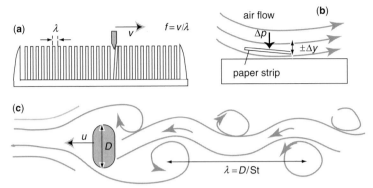

Fig. 9.20. (a) Constant speed and periodic structures. (b) Fluttering of elastic membrane in a steady flow of air. (c) Vortex shedding behind a structure

Periodic motion with sound emission can also be produced by interrupting an airflow. A simple experiment illustrates this effect. Place a strip of paper onto a flat surface and hold it down at both ends. Then blow onto the edge as shown in Fig. 9.20b. The paper strip soon opens up in the front to form a converging channel in which the air speed must increase in x-direction. Now the pressure decreases due to the Bernoulli effect, see Sect. 3.4.1 Since the pressure in the ambient air above the paper is not affected a difference Δp develops between the pressure in the channel and on the outside. This excess pressure closes the gap. The airflow is interrupted. The process repeats itself, again and again. The frequency is high when the mass of the strip is small and the tension stretching the string is high. A similar fluttering of an elastic membrane is obtained when blowing onto a grass blade that is squeezed between the thumbs of both hands, thereby generating a high pitched sound. The human voice, originating from the vocal cords, gets its vibration from a similar periodic processes. The vocal cords of a male open and close typically 120 times per second.

Periodic vibrations in fluids can also be produced by vortex shedding, see Sect. 3.4.3. A cylindrical rod of diameter D imbedded into an air stream of velocity v sheds vortices at the frequency

$$f = \mathrm{St} \cdot U/D, \tag{9.44}$$

where $\mathrm{St} \approx 0.2$ is the Strouhal number.

Example: Consider an object of the dimensions of the vocal cords $D = 2$ mm placed into an air stream of $U = 1$ m/s. The vortex shedding frequency is then $f = 0.2$ (1 m/s/ 0.002 m) $= 100$ Hz. This frequency is typical of a low tone produced by the human voice. Therefore, it is likely that vortex shedding plays a role in voice production.

9.5 Voices

Voices serve animals for communication, to issue warnings, and to give instructions. Voices may express moods, intentions, and social status. Voices work day and night. Beamed voices of high intensity can also be utilized to stun prey, and to illuminate objects for producing acoustic images. This is called sonar. Most animal voices are activated by the breathing apparatus. However, some animals such as grasshoppers and snapping shrimp use legs or other extremities to generate rapid repetitive motion to produce sound, or noise.

9.5.1 The Human Voice

The human voice is produced by several parts of the body, which are all components of the breathing apparatus: The lungs, the vocal cords, the mouth and nasal passages.

The volume between lips and vocal cords can be divided into two sections of length L_a and L_b. Each one of these characteristic lengths has its fundamental open pipe frequency band, called *formant*, at which any tone – however it may be generated – may resonate and hence have an increased amplitude. The mouth cavity with $L_a \approx 11$ cm produces the second formant. The wind pipe section above the glottis with $L_b \approx 5.7$ cm produces the 3rd formant. Both together generate the first formant. $L_a + L_b \approx 17$ cm, with the wavelength $\lambda_1 = 4(L_a + L_b) \approx 0.68$ m. The sound of the first formant is centered about the frequency $f_1 \approx 340$ ms^{-1}/0.068 m = 500 Hz. Since the walls of the oral cavity are soft and irregular the resonances are not sharp, but rather broad bands, called *filter function*. The filter function modi-

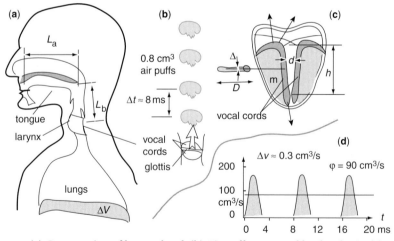

Fig. 9.21. (a) Cross section of human head. (b) Air puffs generated by the glottis. (c) Cross section of larynx with glottis opening area $A_{gl} = d \cdot h$. (d) Time sequence of volume flow φ, and air puffs ΔV

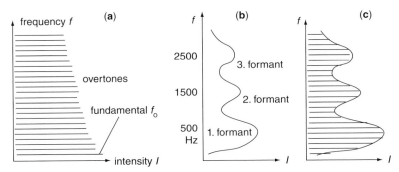

Fig. 9.22. (a) Frequencies produced by the vocal cords with overtones. (b) Formant filter function. (c) Resulting acoustical output

fies the spectrum of overtones of f_0. The spectrum of overtones, filter function, and the resulting sound output, are shown in Fig. 9.22. The oral geometry of humans differs significantly from the mouth of our closest relative, the chimpanzee, who has a straight throat.

The filter function depends on the geometrical dimensions of the throat, the mouth and the nasal cavities, as well as on the position of lips and teeth. During speech and singing these geometrical dimensions are continuously altered, thereby changing the filter function, and hence the frequency spectrum of the produced sound. Opera singers are oral cavity acrobats.

The lungs produce a flow of air. The air stream is shopped by the vocal cords into sequential pulses. A male voice has typically $f_{om} \approx 125$ pulses per second. A female voice has about $f_{of} \approx 250$ pulses per second. The human voice contains f_0 and its numerous harmonics [Sataloff 1992].

Example: A 70 kg male inhales a minimum of about $\varphi_0 = 37$ cm³ air/s, (see Sect. 4.3.3) to maintain his metabolism. Generally the oxygen is not fully extracted so that the average volume flow is larger, say $\varphi \approx 100$ cm³/s. This volume flow scales with body mass M as $\varphi = 100(M/70 \text{ kg})^{3/4}$ cm³/s. The flow must be equivalent to $\varphi = f_{om} \Delta V$, where $\Delta V = \varphi/f_{om} = 100/125 = 0.8$ cm³ is the volume of each air puff. If these air puffs have a duration of 3 ms. The volume flow rate during these puffs is much higher than φ, namely $\varphi_p = 0.8$ cm³/0.003 s ≈ 270 cm³/s $= A_{gl}u$. The puff speed φ_p is the product of average speed u and glottis opening area A_{gl}.

Typical dimensions of the average glottis opening are $h \approx 2$ cm and $d \approx 0.3$ cm, yielding the glottis area $A_{gl} \approx 0.6$ cm². Hence during speech the air escapes with the average speed $u = \varphi_p/A_{gl} \approx 4.4$ m/s. At this speed the air puffs in a male voice of $f = 125$ Hz are separated by the distance $\lambda_p = u/f = 440$ cm s⁻¹/125 s⁻¹ $= 3.6$ cm.

When air flows at sufficiently high speed around obstacles of width D (so that the Reynolds number Re $= UD/v$ exceeds a value of 50) vortices are produced. If $D \approx 6$ mm is the width of the vocal cords, the Reynolds number is Re $= 4.4 \cdot 0.006/1.5 \cdot 10^{-5} = 1800$, well above the onset of vortex shedding. The vortex shedding frequency off cylinders of width D is $f_{vs} = $ St U/D. Typically the Strouhal number is St ≈ 0.2. Vortices of the same sense of rotation are separated by the distance

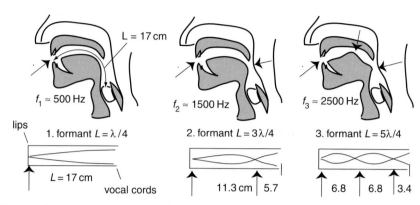

Fig. 9.23. Formants for a person with $L = 0.17$ m

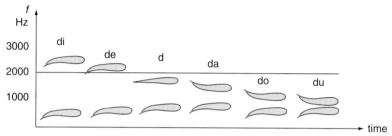

Fig. 9.24. Frequency spectrum of various d-sounds

$\lambda_v = D/St \approx 5D$. In the approximation that the vortex shedding through the vocal cords is similar to the shedding from a cylinder of the same diameter, these vortices would be separated by the distance $\lambda_v \approx 5D = 3$ cm. This distance is close to the puff pulse separation $\lambda_p = 3.6$ cm calculated above: the opening and closing of the vocal cords appears to be in resonance with vortex shedding.

Language consists of groups of vowels and consonants, either pronounced quickly or drawn out in time. Vowels have different pitches. This is illustrated in Fig. 9.24 by the frequency spectrum of various d-sounds.

> **Exercise:** The oral cavity of a child measured from the vocal cords to the lips has a length $L = 8$ cm. Treat this volume as an organ pipe closed at one end and find the fundamental frequency f_0, namely $f_0 = v/4L = 430$ ms$^{-1}/(4 \cdot 0.08$ m$) = 1063$ Hz.

9.5.2 The Frequency Spectrum of Speech

The pitch of voices depends on the shape of the oral cavity which moulds itself to facilitate the music of language. Different nationalities have distinctive different speech spectra. Figure 9.25 shows the frequency distribution in some European languages. As different human races are distinguished by their external body features, different accents are due to specific oral cavity features.

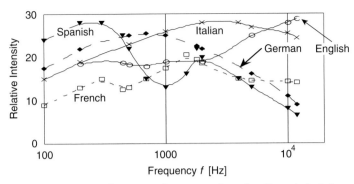

Fig. 9.25. Frequency spectra of European languages, data after Tomatis [1987]

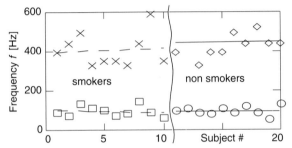

Fig. 9.26. Highest and lowest pitch of voice of a group of ten smokers, and ten non smokers, measured by comparison with a piano [Leung 1999]

The pitch of the human voice not only depends on the dimensions of the voice box but also on the state of the tissue, which in turn can be influenced by habits. In a class project Vivian Leung [1999] investigated if smokers have on average a different pitch of voice than non smokers. A group of ten non-smokers had the average low frequency $f_{low,no}$= 97.1 ± 7.0 Hz and $f_{high,non}$= 450 ± 29.1 Hz. In contrast, a group of 10 smokers had the values $f_{low,s}$= 95.2 ± 8.3 Hz, and $f_{high,s}$= 406 ± 30.5 Hz. The difference of the average low notes falls within the standard devia-

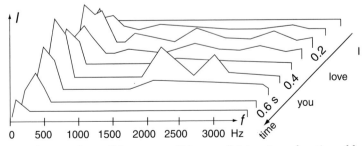

Fig. 9.27. Fourier transform of the sentence "I love you", intensity as function of frequency and time. Signal slices are taken at about 100 ms intervals

tions of the means; however, the average high notes differ by more than one standard deviation of the means. These data suggest that smokers on average have deeper voices than non smokers, probably because the vocal cords are always slightly swollen, and inflamed and therefore more massive than the vocal cords of non smokers.

Sound received by the ears is the superposition of all frequencies that are contained in the sound. If the sound is recorded with a microphone, the signal may be analyzed with a fast Fourier analyzer and converted into a frequency-time histogram. Figure 9.27 shows a human voice, namely the spoken sentence *I love you*. Time runs from back to front. Fourier slices are taken at time intervals of about 100 ms.

9.5.3 The Sound of Frogs

Frogs communicate by voice. If different species live in close quarters they must find ways to communicate without inter-species interference. Somehow they have managed to occupy different sections of the sound spectrum, similar as amplitude modulated (AM) stations on the radio dial. Figure 9.28 shows this spectrum division for 8 species of frogs in a Puerto Rico rain forest, as reported by Narins [1995].

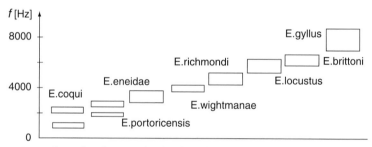

Fig. 9.28. Frogs share the air waves by dividing the frequency spectrum, data after Narins [1995]

9.5.4 The Sound of Snapping Shrimp: Cavitation

The small shrimps Alpheus heterochaelis, called the snapping shrimp, make sharp and very loud noises with their large claw. Fast video camera images, taken by Versluis et al [2000], have revealed that the shrimp produce a water jet of $u \approx 28$ m/s, by shutting their large claw. At this velocity the pressure in the flow is reduced so much by the Bernoulli effect ($p + 0.5 \, \rho u^2 = $ const, see Sect. 3.4.1) that the pressure in the flow falls below the vapor pressure of water, and micro bubbles form. Subsequently the bubbles collapse emitting noise up to 210 dB re 1 μPa, which stuns prey or predators. The collapse of bubbles in fast flows is known as cavitation. Engineers try to avoid this effect, because it causes damage of propeller blades and walls by pitting.

Note that sound levels in water are generally reported with the reference sound pressure $p_0 = 1$ μPa, whereas sound levels in air are referenced to the base intensity $I_0 = 10^{-12}$ W/m². The decibel values in water quoted for the reference value 1 μPa are larger by 64.8 dB than the sound level values referenced to I_0. The conversion between these two reference schemes is described in Sect. 9.23, and Table 9.2. The shrimp noise therefore corresponds to $210 - 64.8 = 145.2$ dB re $I_0 = 10^{-12}$ W/m². This represents, the astonishing intensity $I = 10^{14.5} \cdot I_0 \approx 300$ W/m², which is like the sound of a rifle shot fired off close by.

9.5.5 Insect Sounds

Many insects produce sounds by moving their wings or legs. Particularly loud sounds are made by mole crickets. [Young and Bennet-Clark 1995 and 1998]. These animals actually burrow into the ground and sculpture the entrance of their cave like the horn of a trumpet, to emit the sound at their chosen frequency $f = 2.5$ kHz effectively in all directions. See also problem P 9.8. Some insects that are preyed upon by bats have developed the ability to emit ultrasound to confuse the sonar of the bats [Roeder 1965].

9.5.6 Sperm Whale Sound

Sperm whales are known to generate sound pulses of very high intensity. Levels of up to 215 dB re 1 μPa have been reported. This converts into a sound intensity of $215 - 64.8$ dB ≈ 150 dB re I_0, corresponding to $I \approx 1$ kW/m², an intensity like the sound of a jet plane at takeoff. Sperm whales have a massive organ dedicated to sound production in front of their skull bones. They breathe through one nasal passage. A constriction in the other nasal passage is converted into a sound generator, with two lips in a *clapper* made to flutter when air is sloshed between two air volumes on either side of the clapper. The massive sound organ has a weight of several tons – representing 1/10 of the whales total body mass. It is filled with spermatic oil of very special acoustical properties [Clark 1978] and junk tissue of slightly lower impedance that is organized into cells reminding of lenses. This sound production system is likely a combined lens – amplifier organ, whose exact operation is presently still unknown.

9.6 Information Extracted from Ambient Sound

Passive sound, namely noise emitted by inanimate objects, or sound produced involuntarily by animals contains many different sets of information.

First there is the objective, physical information: (i) distance of sound source, (ii) its direction in space, (iii) is the source coming or going? Then there is the biological content of the information: is the "speaker" the teaching mother, or a potential meal, a mate, or a menace? Last but not least there is the social content of the information such as: does the speaker belong to your social group? Is the message urgent for the listener or the speaker? What is the mood of the speaker? Is there a message of social hierarchy in the voice?

Physical principles help to decode the physical and social content of the acoustic information.

9.6.1 Direction, Echoes, and Shadows

An organism reckognizes its environment if it knows where everything is. Perception is enabled by signals that may be encoded in light or sound waves. Many objects emit their own sound, some respond with sound if there is a wind. Their distance can be judged by the intensity of received sound. Their left-right position, angle ø, can be judged relative to the position of the head by the delay times between the signals received by the left and the right ear, Fig. 9.29. By inclining the head one can also perceive the elevation of the signals.

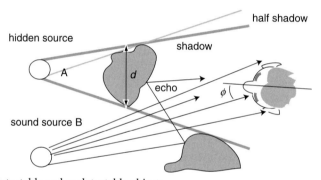

Fig. 9.29. Detectable and undetectable objects

Sound is reflected from objects. Sound can also be obstructed by objects, which cast shadows. Lateral motion of the head may reveal sources hidden behind obstacles.

Similar to light, sound may be diffracted around obstacles (provided the wavelength is smaller than the object width d) so that some sound energy gets into the shadow regime behind the object. This "bending" of the sound beams can be observe with a simple experiment. Close your left ear with the right forefinger. Now extend the left arm fully and rub thumb and forefinger of your left hand: you will hear the high pitched rubbing sound from your left hand in your right ear: The sound is diffracted around the head. Now bring the left hand close to the left ear and rub the fingers again: you hear nothing, because the angle through which the sound waves would have to be bent to reach the right ear is too large.

9.6.2 Signal Spectrum Recognition – Who Is There?

Individual voices differ considerably, and ears and brains are well adjusted to recognize a speaker on the telephone, although one does not see the caller. In addition to these personal traits the voices of people who speak different languages vary in their frequency components. This fact must lead to slight modifications of elements of the vocal tracts that are learned at an early age. For instance Italian has more high frequency sounds in the language, with the result that Italian is well suited to the voices of sopranos. Average frequency distributions of English, French, German, Italian, and Spanish are shown in Fig. 9.26.

9.6.3 Distance and Directions of Sound Sources

Listening with two ears can provide information about distance and direction of noisy objects, [Konishi 1993]. Since the ears are separated by the head diameter d, sound signals from a distant source located in the direction ϕ arrives first at one ear, say the right ear with some intensity $I_R(t)$, and then reaches the left ear with some intensity $I_L(t)$ after a delay time $\Delta t = \Delta s/v$, giving a clue about the direction ϕ of the object from the delay distance

$$\Delta s = v\,\Delta t. \tag{9.44}$$

The extra path length is $\Delta s = d\sin\phi$ can be read from the geometry shown in Fig. 9.30. Knowledge of the delay time Δt yields the angle

$$\sin\phi = (v/d)\,\Delta t. \tag{9.45}$$

We will describe shortly how the brain may measure the delay time. The two sound signals carry two other bits of information, which yield information about the distance of the source: Their absolute intensities I_R, and I_L and the relative difference $\Delta I/I \approx (I_{Rs} - I_L)/I_R$. Let s_R and $s_L = s_R + \Delta s$ be the distances from the sound source to the right and the left ear respectively, and I_0 the intensity of the source at some distance s_0 then these intensities are $I_R = I_0(s_0/s_R)^2$, and $I_L = I_0(s_0^2/(s_R + \Delta s)^2$, so that $\Delta I = (I_R - I_L) \approx I_R/(2\Delta s/s_R + \Delta s^2/s_R^2) \approx 2I_R \Delta s/s_R$, yielding

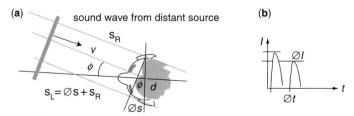

Fig. 9.30. Path differences for waves reaching ears with a delay time Δt

$$\frac{\Delta I}{I_R} \approx 2\frac{\Delta s}{s_R}. \tag{9.46}$$

This intensity information can be used to extract the distance s_R. For remote objects where $d \ll s_R, s_L$ both intensities are practically the same. If the listener can "measure" the absolute intensity I_0 at some reference distance s_0 the unknown distance s_R can be found from the inverse square law $s_R = s_0\sqrt{(I_0/I_R)}$. The brain would do this evaluation automatically by comparing the detected intensity with some "learned" intensity stored in memory. When a sound source is located nearby, the intensities detected by the left and the right ear will differ markedly so that the ears can detect $\Delta I/IR$. Then the distance s_r can be evaluated from (9.44) and (9.49), namely

$$s_R = 2v\Delta t \cdot \frac{I_R}{\Delta I}. \tag{9.47}$$

In this case the distance is derived from a delay time measurement and the signal intensities only, and no absolute reference intensity must be known.

The delay time Δt is generally quite short. For an angle of $\emptyset = 30°$, $d = 0.20$ m, and $v = 340$ m/s in air one finds $\Delta t = 0.29$ ms. This is much shorter than the fastest muscle motion of typically 100 ms. For smaller angles the time interval gets smaller, and if one considers animals with smaller heads, the "stopwatch in the brain" must measure time intervals down to less than 100 μsec [Hartmann 99].

A blind-folded person can determine the direction of a sound source to an accuracy of $\Delta \emptyset \approx \pm 5°$. This feat is accomplished by neurons that fire in coincidence. The delay of the acoustic wave $\Delta t = \Delta s \cdot v$ is compensated by a nerve pulse delay $\Delta t_n = \Delta x \cdot u$ inside the brain. This delay comes about if the nerve pulse propagating at the nerve conduction speed u is forced to travel an extra distance Δx, see Fig. 9.31.

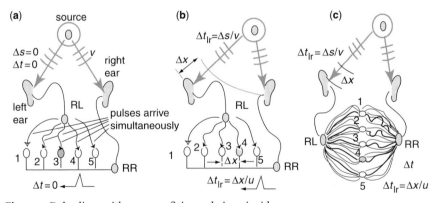

Fig. 9.31. Delay lines with neurons firing only in coincidence

Aquatic mammals, like seals, that mainly live near the surface of the water sometimes use the different travel times Δt of sound in air and in water to guess distances of objects at the water surface. For instance a seal mother can tell at which distance s her baby cries, by listening simultaneously above and below the water surface. In air the travel of the sound is $t_a = s/v_a$. Under water the sound arrives within $t_w = s/v_w$. Therefore there is a delay time between both signals $\Delta t = t_a - t_w = s/\{1/v_a - 1/v_w\}$. This relation can be solved for s yielding

$$s = \Delta t \frac{v_a v_w}{v_w - v_a} = C \Delta t . \tag{9.48}$$

The proportionality constant $C = v_a v_w/(v_a - v_w)$ can be learned by experience, but the delay time must be precisely measured.

9.6.4 Delay Time Measurements

Two different time-delay schemes have been proposed. In Fig. 9.31a the source is located symmetrically in front of the head. From the ears the pulses travel to nodes, which act as relay stations RL and RR, and connect to a row of neurons, each dedicated to a certain direction. These neurons fire only if they receive pulses simultaneously from RR and RL.

The nerves connecting the neurons from RL have all the same length, so that RL stimulates the neurons at the same time. However, the nerves leading from RR to the neurons have different lengths. The path to neuron 1 is the longest, to neuron 5 is the shortest. The path to neuron 3 has the same length as the RL connections, hence $\Delta t_3 = 0$. This particular neuron fires when a sound signal reaches both ears at the same time, namely when the sound source is right in front of or behind the head. Neuron 4 fires at the delay time $\Delta t_4 = \Delta x/u$, see Fig 9.31b. Neuron 5 would fire if the delay time was $\Delta t = 2 \Delta x/u$.

Alternately, direction hearing can be explained as a learned skill. At birth all of the neurons in a detector chain are connected to both ears by two sets of nerve bundles, both bundles having nerves of many different lengths, Fig. 9.31c. Young owls train their directional hearing when they look at a noise making object located in a particular direction. The ears will send signals through the many channels from both nerve bundles, which will all arrive with different time delays, as if there were many noise-making objects located in many different positions. One particular set of two nerves from the two bundles will likely place the object in the direction where the eyes see it, and this is the *correct* set. In Fig. 9.31c such nerve pairs are drawn with heavy lines for all neurons. In particular, neuron number 4 fires for the delay time Δt associated with the extra distance Δs, similar as in the scheme Fig. 9.31b. Training implies that only the correct set of nerves for this direction will be kept in the body, all the other nerves from the two bundles are left to wither and gradually disappear in the body. With such delay lines and coincidence counters, animals can achieve an astonishingly fine acoustical directional

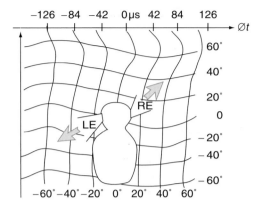

Fig. 9.32. Acoustic field of view of the barn owl. The time delay Δt of 42 µs corresponds to a horizontal angular resolution of 20°. Intensity ratios of $\Delta\beta = 10 \log(I_L/I_R) = 4$ dB correspond to an azimuthal resolution of 20°

discrimination. Humans can determine the direction from which a sound comes to within 5°, while dolphins are reported to achieve resolutions of $\pm 1°$.

The barn owl takes the acoustic resolution of locating sound sources one step further [Konishi 1993]. Its horizontal (left-right) discrimination comes as usual from the delay time between the left and the right ear. A delay time of $\Delta t = t_L - t_R = 42$ µs corresponds to an angle of 20°.

In addition, the barn owl achieves a vertical discrimination from intensity differences between both ears. The right ear is located slightly below the eye level, and points up, as indicated in Fig. 9.32. Sound emitted from the upper hemisphere appears to be louder, than sound coming from the lower hemisphere. A sound source, which moves down in the field of view, appears to become fainter and fainter.

The left ear is located above the eye level and it points downward. A sound source, which moves downwards, appears to get progressively louder. The sound level difference

$$\Delta\beta = 10\log\frac{I_R}{I_L} \tag{9.49}$$

is a measure for the azimuthal position of a source. $\Delta\beta = 4$ dB corresponds to an angular increment of 20°. This resolution is quite sensitive, because $\Delta\beta = 10$ dB corresponds to an intensity ratio of a factor 10.

9.7 Sound Images

The soundscape of the natural surrounding is full of clues about the objects in the vicinity signalling opportunities, or dangers. Animals can read the sound-scape in various ways, either by passive observation or by the active production of sonar search beams, coupled with sophisticated acoustic interpretation of the back-scattered acoustic signals. Many objects emit their own sound, some respond with sound if there is a wind. Their distance can be judged by the intensity of the received sound. Their left-right position can be judged relative to the position of the head by the delay times between the signals received by the left and the right ears. By inclining the head, one can also perceive the elevation of the signals.

However, many objects do not move. They are inaudible. Yet sound will reflect off them. It does not take much energy to produce sound. This is the basis of all sonar systems. First let us see how much energy is actually traveling in a sound wave.

9.7.1 Little Energy Goes a Long Way Traveling as Information

Animals have learned to use sound for finding their way, for detecting opportunities and dangers, and for communication. This is quite a feat, because sound carries very little energy compared to the mechanical energies involved in motion, and other body functions.

Consider the following example: The faint sound from a dripping faucet can be heard, when there are no noises in a room, from a distance of $R = 3.0$ m. Assume the drop of $r = 1.5$ mm radius, and mass $m = (4\pi/3)\rho r^3 = 1.4 \cdot 10^{-5}$ kg, falls from the height of 0.4 m. It acquires the energy: $E_{mech} = mgh = 1.4 \cdot 10^{-5}$ kg $\cdot$ 9.8 ms$^{-2} \cdot$ 0.4 m $= 5.6 \cdot 10^{-5}$ J. A small fraction of this energy is turned into noise with the acoustic energy, E_{ac}. We arbitrarily pick a very small fraction of this mechanical energy, say $E_a/E_{mech} = 6 \cdot 10^{-4}$, to find out how much sound it would produce. Assume that this acoustic energy $E_a = 6 \cdot 10^{-4} \cdot 5.610^{-5}$ J $= 3.36 \cdot 10^{-8}$ J is released during the time $\Delta t = 0.01$ s as the drop splashes into a bucket in the sink. Then this sound has the power

$$P = \frac{E_{ac}}{\Delta t} = \frac{3.36 \cdot 10^{-8}\,\text{J}}{0.01\,\text{s}} = 3.36 \cdot 10^{-6}\,\text{W}. \tag{9.50}$$

At a distance of $R = 3.0$ m this acoustic power generates the intensity $I = P/4\pi R^2 = 3.36 \cdot 10^{-6}$ W$/(4\pi (3$ m$)^2) = 2.97 \cdot 10^{-8}$W/m^2, which represents a power level of $\beta = 10 \log_{10}(I/10^{-12}) = 10 \log_{10}(2.97 \cdot 10^{-8}/10^{-12}) = 44.7$ dB. This sound can very well be heard. If one takes the threshold of hearing as 4.7 dB, then a sound with an intensity 1/10,000, could just be perceived. For the above example this implies $E_a/E_{mech} = 6 \cdot 10^{-8}$.

While the energy carried in sound is very small, the metabolic effort to produce it may be quite significant [Speakman et al. 1989]. A way to reduce these energetic costs is to generate sound in phase (resonance) with the abdominal muscle activities, as observed for bats by Lancaster and Speakman [2001].

9.7.2 Delphinid Acoustic Apparatus

Aquatic mammals can generate a wide range of sound frequencies. Low tones below 20 Hz are used to scan the seascape to recognize islands and other large-scale features of the terrain. High tones are used for communication and echolocation of prey. The voices are matched by very sensitive ears [Hemillä et al 1999].

Essentially dolphins produce three distinct types of calls: echolocation clicks, pulsed calls, and whistles. The first are sonar calls that are used for orientation and to find prey in murky waters. The second are issued for communication, and the third convey information about the identity of the caller as well as the emotional state of the animal.

The mammalian voice was developed for sound production on the land, where the expelled breath from the lungs carries the sound through the open mouth directly into the air. Aquatic mammals had to modify the vocal system in several ways. First, they emit sound with closed mouth. Second, their sonic vibrator is in good acoustic contact with the surrounding water. Third, they have developed a sound concentrating system. Its main component is the melon, a fatty lens-shaped tissue segment, Fig. 9.34, which is connected to the sound source by low impedance sound channels. The sound is generated by phonic lips, thick vocal cords, which are in good acoustic contact with the sound concentrator [Huggenberg et al. 2001]. The vocal cords are excited by air that moves between vestibular air sacks in region (1), just below the nostrils and, and the lower regions of the windpipe, location (2), Fig. 9.34 a.

Sound for communication is emitted in all directions. The echo location clicks contain broad band noise with very short rise times, focused partly by reflection from the bone of the skull and partly by refraction through the lens shaped melon. Fourier analysis shows that these sounds contain energy at all frequencies but biased to very high frequencies. Pulsed calls resemble human vocalization, except that no air escapes from the animal. Whistles are narrow band calls that change pitch in a unique way for each animal. This personalized music score likely carries messages over much wider distances than the clicks and the pulsed calls. In fact when the oceans where not full of the background noises from human activities

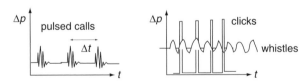

Fig. 9.33. Pulsed calls, whistles, and clicks, used by cetaceans for communication and echo location

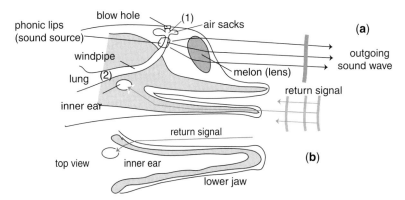

Fig. 9.34. (a) Path of sound generated by dolphin, and sound received through lower jaw bone. (b) Path of sound reaching ear of dolphin

whales could detect the songs of their kin over distances from Hawaii to Alaska, provided the sound was traveling in the Sofar channel.

Sound can reach the inner ear thought the external auditory orifice as well as through the jaw bone, Fig. 9.34b. In fact the tip of the jaw of bottle-nose dolphins has a round bulb which looks like the hand held microphone used by jazz singers.

Blue whales use sounds of down to 5 Hz for their echolocation. Such low frequencies allow them to perceive structures down to

$$\Delta x \approx \lambda = \frac{v}{f} = \frac{1500 \text{ m/s}}{5 \text{ s}^{-1}} = 30 \text{ m} . \tag{9.51}$$

This is the size of swarms of krill from which they feed. Also they can make out islands and undersea mountains to find their way on their migratory routes.

9.7.3 Echo Location of Bats

Sound, emitted by an animal, can be used for echolocation or sonar provided the sound is reflected off objects, and enough intensity reaches the ear of the animal. The emitted signal should have distinct frequencies and sharp rise or fall times. Sonar has some very nice features but also some shortcomings.

Distance can be assessed from the delay time between outgoing and reflected signals. The size of objects can be determined from the magnitude of the reflected signal and the object distance that was judged from the delay time. The Doppler shift of the reflected signal yields information about the speed of a moving object. The shortcomings of sonar are its limited resolution $\Delta x \approx \lambda$, which is determined by the wavelength $\lambda = v/f$ of the sound. Objects smaller than the wavelength of sound do not show up. In addition, sonar can be jammed by noises of similar frequencies. Some beetles have learned to fool bats with this ruse [Roeder 1965].

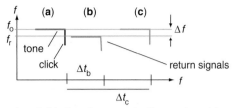

Fig. 9.35. (a) Outgoing signal. (b) Signal returned off a moving object nearby, (c) Signal returned off a more distant stationary object

A suitable sonar signal is a single frequency tone followed by clicking noise, which is really a superposition of many different frequencies [Suga 1990], Fig. 9.35a. The emitted signal returns at the same f frequency and with the delay time Δt_c if it is reflected off a stationary object.

The click is useful to determine the echo return time Δt with good definition. The distance d to the object is

$$d = v \cdot \Delta t / 2 . \tag{9.52}$$

When the signal is bounced off an object moving at the speed u the return signal comes back with a shifted frequency $f_r = \pm \Delta f$, and the delay Δt. If the velocity u of the moving object is small compared to the speed of sound v, the frequency difference is $\Delta f \approx f_0 2u/v$. It arises from the Doppler effect associated with the velocity of a moving object, which is discussed next.

9.7.4 How Bats Know the Speed of Their Prey

The body of a hovering moth acts like a stationary mirror, the distance of which is known to the bat by the travel time of the return pulse. The carrier frequency f of the sound is set by the *sender*, namely by vocal cords of the bat, which move somewhat like the membranes of a speaker. Sonar relies on the physical effect that sound is reflected off objects. Therefore, one would expect that the bat exactly hears the frequency f which it has initially produced. However, there are situations where the bat will hear a different frequency, namely if either the bat itself moves (towards, or away) from the reflecting object, or if the reflecting object moves. In either case the frequency f will be shifted by some amount Δf, by a phenomenon known as the Doppler effect. The change of frequency is called the Doppler shift.

Let v be the velocity of the sound wave, u_s the velocity of the source, u_r the velocity of the receiver, and f_s the frequency emitted by the sender. The frequency observed by the receiver is

$$f_r = f_s \cdot \frac{v \pm u_r}{v \mp u_s} . \tag{9.53}$$

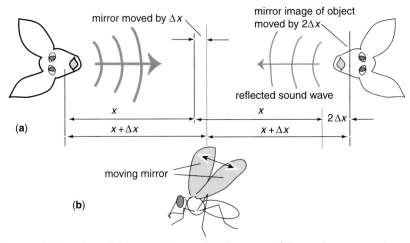

Fig. 9.36. (a) Location of object and image. (b) The wings of the moth act as moving mirrors producing a Doppler shift

The upper signs in numerator and denominator hold for mutual approach, namely $+u_r$ if the receiver moves towards the source, and $-u_s$ if the source moves towards the receiver.

Doppler shifts also occur when a wave is reflected off a moving object. This shift is used in order to assess the speed of their prey. The moving object acts like a moving mirror. The mirror image, namely the sender of the sound Fig. 9.36 (the bat itself that emitted the source frequency f_s) appears to be moving at twice the mirror velocity

$$u_s = 2\, u_m .\tag{9.54}$$

Of course the moth acts like a mirror in this case. Due to the motion of the moving mirror the reflected sound waves carry the frequency

$$f_r = f_s \cdot \frac{v \pm u_r}{v \mp 2u_m} .\tag{9.55}$$

If both, the mirror velocity u_m and the receiver velocity u_r are small compared to the sound velocity v the received frequency can be given as a linear function of velocities. Take the first term of the fraction $v/\{v-(\pm 2\, u_m)\}$ and expand it to get $1 \pm 2\, u_m/v$. Then take the second term of the fraction in (9.55) $\pm u_r/\{v-(\pm 2u_m)\}$. Here we neglect the term $2\, u_m$ in the denominator to get the first approximation $\pm u_r/\{v-(\pm 2\, u_m)\} \approx \pm u_r/v$. Together these approximations amount to $f_r \approx f_s(1 \pm 2\, u_m/v \pm u_r/v)$. In this approximation the frequency shift $\Delta f = f_r - f_s$ becomes

$$\Delta f = f_r - f_s \approx \pm f_s \cdot \frac{u_r + 2u_m}{v} .\tag{9.56}$$

(a) **(b)**

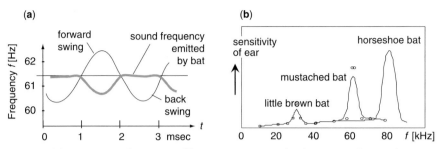

Fig. 9.37. (a) Frequency adjustment of bats to compensate for their own velocity after Suga [1990], (b) sensitivity of ears of bats

For a stationary observer the frequency shift reduces to the simple relation $\Delta f/f = 2\,u_m/v$. An insect hovering in the air with flapping wings will act like a moving mirror that moves back and forth, approximately like a sine wave $u_m = u_{mo}\sin 2\pi f_w t$. Then the received signal will oscillate around the carrier frequency. From the amplitude u_{mo} of the frequency fluctuations the bat can obtain clues about the size of its prey. The intensity of the reflected signal together with its delay time contains information about the size of the target. If the target approaches, u_m is positive and the returned signal has an increased frequency.

A frequency increase could however also be the consequence of the forward speed $+u_r$ of the receiver itself. To avoid confusion, bats automatically lower the pitch of their voice in proportion to their own (forward) flying speed. This effect was detected by Suga [1990] when measuring the voice frequency of a bat placed in a cage on a swing, Fig. 9.37a. The animal lowered its pitch during the forward swing only, because in real life bats do not hunt flying backward.

Bats use frequencies that are matched to the size of their prey. In order to be "seen" the wavelength must be smaller than the prey dimension. To observe a 5 mm bug the wavelength should be $\lambda \leq 5 \cdot 10^{-3}$ m, then

Different species of bats have learned to use different frequencies, Fig. 9.37b. Their ears are only sensitive to their hunting frequency. For instance the horseshoe bat uses $f = 80$ kHz.

Example: With this information one can estimate the frequency variations observed by a mustached bat emitting a signal of $f = 60$ kHz, when acoustically *seeing* a loscust which flaps its $L = 5.0$ cm long wings from $\phi = +45°$ to $-45°$ at the frequency $f_{wing} = 50$ Hz. The maximum *Doppler width* Δf_D of the return signal of the flapping wings is found as follows; At an opening angle of 45° the wing tips have an oscillation amplitude $A = L\sin\phi = 0.05 \cdot \sin 45° = 0.036$ m. Their maximum speed is $u_m = A\,\omega = A\,2\pi f = 2\pi \cdot 50 \cdot 0.036 = 11.1$ m/s. This corresponds to a Dopper shift $\Delta f_+ = f \cdot 2\,u_m/v = 60{,}000$ s$^{-1} \cdot 22.2$ ms$^{-1}/340$ ms$^{-1} = +3.9$ kHz at the maximum approach velocity. Then the received frequency is $f_+ = f + \Delta f_+ = 60$ kHz $+ 3.9$ kHz $= 63.9$ kHz. When the wingtips move away from the bat the frequency will be $f_- = f - \Delta f = 56.1$ kHz. The Doppler widths of the return signal is $\Delta f_D = 63.9 - 56.1 = 7.8$ kHz.

9.8 Sound, the Social Sense

Sight and sound provide quite different information. Sight gives fine details of structures and objects, instantaneously and in parallel. Much of the pre-processing is done by the eye. Sound arrives sequentially, much of the processing is left to the brain. But sound reveals activities, because sound is made by things that move, by events that happen, and by animals that act, all of which imply threats or opportunities. Hardly any social behavior, or learning has appeared on earth without sound. Sound activates the brain and makes it amenable for higher life forms. Sound is the basis of all higher forms of language, it enables social interactions. Apparently people who lose the ability to see remain more socially integrated with the rest of the population than people who lose their hearing. Both sound and light can produce images of the surrounding. However, in murky, and turbulent water it is impossible to *see* with either light or sound.

9.8.1 Comparison of Light and Sound Images

Light and sound are wave phenomena with very different frequencies, phase velocities, and absorption constants. Therefore both phenomena facilitate different applications in signal detection, production, and communication, as indicated in Table 9.4.

Information encoded in sound waves travels fast and it is very energy effective. Acoustic powers emitted by sound making organs, are small fractions of Watt, whereas moving a limb or the whole body involves mechanical powers of the or-

Table 9.4. Comparison of light and sound images

parameter	light	sound
phase velocity	$c_0 = 3 \cdot 10^8$ m/s, water $c = c_0/n$, $n = 1.33$	in air $v \approx 340$ m/s, in water $v \approx 1500$ m/s
f	small range around $f \approx 6 \cdot 10^{14}$ Hz	large range $f = 10 - 10^5$ Hz
λ	400 –700 nm in air $\lambda \approx 3$ mm –30 m	
1/e range	in air $\approx \infty$, in water ≈ 1 m	in air ≈ 10 km, in water ≈ 1000 km
smallest image transmitted	≈ 10 μm	in air, bats ≈ 1 cm, in water, whales ≈ 1 km
information	point images	phase & intensity imaging
image analysis	images (parallel processing)	phase and fourier spectrum
superposition	light from different sources does not interfere sources	all sound waves from different add up
other limits	daylight needed	time resolution of brain
problems	biological search beams do not exist	air – water interface transmits only 1/3400 of intensity
3D images	yes, with 2 eyes and brain	very limited
use for communication	limited	excellent, leads to language, learning, social behavior

der of 10^2 W. Acoustical signals must be read sequentially, which takes some time. In contrast optical signals are received simultaneously, instantaneously.

Information has an influence on the survival of animals only if the recording and analysis of data leads to intelligent response, namely seizing opportunities or averting dangers.

The properties of light and sound lead to different forms of intelligence. The simultaneous data of optical signals leads to images, imagination, and recognition of beauty. The sequential data of sound alerts the organisms to sequences, of cause and effect, which points towards logical thinking. The ease of producing and modulating sounds lead to language and education. These two components are essential for social behavior, civilization, and culture.

9.8.2 Why Sound Images Are Not Always Good Enough

Sound, used passively, unveils clues regarding happenings in the environment. Sound used actively, namely sonar, reveals objects even if they are not moving. Cetaceans and bats have learned to actively use sound to map their surrounding, and their prey. Sound is also used very effectively for communication by a great many species.

Unfortunately sound is

- useless for orientation in noisy environments,
- of limited value in muddy waters,
- unsuitable in dense vegetation, where multiple scattering occurs,
- blown away by the wind,
- not good for talking over large distances, because of a slight dispersion: different frequencies travel at different speeds,
- sonar cannot see objects that are smaller than the wavelength λ, (sub millimeter scale), and ultra high frequencies are hard to make, and
- the noise of the sonar beam can well alert the object of the sonar scan.

Many of these disadvantages are overcome by the optical sense. Is there a way of observing the surrounding without giving away the presence of the observer? Light can do so, because during the day the light is always there. On the other hand light is absent during night time, underground, and deep down in the water, and bioluminescence only shines with very low intensity, insufficient to illuminate a scenery. Both light and sound become useless in turbid, muddy waters, where an *electric field sense* may yield alternate information about the surrounding.

Table 9.5. Frequently used variables of Chap. 9

variable	name	units	name of units
f	frequency	$\mathrm{Hz} = \mathrm{s}^{-1}$	Hertz
I	intensity	$\mathrm{W/m^2}$	Watt per square meter
k, k_H	spring constant	$\mathrm{N/m}$	Newton per meter
k_z	wave number, or spatial frequency	m^{-1}	
n	mode number		
Δp_0	sound pressure amplitude	$\mathrm{N/m^2} = \mathrm{Pa}$	Pascal
s_0	displacement amplitude	m	meter
T	period	s	second
u_0	oscillation velocity amplitude	$\mathrm{m/s}$	
v	phase velocity		
Z	impedance	$\mathrm{kg\,m^{-2}s^{-1}}$	
β	sound level	dB	decibel
γ	adiabatic exponent		
λ	wavelength	m	meter
ω	angular velocity of oscillation	$\mathrm{rad/s}$	

Problems and Hints for Solutions

P 9.1 Long Distance Whale Talk

Assume that whales produce sound at a level of 100 dB re $I_0 = 10^{-12}$ W/m². Suppose they talk in the Sofar channel between Alaska and Hawaii. a) What is the intensity that reaches the other end (assume cylindrical spreading). b) How long does it take for a message to go from one end to the other? c) If there is background noise of 40 dB, and the signal must be 3 dB above the noise level, over which distance can they talk to each other?

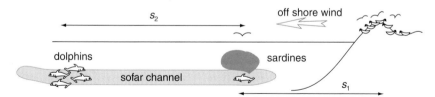

Fig. 9.38. Sea gulls and dolphins

P 9.2 Phase Velocities, and Travel Times

A swarm of sardines is discovered by a roaming seagull, and by a dolphin. Simultaneously both scouts call their friends: there are other seagulls sitting on the shore 2 miles away, and a school of dolphins is playing 3 miles out to sea. a) How

long does it take for both groups of animals to hear the signals of their scouts?
b) If both predator groups depart to the new feeding ground immediately, which
arrives first, and by how much? (To answer this question you must find the average speed of a gull and a dolphin.) c) How does the race go if there is a 30 miles/h
off shore wind?

P 9.3 Seal Mother Calling

A mother seal can measure the distance to her baby swimming at the surface by
listening to its cry under water and above the surface. Sound travels faster in water than in air, thus the cry is first heard in the water and later in the air. By detecting the time difference between sound heard above and below the surface the distance can be determined. a) If this delay time is 115 ms, how far is the baby away?
Speed of sound in air is $v_{air} = 340$ m/s, in water $v_{water} = 1500$ m/s. b) Could she be
confused by any sound that travels halfway through the air and then travels the
rest of the way under water thus having an average velocity somewhere between
the sound velocity of air and water? Explain. c) Is she likely confused by other
sounds? Make a list of typical sounds, created at the water surface that could be in
the natural environment of the seals.

P 9.4 Helmholtz Resonators and Organ Pipes

a) Obtain the typical dimensions of a frog's mouth and its throat volume while the
animal croaks. Calculate the frequency of its voice assuming that the vocal organ
can be modeled as a Helmholtz generator. b) Estimate the frequencies that could
be produced with the air sacks in the dolphins breathing apparatus, if these cavities could be considered as Helmholtz resonators or as closed organ pipes.

P 9.5 Fundamental Voice Frequency, and Vortex Shedding

Objects of lateral width d, which are imbedded in a steady flow of air, shed vortices
at the rate $f_{vs} = St \cdot U/d$, where U is the flow velocity around the object. St is the
Strouhal number, which has the typical value St= 0.2. Model the opening between
the vocal cords (height h and width d) as a vortex shedding object, and derive the
vortex shedding frequency as function of the body mass M and the vocal cord dimensions h, and d. Remember that a person of $M = 70$ kg needs to breathe
2.2 l/min to support the resting metabolic rate ($\Gamma = 3.6 M^{3/4}$) if all the oxygen is entirely extracted. (Actually the typical breathing rate for a male is 6 l/min, explain
why.) Assume a typical $h = 2$ cm, and a lateral dimension $d \approx 0.5$ cm of the vocal
cords, and assume that the average opening of the vocal cords is $d \approx 0.3$ cm. Calculate the vortex shedding frequency for a 50 kg female. Show how f_{vs} scales with
body mass M and opening dimension d and h.

P 9.6 Elastic Vocal Cord Oscillator

The vocal cords are pulled lengthwise by muscles and they vibrate when we speak,
Fig. 9.21. Assume that they change their length periodically as they vibrate. Treat a
vocal cord as an elastic object of length $h = 2$ cm, width $D = 0.3$ cm and thickness

$\Delta = 0.3$ cm with a Young's modulus of $Y = 10^8$ N/m², and determine its elastic oscillation frequency f_{el}.

P 9.7 Exercise: 1D versus 2D Sound Orientation

The human ear is set up to determine the left-right direction from which a signal comes. An individual turns the head to ascertain the sound direction. This is a good enough technique for animals that hunt on the ground, and thus only have to worry about 2 dimensions. Birds need to know the left-right and up-down direction. Therefore night hunters like owls must turn their head around a vertical axis to find the left-right direction of a sound source and must turn the head around a horizontal axis to ascertain the up-down direction. Explain why.

P 9.8 The Mole Cricket

Waldo the mole cricket believes that he makes sound with a Helmholtz generator. His vocal instruments are tuned to a frequency $f_0 = 2500$ Hz. He has dug out a hole of $V = 4$ cm³. He sits in a narrow passage of $L = 3.0$ cm length.

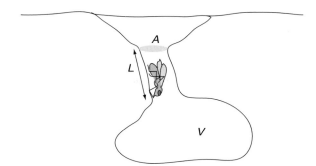

Fig. 9.39. Mole cricket in his burrow. The volume V and the bottleneck, of length L and cross section area A form a Helmholtz generator

a) If he wants to tune the resonance frequency of his Helmholtz-burrow to its instrument frequency f_0, how large should he make the bottleneck area A? b) If the cricket digs the channel cross section area A just 10% larger than that calculated in part a), would the Helmholtz frequency go up or down? c) By what % would the frequency change? d) If the cricket sound spreads out uniformly into the hemisphere above his burrow and this sound has a sound level of 40 dB at a distance of 60 m what is the power of the sound source? e) How large is the displacement amplitude s_0 of the air molecules at this distance?

P 9.9 Blue Whale Voice

The lung of a blue whale has about the size and volume of a Volkswagen Beetle. It is connected to the blowhole through a windpipe of about $d \approx 30$ cm diameter, one meter long. Estimate the lowest frequencies that this system could generate using any of the principles described in this section?

P 9.10 Shark Attack

A diver makes a deep dive wearing a facemask that covers eyes, nose and mouth. At 35 m depth he suddenly notices a shark cruising towards him, he lets out a short shriek at the top of his voice with an intensity of 95 dB. What is the intensity of his voice transmitted into in the water just outside his face mask. Just consider the transition of sound from the air to the water. First find the pressure and density at that depth. Look at Sects. 33.1 and 32.2.

Sample Solutions

S 9.3 a) Sound signal travel time in air, $t_a = \Delta s/v_a$. Travel time in water $t_w = \Delta s/v_w$. Difference $\Delta t = t_a - t_w = \Delta s (1/v_a - 1/v_w)$. Distance of source $\Delta s = \Delta t(v_a \cdot v_w)/(v_a - v_w) = 0.115 (340 \cdot 1500)/(1500 - 340) = 50.56$ m. b) Only the very small fraction $(\rho v)_{air}/(\rho v)_{water} \approx 0.0003$ of the impinging sound intensity will be transmitted from air into the water, therefore the other two signals clearly stand out.

c) Sounds of waves, slapping sound of seals hitting the water with their flippers, noise of seagulls, Further there is the noise of the 5 cm long snapping shrimp. They produce water jets of speeds up to $u = 30$ m/s with their claws. At this speed the jets lower the pressure around micro bubbles in the flow below the vapor pressure of water. The bubbles expand up to a few cm in diameter, and subsequently collapse within 0.1 ms producing a loud pop as they cavitate (reported in *Science* Sept. 22/2000).

S 9.5 Hint Assume that the volume flow rate of air into the lungs, φ is a factor of $2.7 = 6$ l/min/(2.2 l/min) times larger than the volume flow ($\varphi_0 = 2.2$ l/min) required to sustain the minimum metabolic rate φ_p. This excess arises because the oxygen is not entirely extracted in the lungs. The rate φ_p is tied to the metabolic rate $\Gamma = 3.6\, M^{3/4}$. Assume that the ration of opening time τ to cycle time T of the vocal cords is a constant $C = \tau/T = \tau f$. Then $f = C \cdot \text{St}\, \varphi \cdot (M/70)^{3/4}/(h \cdot d^2) = \text{const}\, M^{3/4} \cdot h^{-1} \cdot d^{-2}$.

S 9.8 a) Solve the relation for the Helmholtz frequency for the area $A = LV(2\pi f/v)^2 = 0.03 \cdot 4 \cdot 10^{-6} \cdot (2\pi \cdot 2500/340)^2 = 2.56 \cdot 10^{-4}\, m^2$, where a sound speed $v = 340$ m/s has been assumed. b) The sound frequency will increase. c) By differentiation of the Helmholtz frequency relation (9.41) one finds $df/f = \frac{1}{2}\, dA/A = 5\%$. d) Solve $40\ dB = 10 \log_{10}(I/10^{-12})$ for $I = 10^{-8}$ W/m². The sound is emitted into the semisphere with the surface area $A = 2\pi R^2$. Solve $I = P/A$ for $P = I \cdot A = 10^{-8} 2\pi\ 60^2 = 0.226$ mW. e) $s_0 = (\frac{1}{2}\pi f) \cdot (2I/\rho V)^{1/2} = (1/5000\,\pi) \cdot (2 \cdot 10^{-8}/1.29 \cdot 340)^{1/2} = 0.43$ nm.

10. Body Electronics and Magnetic Senses

The molecules that came together
In early forms of life possessed a double nature:
They carried energy as well as charge in their design.
The charge may well have been unwanted ballast.
However, charge allowed to carry signals,
Which now enable action, thinking, pain.
These are the basics for all higher organisms
And are the tools of homo sapiens.

Electrical Effects Appear on Several Levels

Electricity and magnetism, the twin sisters of the electric and magnetic field have separate utilities for animals. Electrical effects act inside and between cells, and thus enable body functions. They provide external information, and electrical language. Magnetic fields help with navigation.

Electrical effects mediate ionic reactions in the cells. By keeping wanted ions inside, and unwanted charge carriers outside, all cells become charged capacitors – little Leyden jars. Electrical charge – waves carry the signals along nerves, the hard-wired cables of the body, which enable all the controls and senses. For instance, electrical effects relate information from the eyes, ears, nose, and fingertips to the brain, and send instructions back to the muscles.

Further, electrical effects facilitate the *seventh sense*, the electrostatic perception of sharks, tropical fish, and platypus, which primates never acquired. Magnetic effects are the basis for the *eighth sense*, the magnetic sense of bacteria, turtles, pigeons, and many other animals, to detect north and up-down on the globe, and provide clues about the location. These two electric/magnetic (e-m) senses complement the major six senses of seeing, hearing, balance, touch, smell, and taste.

The e-m senses are true field senses – they act over distance. The magnetic field of the earth is present everywhere, so the magnetic sense is global. Electric field senses on the other hand are limited to close range, since the biological electric fields, to which they respond, are generally small. The electrostatic and magnetic senses are important in settings where light and sound fail. They give important information in locations that are bare of physical orientation points: muddy waters, foggy skies, the open seas, and generally noisy environments.

Living cells generate small electric fields and some animals have learned to *see* with their electrostatic sense a meal hidden under the sand, or in turbulent waters.

10.1 The Electrical Machinery of Life

From the very beginning, electrical effects have been important to build organisms and to maintain life functions. Eons ago chain molecules with polar ends assembled themselves to form lipid bi-layers, creating little bubbles that would be-

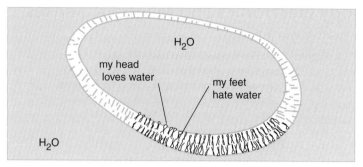

Fig. 10.1. Lipid bi-layer forming a closed vesicle that may harbor life

come the cell walls of the first single cell organisms. Cell walls must have the ability to place molecules that are *good* for life inside, and *bad* ones outside. It so happens that the inside of a living cell contains more negative charge. If charges cannot freely go from inside to the outside the cell wall has insulating properties. An insulating vessel filled with charges acts like a capacitor. By adding switches to allow charge to flow occasionally, an electrical discharge system was invented which would become much later in the evolution the basis of senses, nerve conduction and brain functions.

Many life functions depend on the charging and discharging of living cells: detecting signals of sound, light, and heat with sensory cells, conducting nerve impulses, switching cells in the brain, activating muscles, giving lectures, expressing political opinions, and writing exams.

10.1.1 Life Started in Leyden Jars

The basic building blocks of big bodies are the cells: assemblies of organic molecules and various inorganic ions enclosed by a membrane. The membrane, typically $d=7$ nm thick, is bridged by several different types of large organic molecules. Some of these act as ion pumps that move certain ions to the inside and others to the outside. Other large organic molecules act as valves or gates that let some of these ions go to the other side, Fig. 10.2. A living cell requires certain concentrations of these positive and negative ions, which in total amount to a net negative charge inside the cell. Therefore, each cell acts like a small capacitor, similar to a Leyden jar: the cell has a potential difference (voltage) between the inside and the outside, which changes if charges are allowed to flow through the membrane.

In order to support life functions, certain ions such as Na^+, K^+, H^+, Cl^-, bicarbonate molecules, and organic molecules, have to be either outside or inside the cell. Typical ion concentrations are given in Table 10.1.

Due to the concentrations of these ions, the inside of the cell is negatively charged at −70 mV. The voltage difference of $V=-0.07$ V across the $d=7$ nm thick cell wall generates an electric field E,

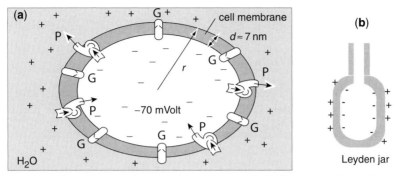

Fig. 10.2. (**a**) Cells, radius r with pumps (P), and gates (G). (**b**) "Leyden Jar" capacitors that store charges

$$E = V/d = 0.07\ V/7\cdot10^{-9}m = 10^7\ V/m\ . \tag{10.1}$$

Note that this field is stronger than the minimum field $E=3$ kV/mm that causes electrical breakdown in air[1]. There are no sparks across the cell wall because sparks arise only after there has been an avalanche of electrons through a certain distance of space, say 50 μm. All avalanches need a distance to develop. A snow avalanche in the mountains, which has run for only a few feet does not have much punch. More will be said about this in the section on nerve conduction.

10.1.2 Forces Created by Electric Fields

An electric field E is created between positive and negative charges. The field exerts forces on charges q that happen to be located inside the field. Positive charges are pushed away from the +side of the field whereas negative charges are attracted to the +side of the field. Charge is measured in the unit coulomb Cb[2]. Such electrostatic forces are the basis for the generation of muscle forces.

The local direction of the force can be illustrated by field lines that go from the positive charge to the negative charge. At any point in space an electric field E has a *magnitude* and a *direction*. The electric field is a vector measured in the unit Volt/m. If there is an electric field E any charge q will experience the force

$$F = q\cdot E\ . \tag{10.2}$$

---

[1] Typical electical field strengths in household appliances are 10^3 V/m. For instance if a bare electrical wire of an 110 V line is located 1 cm away from the ground, the electrical field is $E=110$ V/0.01 m$=1.1\cdot10^4$ V/m. The air in a spark gap breaks down in atmospheric air if a field strength of 3 kV/mm$=3\cdot10^6$ V/m is applied. This breakdown occurs at the minimum $(p\cdot d\approx6\cdot10^{-3}at\cdot mm)$ of the Paschen curve that plots breakdown voltage versus the product pressure p times gap distance d.

[2] Capacitance has the unit Coulomb with the symbol Cb, which is chosen so that it cannot be confused with the parameter for capacitance C.

Consider a simple example: an electron with the charge $q=e=1.9\cdot10^{-19}$Cb inside the cell membrane *sees* the field $E=10^7$ V/m and experiences the force

$$F=eE=1.6\cdot10^{-19}\cdot10^7 \text{ V/m}=1.6\cdot10^{-12}\text{ N}=1.6 \text{ pN} . \tag{10.3}$$

The unit pN is called a pico Newton. The word pico represents the factor 10^{-12}. This force is approximately ½ the force that an individual cross-link pushes the *myosin* head parallel to the *actin* in a muscle fiber.

10.1.3 Moving Charges into and out of Cells

Cells have the electrical property of capacitors, which are characterized by their capacitance C, measured in the unit Farad F

$$C=\frac{\varepsilon_0 K A}{d} . \tag{10.4}$$

The permittivity constant of a vacuum $\varepsilon_0=8.85\cdot10^{-12}Cb/Nm^2=8.85\cdot10^{-12}F/m=$ $8.85\cdot10^{-12}$ pF/m has the unit pF/m pronounced picofarad per meter. The dielectric constant K describes the material between the surfaces, which hold the charge. The value of K indicates how much more charge the capacitor *with* the cell wall material can hold than vacuum ($K_{vac}=1$). $A\approx4\pi r^2$ is the surface area of a cell, and $d\approx7$ nm is the thickness of the cell wall. Myelinated nerve cell membranes have a typical value of $K\approx8.8$, and a typical specific capacitance of $c=C/A\approx1$ μF/cm^2. The basic electrical phenomena associated with life are the charging (actions of the *pumps*) and the discharging (action of the *valves*) of these elementary capacitors.

Body fluids contain various positive and negative ions. Typical values for blood plasma are given in Table 10.1. Charged particles in motion become part of an electrical current. A moving charge q represents a current element. If n charges per m^3 move at a speed u, there is a the current density j

$$J=nq\cdot u \text{ Cb/(sm}^2) . \tag{10.5}$$

Table 10.1. Typical concentrations for mammalian blood plasma, and author's blood values in μmol/ milliliter. Conversion: 1 μm/ml$=6\cdot10^{17}$ part/cm^3

species	H_2O μmol/mL	Na$^+$	K$^+$	other$^+$	H$^+$	total$^+$	Cl$^-$	bicar bonate	organic$^-$	total$^-$
external	$5.56\cdot10^4$	145	4	5	$3.8\cdot10^{-5}$	154	120	27	7	154
internal	$5.56\cdot10^4$	12	155		$13\cdot10^{-5}$	167	4.1	8		167
ext/int		12	1/39		1/3		29	3.4		
author's		139	3.9				103	27		

The total electrical current J is the sum of all current elements $J = \int jdA$. The basic electrical phenomena associated with life are the charging (action of the pumps) and the discharging (action of the gates) of the elementary "cell capacitors". The flow of current is described by Ohms law, which connects current J, measured in Ampere, electrical resistance R, measured in Ohm Ω, and voltage V

$$V = RJ. \tag{10.6}$$

If there is an electric field E, and charges q are present, these charges will be forced to move. They become part of a current that is impeded, or *slowed down* by the resistivity ρ of the material. This law can be immediately applied to describe the flow of charges through a valve in a cell wall, or through some other section of organic material. The rate at which the charges flow through such a conductor depends on its electrical resistance R. Consider an object of length l and cross section A. The magnitude of its resistance R is related to the specific resistivity ρ of the conducting element, its cross section area A, and its length l:

$$R = \frac{\rho l}{A}. \tag{10.7}$$

Typical values are shown in Table 10.2. Materials with high resistivity are called insulators. When the electrical resistance is small one calls the material a "conductor". Electrical charges can move freely in a good conductor. The electrical conductivity σ is defined as $\sigma = 1/R$

Table 10.2. Resistivities of materials. Data after Denny[1993]

material	iron	carbon	sea water	fresh water	organic material	blood	fat	wood	glass	hard rubber
ρ in $\Omega\cdot m$	$9.7\cdot 10^{-8}$	$3.5\cdot 10^{-5}$	$2\cdot 10^3$	$9\cdot 10^9$	$5\cdot 10^7$	$1.5\cdot 10^7$	$25\cdot 10^7$	10^8 to 10^{13}	10^{10} to 10^{14}	10^{13} to 10^{16}

Organic materials are good insulators. They may have a typical resistance $\rho \approx 5 \cdot 10^7\,\Omega m$. A section of nerve axon, $L = 1$ cm long, typically has an electrical resistance of $R = 2.5\cdot 10^8\,\Omega$, [see Pflegl Nichols 1991].

When voltage differences drive currents through a network of electrical resistors R_1, R_2, ... the currents and voltages will split up in a way that is governed by some simple rules. These are identical to the rules for thermal currents and thermal resistors driven by temperature gradients. Currents passing through resistors mounted in *parallel* (see Fig. 10.3a) split up, and the voltage drop V across the resistors is the same.

$$J = V/R_p = J_1 + J_2 \quad \text{(a)}, \quad \text{and} \quad V = J_1 \cdot R_1 = J_2 \cdot R_2 \quad \text{(b)} \tag{10.8}$$

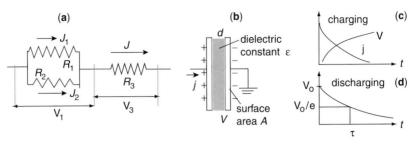

Fig. 10.3. (a) Series and parallel circuits. (b) Capacitor. (c) Current and voltage during charging, and (d) discharging of the capacitor

As a consequence the resistors R_1, R_2, mounted in parallel in Fig. 10.3 combine to the parallel network resistance R_p as

$$1/R_p = 1/R_1 + 1/R_2 .\tag{10.9}$$

If current flows through a network of resistors in *series*, Fig. 10.3a, the current I is the same in every resistor, and the voltages V_1, V_3 across the resistors add up $V = V_1 + V_3$, so that the total resistance of the network is $R = R_p + R_3$.

Due to the action of pumps, the cells in a body are generally charged up. Electric charges start to flow when ionic channels are opened. This process may be compared to the charging and discharging of a capacitor. If a current is made to flow onto the plates of a capacitor, the voltage on the capacitor plates gradually grows. When a capacitor C is discharged by connecting the two plates through a resistor R, the voltage at the capacitor plates decreases gradually from the initial value V_0 according to the value $V(t)$

$$V(t) = V_0 \cdot e^{-t/RC} .\tag{10.10}$$

Within the decay time τ, sometimes called the *e-folding time*

$$\tau = RC\tag{10.11}$$

the voltage decreases from the initial value V_0 to the value $V(\tau) = V_0/e = V_0/2.718 = 0.368\ V_0$.

Example: Take $R = 1.5\ \mathrm{k\Omega}$, $C = 4.3\ \mathrm{\mu F}$, $\tau = 1.5 \cdot 10^3 \cdot 4.3 \cdot 10^{-6} = 0.00645\ \mathrm{s} = 6.45\ \mathrm{ms}$.

Typically the voltage at the end of a nerve cell, called an axon, has a decay time of $\tau \approx 4$ msec. This scenario sets the stage for nerve conduction, sensory organs, brain, and thinking. It also forms the basis for all the active uses of electricity for electro-static sensing, and for electrical warfare (electric eels).

10.2 Conduction of Nerve Pulses

Nerve cells in an organism are integrated networks of cell bodies, dendrites, axons, and terminal branches. A good basic introduction to nerve conduction is given by Pflegl Nichols [1991]. Nerve pulses are generated by charges that move across cell membranes, or that cross gaps between nerve segments.

10.2.1 Gates in Cells

A nerve cell looks like a brittle star fish with many short spiny arms and one very long arm called the axon. Impulses arising from a cell body always flow into the axon and from there to the terminal branches. Axons are the long tail of nerve cells that connect to other nerve, or brain cells. Before looking at the process of conduction, one should have an understanding of the electrical elements that are common to most cells.

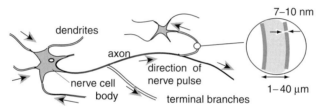

Fig. 10.4. Nerve cell branching

Nerves have an insulating cell wall that is typically 7–10 nm thick, containing pumps that move ions across the cell wall, and gates that allow them to flow in the opposite direction. A nerve axon has typically a diameter of $d=1$–40 μm, and a wall thickness of 7–10 nm. Through the action of pumps, Na^+ ions are accumulated on the outside and K^+ ions on the inside of the cell membrane. This leads to a charge of –70 mV on the inside of the cell membrane. Similar as other cells, the nerve membranes with a thickness of $\delta \approx 7$ nm support an electric field of $E \approx 0.07$ $V/7 \cdot 10^{-9} m = 10^7$ V/m.

The electric field points into the cell. The field would drive the positive sodium ions Na^+ into the cell, if they could flow freely. However, the nerve membrane is, of course, normally a very good barrier to prevent such migration. The *pumps* are big molecules sticking though the nerve walls at many places. They constantly force positive ions to accumulate on the outside, like charges on one plate of a capacitor. Hence, one may consider a section of cell wall as a charged capacitor. If charges were free to flow to the other side, the voltage would first go to zero and then the capacitor would charge up in the opposite direction; the inside would become positive because charges tend to overshoot a neutral equilibrium.

Imbedded in the cell wall besides the pumps are *gates*, which allow Na^+ ions to pass through the membrane if they are open. The gates may be opened by apply-

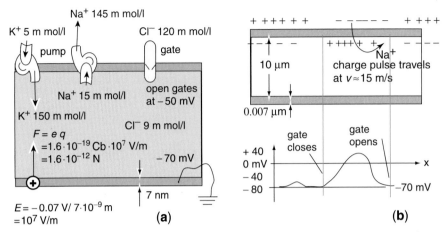

Fig. 10.5. (a) Pumps and gates in a nerve. **(b)** Snapshot of traveling electric pulse as function of position x on axon

ing a voltage, or by mechanical stimulus. While the pumps are the elementary electro-motoric forces, which constantly try to maintain the negative charge on the inside, the gates are the active elements, which can suddenly open and upset the charge distribution. Pumps and gates represent the active elements that lead to nerve conduction and drive all biological electrical effects.

It is easy to imagine a hydraulic analog to this discharging and recharging mechanism. Consider two water reservoirs (a), and (b), which are connected by a pump that continuously lifts water from (a) to the (b), and a valve and large diameter pipe that allows water to flow back to (a). The action of the pump raises the water in (b) to a level determined by the pressure head of the pump, and lowers the water level in (a). When the valve is opened the water from (b) rushes back into (a) gaining speed in the connecting pipe, so that the water in (a) overshoots the equilibrium position. After the valve has been closed the action of the pump gradually restores the higher water level in (b).

The negative voltage at one place inside the cell walls suddenly turns positive when a nerve is triggered to fire. This reversal of polarity causes the nearest gate to open, letting all the nearby Na^+ ions in. If enough charges flow to the inside very

Table 10.3. Comparison of different axon types of identical axon diameter, $D=10$ μm. Data adopted from Davidovits [2001]

axon properties	non myelinated	myelinated axon
resistance per m of fluid inside and outside	$6.4 \cdot 10^9$ Ω/m	$6.4 \cdot 10^9$ Ω/m
conductivity of axon membrane	$1.3 \cdot 10^{-4}$ moh/m	$3 \cdot 10^{-7}$ moh/m
resistivity	$7.7 \cdot 10^3$ Ω m	$3.3 \cdot 10^6$ Ωm
resistance per m of axon length (gates closed)	$\approx 10^{14}$ Ω/m	$4.2 \cdot 10^{16}$ Ω/m
capacitance per unit length of axon	$3 \cdot 10^{-7}$ F/m	$8 \cdot 10^{-10}$ F/m
RC time	$3 \cdot 10^7$	$3 \cdot 10^7$

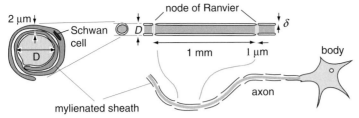

Fig. 10.6. Nerve axon with myelinated sheath

quickly two things will happen. (i) The inside turns positive, reversing the direction of the electric field. (ii) The reversed voltage causes the gate to close again, so that no further charges can flow.

The pumps are continuously operating but initially they cannot keep up with a rapid inflow of Na^+ ions. However, the pumps are working at a steady rate. After the gates have been shut again the pumps gradually restore the initial charge distribution, Fig. 10.5b.

An impulse can be conducted by a nerve axon because each local group of charges that flows through one gate will open the next gate down the line when the inside has become positive enough. The amount of charges available to flow depends on the resting voltage, which in turn depends on the electro-motoric force of the pump and the local capacitance. This next gate down the line will in turn let charges through which subsequently open the adjacent gate. Thus the charge pulse travels along the axon. The speed of the electrical pulse depends only on the rate at which the charges flow in and on the speed at which gates open in response to the changed electrical environment.

The nerve conduction velocity depends on the type of nerve and its diameter. Nerve conduction is faster if the axon is encased in a myelin sheath. The bodies of brain cell nerves are mainly located in the gray matter. They are interconnected by myelinated axons, located mainly in the white matter. Various nerve phase velocities are shown in Table 7.8, and are plotted in Fig. 7.12. The signal conduction velocity v_{nerve} increases steadily with nerve diameter D. The data may be approximated by the relation

$$v_{nerve} = 2.8\ D^{1.3}\ \text{m/s} , \qquad (10.12)$$

where D is given in μm. Thick nerves are fast. The octopus intrinsically knows why it has a thick nerve connecting its ink jet to the brain. This nerve's big diameter of $D=500$ μm [see Davidovits 2001] guaranties a quick response when the animal wants to become invisible in a hurry.

10.2.2 Moving Charges by Mechanical Stresses: Piezo Effect

Some materials such as bones show the piezo effect: when a cylinder of certain materials is stressed in axial direction, one end charges up positively, and the opposite end charges up negatively. This creates an electrical field E. The more stress is applied, the larger the field becomes.

When animals move around their bones are exposed to stresses of varying magnitude. The piezo effect translates these stresses into voltage signals. Apparently bones are reinforced in response to these local voltages: bones are built up most where the largest stress appears. Thus, the piezo effect is a simple feedback mechanism for building skeletons that are strong only where strength is needed. Similarly the piezo effect may play a role in soft tissue healing.

10.2.3 Electrical Signals of Muscle Activities

Muscle activities are always associated with electrical signals. This is well known from every electro cardiogram (ECG). The electro-motoric force *emf* of the muscle is a repetitive function of time, with signal strength of the order of milli volts. The features of the voltage trace of an ECG are associated with parts of the heart activity, Fig. 10.7. These features are generally not exactly reproducible for every heart-beat.

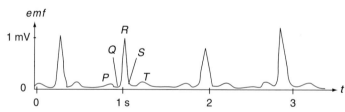

Fig. 10.7. Typical electro cardiogramm (ECG) with atrium polarization (P), ventricle polarization (QRS), and re-polarization (T)

10.3 Passive Use of Electrical Fields

Living cells produce weak electric fields. In turbulent and muddy waters and during the night, neither light nor sound can be used to find the way. Then the detection of biological electrical fields can help an organism to find a meal. Indeed, several aquatic animals possess the ability to detect electric fields: Sharks [see Bastian 1994], fresh water fish, star nosed moles [see Zimmer 1993], and platypus. They have developed this ability since their prey inadvertently generates small electric fields.

10.3.1 The 6th Sense: Electrical Detection of Prey

To see how organic prey may be found, consider the electric fields which are produced by a living organism. Let the animal, which is buried in the sand, be represented by a battery.

Body functions generally involve weak electrical potentials. Some of these biological electrical fields leak out of the body. For instance, the gills of animals buried under sand emit a small electric field, and a shrimp [see Schleich 1973] may produce a field of $E=1\,\text{mV/cm}$ at a distance of $r=8$ cm.

Electrical fields spread out in space through non-conducting materials such as sand and water. A shark can detect electric fields as small as $E=5\cdot10^{-9}\,\text{V/cm}=5\cdot10^{-7}\,\text{V/m}$. Platypus responds to fields of $E=5\,\mu\text{V/cm}=5\cdot10^{-3}\text{V/m}$. This 6th sense, with sensors to detect very weak electrical fields, enables such animals to locate prey buried under sand and in muddy waters.

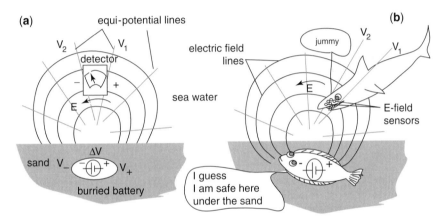

Fig. 10.8. (a) Electrical field of battery buried under sand. (b) Shark and flounder

The principle of detection is shown in Fig. 10.8a. A *battery* sets up an electric field, which extends through the surrounding medium, such as wet sediments and water. Shown are some field lines, and equi-potential lines. Along the field lines the voltage decreases by the amount $\Delta V=V_+-V_-$ because the voltage must decay from the value V_+ at the plus pole of the battery to the value V_- on the minus pole on the battery. The electric field produces the potentials V_1 and V_2 at the location of the detector. The voltage difference $\Delta V=V_2-V_1$ can be measured.

The biological application is illustrated in Fig. 10.9b. The shark has special organs to detect very small voltage differences. In order to sense the location of the voltage source the animal must be able to detect the direction in which the electric field increases. A similar system would not work in air, because the electrical resistance of air is so high that voltage differences created by animal tissue between detectors, which are separated by fractions of a meter, would be much too small to be picked up by any biological system.

10.3.2 Electrical Detectors of Fish

The electrical field detectors of aquatic animals look roughly like bottles with long and narrow necks that stick like pores through the skin, Fig. 10.9. The bottles have a relatively high electrical resistance. The bottles are filled with *lumen*, a fatty tissue of lower resistance.

The electrical transducers that convert an electrical field into a nerve signal are big cells located at the bottom of the detector bottle. The sensory cells act like leaky condensers: if more current flows in than diffuses out of the cell, the cell charges up. The state of charge is detected by the sensory neurons. In the presence of electric fields Ca^{2+} ions migrate into the cells, and K^+ ions move out. This causes some net charges to appear at the connected nerve ends, thereby sending off a nerve signal. Saltwater animals respond to low frequency fields. Some fresh water fish can sense oscillating fields with frequencies in the kHz range.

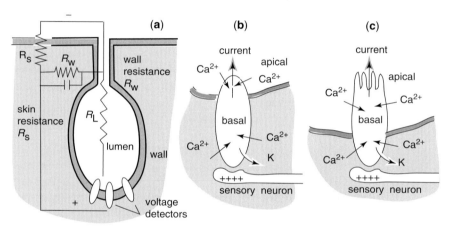

Fig. 10.9. (a) Electrical detectors imbedded in skin after Bastian [1994]. (b) Low frequency marine detectors, (c) high frequency detectors of weakly electric fresh water fish

Sharks carry electrical detectors in the front and on the top of the back of the head. They use these sensors in close vicinity to their prey. They shut their eyes during an attack. Divers make use of this fact when they work inside a metal cage. The metal mesh acts like a Faraday cage, which prevents any electric fields on the inside from leaking out. Therefore, the shark cannot detect the biological field of the diver when approaching the cage, and turns away. Of course a sturdy metal cage is also a good mechanical barrier.

10.3.3 Platypus

The platypus feeds in murky waters of rivers and streams often at night, with ears, eyes, and nose closed. It has a flat rubbery bill similar to that of a duck. The animal has about 13 000 electro receptors located on the outside of the bill. These gland

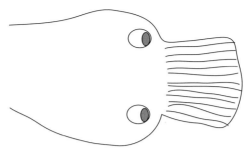

Fig. 10.10. Platypus head with beak and gland duct receptors

duct receptors have a sensitivity threshold of about 5 µV/cm, see Schleich [1986]. The sensors are arranged in a striped pattern.

Some of the sensors also respond to pressure pulses, Pettigrew et al. [1972]. In the search phase the platypus scans its environment by moving its head periodically from side to side in quick and jerky motions called saccades lasting $T = 1/3$ to ½ second. When coming closer to a potential prey the platypus may increase its sweep rate from $f \approx 3$ to about 14 Hz. Apparently the animal senses the direction in which the electric field increases.

10.4 The Active Use of Electric Fields

We all know the happy sound of birds in fields and forests. A similar variety of *electrical* noises are produced by electrical fish in tropical waters. Many tropical fish have very distinct electrical melodies. These e-songs may indicate territorial claims or inter-species communication. Some fish produce continuous wave signals, like the chirping of crickets. These are called *wave* fish. Others burst out in short electrostatic blips, like the twittering of chickadees. Such fish are called pulse fish. Generally these electric fields are very weak, with magnitudes of a few millivolt per meter close to the animal. These weak fields drop off rapidly over a few body lengths.

There is a darker side to the use of electrostatic pulses. Some fish are able to produce voltage pulses of a few hundred volts and deliver currents of the order of 1 Ampere with a duration of a few milliseconds. They use electrical discharges as weapons.

10.4.1 Fields and Signals

How do such signals come about? Any animal generates voltage pulses as part of its ongoing metabolism. Humans typically have variations of $\Delta V \approx 100$ mV, caused by gates and pumps, see Sect. 10.2. By switching n cells in series it is theoretically possible to produce voltage pulses of the magnitude $\Delta V_n = n \cdot \Delta V$. It is then only a small step to generate arbitrary sequences of electrical pulses or single-frequency-

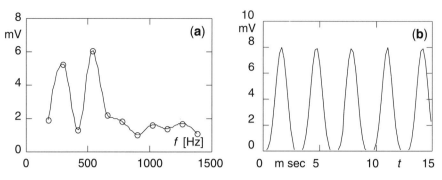

Fig. 10.11. (a) Fourier component of a *pulse fish* signal. (b) Typical *wave fish* signal

waves. Electrical fish have learned such tricks. These animals are often quite small.

Some of these fish generate pulses, others emit continuous waves. Pulses can be thought of as a superposition of different frequencies. In a classroom experiment, a small African elephant nose fish of the species *gnathonemus petersii* was observed in a tank equipped with electrodes connected to a recording system with a Fourier analyzer. The frequency components for this pulse fish are shown in Fig. 10.11a.

An example of a wave fish is the black ghost knife fish, *apteronotus albiforns*. Two specimens were studied during a lecture demonstration. Figure 10.11b shows a typical waveform. The electrical signals seem to express an identity for each animal. We observed signals of two specimen in separate tanks. Each chirped at the fundamental frequencies $f_2 = 1388$ Hz, and the first overtone $f_3 = 2088$ Hz. Then both fish were placed into the same tank. Now one of the individuals changed its pitch: we then recorded an additional song at the frequencies $f_2' = 1448$ Hz and $f_3' = 2178$ Hz. Note that the ratios of the frequencies remained the same: $f_3/f_2 = f_3'/f_2' = 3/2$.

Occasionally it is observed that two fish sing at very close frequencies, so that the resulting electrostatic waves produce a beat frequency.

The electrical languages, namely the voltage-time traces, and the frequency spectra of pulse and wave fish are quite diverse, and are distinct for each species, like bird songs in a forest.

Electrical fish are only found in fresh waters of the tropical regions of the globe. Fresh water has a high resistivity ρ_f, so that an electric fish only has to generate a small current $J \propto V/\rho_f$ in order to maintain its body voltage V. In contrast the resistivity of salt water is about 3 orders of magnitude lower, see Table 10.2. Therefore, in salt water an electric fish would have to support about 1000 times larger currents. This would be too large a burden on the metabolism of the animal.

10.4.2 Living Batteries and Voltage Sources

The bio-electrical signals are generally in the milli-volt range, and the electrical organs are often restricted to a small tail-section of the body. However, there are exceptions, such as the electric eel, Fig. 10.12a. Its electrical organs are composed of $N_{la} \approx 120$ lamellae which in turn contain up to $N_{el} \approx 2500$ flattened cells called electrocytes. Two smaller organs are used for navigation, and the big unit is its attack weapon. The electrocytes are individually charged by pumping Na^+ ions to the outside. This process takes typically $\Delta t_{ch} \approx 25$ s.

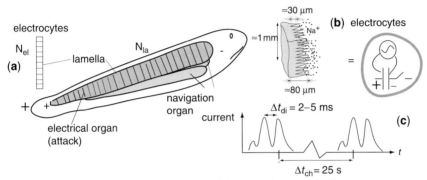

Fig. 10.12. (a) Electric eel. (b) Electrocyte model. (c) Discharge pulse form

When fully charged the electrocytes have a potential of −90 mV, due to the concentration of K^+ ions inside and Na^+ ions outside the electrical cells. The electrical discharge is initiated when a nerve pulse originating in the brain liberates the neurotransmitters acetylcholine from the nerve terminals on the electrocyte's caudal (tailward) surfaces. The released acetylcholine binds to the electrocytes, opening chemically gated sodium channels. This results in an influx of positive ions. As a consequence, the potential inside the electrocytes briefly goes up to +50 mV, resulting in a voltage spike of $V_0 = 0.05$ V−(−0.09 V)=+0.14 V and a discharge current of about $I_0 \approx 10^{-4}$ A. When the N_{el} electrocytes in one lamella are discharging in parallel they generate a current $I = N_{el}I_0 \approx 1$ A that lasts for typically $\Delta t_{di} \approx 2$–5 ms.

Since the sodium channels are only located on the caudal side of the cells the sodium flux is unidirectional in the caudal to rostral direction. Insulating tissue around the organ prevents internal short circuits so that a voltage pulse up to $V = N_{la} \cdot V_0 \approx 600$ V arises when the lamellae fire in series.

10.5 Navigation by Magnetic Fields

The magnetic field of the earth, weak as it is, presents a bearing by which animals can find directions (compass sense), or changes in position otherwise undetectable such as those found in the open sea, or high in the air (map sense). This opportunity has been realized by dolphins, whales, turtles, sharks, alligators, algae, mollusks, pigeons, nightingales, thrushes, honeybees, blind mole rats, hamsters, newts, salamanders, and bacteria. The magnetic sense uses the inclination and/or intensity of the Earth's magnetic field to detect compass North. Further, the motion of charged particles at some velocity u across magnetic fields B induces electric fields of the magnitude $E=u\times B$. Such induced electric fields may give sharks navigational clues.

10.5.1 The Origin of Magnetic Fields

Magnetic fields B are always there when currents J flow steadily, or when displacement currents $\varepsilon_0\partial E/\partial t$ are suddenly initiated. Currents are surrounded by magnetic fields. For a steady current of the current density j (measured in Ampere/m²) the magnitude of B can be found from Ampere's law.

$$\int B dl = \mu_0 \int j dA \tag{10.13}$$

The right-hand side of this equation contains the total current $J=\int j dA$, and the left-hand side is the line integral of the magnetic field encircling this current, see Fig. 10.13a. For instance, at the edge of a cylindrical conductor of radius R the magnetic field has a value $B_0=\mu_0 J/2\pi R$, where $\mu_0=4\pi\cdot 10^{-7}$. The field drops off with radial distance r as

$$B(r)=\mu_0 J/2\pi r. \tag{10.14}$$

Three different units for magnetic fields are in common use: Tesla T, Gauss G, and gamma Γ. They differ by factors of ten. $1\,T=10^4\,G=10^9\,\Gamma$. The direction of the

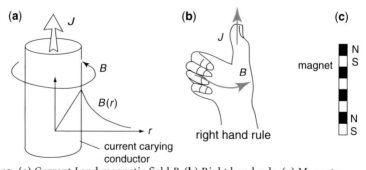

Fig. 10.13. (a) Current J and magnetic field B. (b) Right hand rule. (c) Magnets

magnetic field of a current J is given by the right hand rule, illustrated in Fig. 10.13b.

Magnets line up in magnetic fields. A magnetic field exerts a force acting onto magnets. A magnetic North pole N always attracts a South pole S. This effect lines up magnets in alternating sequences: North – South – North – South, Fig. 10.13c.

A charge q moving at a velocity u through a magnetic field B experiences the force[3].

$$F = q \cdot u \times B = j \times B \tag{10.15}$$

If the charge is attached to a particle, which is embedded in a fluid flow with the velocity u, an electric field E_{ind} is induced. It has the magnitude

$$E_{ind} = v \times B . \tag{10.16}$$

Such fields were first discussed by Faraday, but sharks have used them long before: they can infer the local water flow velocity v by detecting with their electric field sensors the induced field E_{ind}.

10.5.2 The Earth's Magnetic Field

Several animals like pigeons, trout, and honey bees, have acquired the ability to navigate by means of internal magnetic *compass needles*: magneto-tactic bacteria.

The bacteria are part of the magnetic sense of these animals, which helps them find their way using the magnetic field of the earth B_e. It has the average magnitude

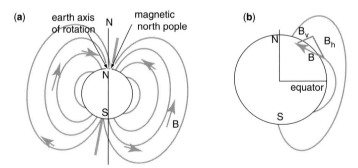

Fig. 10.14. (a) Earth's magnetic field. **(b)** Field lines dipping in vertical and horizontal direction

---

[3] The cross product symbol $\times$ indicates that the force F points – by the right hand rule – into a direction at right angle to both the direction of the velocity vector u and the direction of the magnetic field vector B.

$$B_e \approx 5 \cdot 10^{-5}\,\text{T} = 0.5\,\text{G} . \hspace{3cm} (10.17)$$

On the northern hemisphere the magnetic compass needle points towards the earth's magnetic north pole, which is located in the North West Territories of Canada. The magnetic north pole has no fixed location. It moves presently at a rate of ≈ 40 km/year in eastern direction [Jacob 2002]. Apparently the magnetic field of the earth changes its direction reversing North and South with a cycle time of 500,000 years. The earth's magnetic field has a horizontal B_h and a vertical B_v component, see Fig. 10.14. The downward direction is important for bacteria. North South is important for sailors, and large animals.

10.5.3 The Magnetic Sense

Magnetic fields help animals find direction (map sense), and location (map sense). The magnetic sense likely involves small biomagnetite particles found in the above animals' heads. These particles range in diameter from < 40 nm, to 50–100 nm, to 1–3 μm. They are found in cells whose associated nerves respond to changes in induced magnetic fields. Biomagnetite is essentially paramagnetic, in other words it has randomly-oriented fields that can align in an induced magnetic field. Also, pigeons have clustered 40 nm particles in 1–3 μm diameter clusters, arranged in chains 150–200 μm long. Sensitivity thresholds for a list of animals with known magnetic senses are shown in Table 10.4.

There are several principles by which magnetic fields could stimulate sensory response. Biomagnetite crystals may exert a torque on intracellular filaments that is large enough to trigger a sensory neuron. This mechanism would be the basis for the compass sense. The migrating loggerhead turtle is thought to detect the

Table 10.4. Magnetic abilities for some species

animal	compass sense	map sense	biomagnetite characterization	sensitivity to B (lowest threshold)
pigeon	c	m	50–100 nm particles in chains, on brain surface and in beak	10–20 nT inferred from experiments
dolphin	c	m	in head tissue	< 2000 nT
salmon	c		50 nm particles in snout	
rainbow trout		m	50 nm particles in cells 10–15 μm diameter, 2–3 cells/250 μm³	< 7500 nT
magnetotactic bacteria	c		40–100 nm particles in a chain 1200–1600 nm long	30–45 μT
honeybee	c		50 nm particles	260 nT
thrush, nightingale	c	m	?	< 7200 nT
seal, turtle	c	m	?	< 9000 nT

difference in the earth's magnetic field at two separate points on its migration path, thereby providing a clue about the location.

From studies on rainbow trout it is known that changes in the magnitude of an induced magnetic field stimulate sensory nerves adjacent to cells containing biomagnetite. The exact mechanism has yet to be experimentally determined theoretically, the use of torque is a plausible mechanism.

10.5.4 Magneto-Tactic Bacteria with Ideal Compass Needles

The basic elements for magnetic navigation are small crystals of magnetite Fe_3O_4 (lodestone). An interesting optimization is found by magneto-tactic bacteria. These small organisms of length $L \approx 1$ μm, volume V, and density ρ, live in sediments of density ρ_o, where *up* is bad (too much poisonous oxygen) and *down* is good (lots of nutrients) [Blakemore and Frankel 1981].

With their small size they are subject to Brownian motion, which knocks them about, and tumbles them around. Gravity is not large enough to tell them where down is, because the energy E_{gr} gained by falling through the height $h \approx L$

$$E_{gr} = (\rho - \rho_o)VgL = 0.1 \cdot 10^3\ kg/m^3 \cdot 10^{-18}m^3 \cdot 10^{-6} \approx 10^{-22}\ J \tag{10.18}$$

is much smaller than the energy of Brownian motion $E_{br} \approx k_B T \approx 4 \cdot 10^{-21}$ J.

However, since the bacteria have a built-in compass needles, the earth's magnetic field B_e applies a torque and aligns the organisms in the field lines, which have a downward component. By swimming along this direction the bacteria can move downwards to food and safety. The magnetic moment of the internal *compass needle* $M \approx 6 \cdot 10^{-20}$ J/G, in the earth magnetic field $B_e = 0.5$ G generates the energy

$$E_{mag} = 6 \cdot 10^{-20} J/G \cdot 0.5\ G = 3 \cdot 10^{-20} J > E_{br} > E_{gr}. \tag{10.19}$$

This inequality indicates that the torque generated by the magnetic field easily overcomes the random Brownian motion. The magnetic needle is made up from l

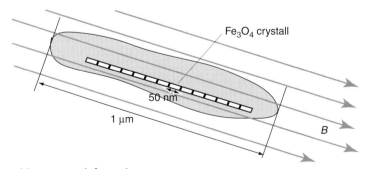

Fe$_3$O$_4$ crystall

50 nm

1 μm

B

Fig. 10.15. Magneto-tactic bacterium

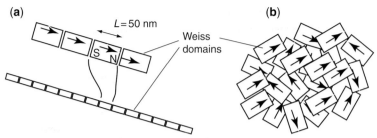

(a) $L = 50$ nm   Weiss domains   **(b)**

Fig. 10.16. Fe_3O_4 crystals. (a) $L \approx 50$ nm: ferromagnetic single crystals lined up. (b) $L >$ 100 nm, multiple Weiss domains with random magnetization

≈ 50 nm long Fe_3O_4 crystals. About 20 of these small crystals arrange themselves in one or two magnetic chains in the cytoplasm parallel to the long direction of the cell, Fig. 10.16a. Each small magnet is surrounded by a sheath that is always adjacent or contiguous with the cytoplasmic membrane. There are *north-seeking* and *south-seeking* bacteria species.

Nature has optimized the dimensions of the small Fe_3O_4 magnets, to maximize the magnetic strength, [Blooom, and Mouritsen 1995]. Crystals of less than 40 nm length are super-paramagnetic. Crystals of length l from 40 nm to 100 nm are ferromagnetic single domain structures. Fe_3O_4 crystals of over 100 nm length contain more than one Weiss domain, with statistically disoriented magnetic vectors, see Fig. 10.16b. In order to act like strong magnets these structures would have to be magnetized from outside so as to line up all the Weiss domains. By keeping the size of the Fe_3O_4 crystals to about 50 nm, the organism obtains ferromagnetic structures, which need not be magnetized, because they automatically line up head to toe, a South Pole seeking out the North Pole of the next unit crystal.

Magnetic bacteria that became fossilized, retained their magnetic properties, and have become magnetite. In spite of many years of research on magnetic storage for computers, humans have not managed to produce magnetic materials as homogeneous as found in *magnetite* from ancient bacteria. The magnetite crystals assembled by bacteria are perfect and free of any defects. This structure remains if such bacteria collect at the bottom of a lake or ocean and become part of the forming magnetite layer. Such pure crystals differ strikingly from any magnetic mineral that the entire recording industry for computer discs and magnetic tapes has produced in the last 50 years.

10.5.5 Pigeons Trout, Turtles, and Dolphins

Many animals can detect the earth magnetic field, small as it is at $B_e \approx 0.5$ Gauss [Keeton 1974], and they navigate with built in "compass needles". Examples are pigeons [Winklhofer et. al. 2001], [Walcott et al. 1979], trout, turtles [Lohmann and, Lohmann 1996], and dolphins [Zoeger et al. 1981].

The direction of the magnetic field is the first coordinate for orientation. Superimposed on this average field are local perturbations, which vary more rapidly, and may therefore yield a second coordinate for orientation [Walker et al. 2000]. The average earth magnetic field may vary in magnitude and direction from year to year by $4 \cdot 10^{-4}$ to $10^{-3}\,\Gamma$. Often homing pigeons react with variations of their flight path, in response to these fluctuations of the earth's field. Therefore, it is believed that the sensitivity threshold of the pigeon *compass* is $\Delta B \approx \pm 10^{-4}\,\Gamma$. The slow variation of the earth's magnetic field apparently does not diminish the magnetic navigation abilities of these animals.

10.5.6 Life on Mars?

In 1984 scientists discovered a meteorite on Allan Hills, Antarctica that was later identified with high probability as a sedimentary rock fragment coming from our neighbouring planet Mars. Very pure magnetite crystals were found recently on this specimen, which are entirely free of any defects. Scientists from NASA [Clement 2000] concluded that these crystals must have been assembled by magnetic bacteria. This presumed bio-magnetic origin is the first tangible evidence for life on Mars.

10.5.7 Orientation by $u \times B$ in the Earth Field

When a fluid carrying charges flows at the velocity u across a magnetic field B, an electric field E is generated at right angles to both the flow and the magnetic field.

This effect is used in magneto hydrodynamics generators, called Faraday generators, and it has a practical application for animal navigation. The strength and direction of this field are given by the vector equation

$$E = u \times B,\tag{10.20}$$

where B is in Tesla T, u in m/s, and E in V/m. Weak electric fields are generated by this effect when water currents with ionic charges, flow across the earth magnetic field.

Faraday tried to measure this effect across the Thames River under London Bridge. He concluded that the Thames carried enough ions and the earth magnetic field would be strong enough to yield a small voltage difference between the electrode placed on the right and the left river bank, at the locations (a) and (b) on Fig. 10.17. The total voltage across the river of width L would be

$$V = E\,L = L \cdot u \times B.\tag{10.21}$$

Taking the earth magnetic field $B \approx 5 \cdot 10^{-5}$ T, a velocity of $u = 1$ m/s, and $L \approx 100$ m, one has $E = 1\ \text{m/s} \cdot 10^{-4}\ \text{T} = 50\ \mu\text{V}$ and $V = L \cdot E = 0.005$ V. Faraday failed to

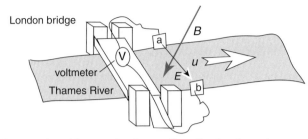

Fig. 10.17. Voltages induced in a conducting medium flowing through a magnetic field

measure the effect, because he could not measure precisely enough the voltage difference between the two terminals a and b.

Similar small fields are generated as ocean currents move through the earth magnetic field. Sharks reportedly can detect fields down to $E=500$ nV/m $=$ 5 nV/cm. Such a field is generated when the potential difference of a 1.5 V battery is applied across the Atlantic. When a shark swims with the velocity u at right angles to the earth's magnetic field, it will experience the electric field $E=uB_e$. At the velocity $u=1$ m/s the induced field is 50 µV/m, as calculated above. This field is 100 times stronger than the shark's detection threshold. Therefore, it is likely that sharks use these fields for navigational clues on their travels. The direction of the fields would help them to maintain a certain course.

Table 10.5. Frequently used variables of Chap. 10

variable	name	units	name of unit
B	magnetic field	T, or G, or Γ	Tesla, Gauss, gamma
C	capacitance	F	Farad
E	electric field	V/m	Volt/meter
j	electric current density	A/m²	Ampere/m²
J	current	A	Ampere
K	dielectric constant		
q	electric charge	Cb	Coulomb
R	electrical resistance	Ω	Ohm
ρ	electrical resistivity	Ωm	Ohm-meter

Problems and Hints for Solutions

P 10.1 Shark Tachometer

Sharks are known to detect electric fields down to 0.5 μV/m. When they swim at the velocity u through a magnetic field B an electric field, sometimes called the Faraday field, $E = u \times B = u \cdot B \cdot \sin\phi$ is generated, where ϕ is the angle between u, and B. Assume a shark (width at the head $d = 0.15$ m) swims at the velocity $u = 8$ m/s at right angle through the earth's magnetic field, $B = 5 \cdot 10^{-5}$ T. a) Determine the Faraday field generated in this motion. Can a shark detect such a field? b) What is the voltage difference $V = d \cdot E$ between detectors located at the left and the right side of the head of the animal? c) By measuring this electric field the shark actually detects its own velocity. What is the smallest speed that the shark could just barely measure given the resolution ±0.5 μV/m of its electric sense?

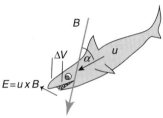

Fig. 10.18. Shark tachometer

P 10.2 Currents in Nerves

(a) Calculate the averag electrical resistance R of an 8 cm long section of a non myelinated afferent nerve of $D = 17$ μm. a) How large is the cross section area A_n of the nerve? b) How much induced current J_i (measured in A) will flow through this nerve section if an electric voltage difference of 1.5 V is applied to both ends? What is the current density j_i A/m²? c) Compare the induced current J_i to the current J_p produced by the pumps of a typical cell membrane of an area equal to the nerve cross section area A_n.

P 10.3 Zapping a Meal

Electric eels kill their prey with electrical pulses of up to 600, and typically 1 A lasting for about $3 \cdot 10^{-3}$ s. These pulses are generated by sodium ions Na^+ that are suddenly admitted through channels in the walls of the specialized electrocyte cells. The cells have finger-like structures of typically 4 μm radius and 30 μm length that are arranged about 10 μm apart, so that 10,000 of these fingers protrude side by side on a surface area of 1 mm × 1 mm.

Assume that the sodium pumps in the cell walls maintain a steady state concentration of 140 mmol/l. a) How many sodium ions must diffuse into the cells to support a current pulse of $I = 1$ Amp. b) If all the Na^+ ions of a $\Delta s = 1$ μm thick layer outside the cell walls diffuse into the cells during the $\Delta t = 3$ ms long current pulse, how much cell surface area A is needed to support the pulse? c) Suggest a reason why the cells have finger-like protuberances. d) Are there enough Na^+ ions between the fingers to support the current pulse?

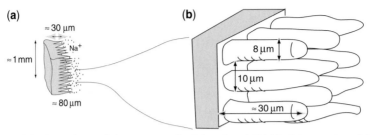

Fig. 10.19. (a) Dimensions of electrocyte cell of electric eel. (b) Model geometry of fingers

Hints for Solutions

S 10.1 The Cross product $E=u \times B=u \cdot B \cdot \sin \phi$ gives the field strength E in V/m if B is given in Tesla and u in m/s. a) In numbers $E=5 \cdot 10^{-5}\text{T} \cdot 8\text{m/s}=4 \cdot 10^{-4}\text{V/m}=4 \cdot 10^{-6}\text{V/m}=4\,\mu\text{V/m}$. This is a factor $4\,\mu\text{V}/0.5\,\mu\text{V}=8$ times larger than the resolution of the electric sense. b) The voltage difference between both sides of the head is $\Delta V=d \cdot E=4 \cdot 10^{-4}\text{V/m} \cdot 0.15\ \text{m}=6 \cdot 10^{-5}\ \text{V}=60\ \mu\text{V}$, therefore, not much is needed.

S 10.3 a) A current pulse of 1 Ampere lasting 3 ms requires 0.003 Cb$=n \cdot e$. The charge of one sodium ion Na$^+$ is $e=1.6 \cdot 10^{-19}$ Cb. Thus $N=0.003$ Cb$/1.6 \cdot 10^{-19}$ Cb$=1.88 \cdot 10^{16}$ Na$^+$. b) 140 mmol/l corresponds to $n=(140 \cdot 10^{-3}\ \text{mol/l}) \cdot (6 \cdot 10^{23}\ \text{atoms/mol})=8.4 \cdot 10^{22}\ \text{atoms/l}=8.4 \cdot 10^{16}\ \text{atoms/mm}^3$. A surface layer of $a=1$ mm^2 and 1 μm thickness has a volume of $V_{sl}=0.001$ mm^3, and hence contains $n_{sl}=n \cdot V_{sl}=8.4 \cdot 10^{13}$ Na$^+$ ions. To generate a current pulse of $1.88 \cdot 10^{16}$ Na$^+$, a total cell surface area of $A_t=N/n_{sl}=224$ mm^2 is required. c) The finger geometry permits more surface area to be packed into a small volume. d) The number $N_f=V \cdot n$ of Na$^+$ ions surrounding the fingers is found by first determining the fluid volume V between the fingers, and then multiplying this volume by the number density $n=8.4 \cdot 10^{16}$ atoms/mm^3.

11. Better Physics: The Trifle of Difference

In the struggle for existence… the merest trifle (of difference)
would give the victory to one organic being over another…
but probably in no one case could we precisely say
why one species has been victorious over another.

Darwin

… Yet it is life itself
That makes the differences
From which the actions flow.

Physics Enables and Sets Limits

Throughout this book it has become apparent how intimately Physics is intertwined Zoology. Physics deals with dead concepts and inanimate parameters such as temperature mass, position, time, velocity, and energy, which are often expressed as equation linking various parameters. Zoology, in contrast, is seen as a living subject dealing with the complexity of design and the actions of animals. When talking about Zoology one thinks of pictures of eagles circling high in the air, of dolphins squeaking in the aquarium, of ant hills in the forest, and of the multitude of strange animals seen in zoos. And yet the physics laws rule supreme and permit the living world to perform certain actions with grace, and efficiency. The physics models thus help to understand some of the actions and the body design of animals.

Physics concepts are used here to interpret, and sort out biological facts. In this perspective, physics becomes the grammar in the book of life that displays and explains how animals function. Like other forms of philosophy, physical models are non-exclusive. There are often several different ways to arrive at the same conclusion. The physics perspective, which focuses on a few effects only, totally ignoring others, often clears the view to see a larger picture. Thus, one may hope to gain an embracing perception of some principles of life, which are not obvious from biological data. Yet at the same time one must be aware that this biased focus may dull the view of recognizing multiple use of certain organs. Competing demands may force organisms not to operate at the single-use optimum.

The previous chapters have provided many examples of body design and animal actions where the exploitation of *physics principles* within the framework of *biological constraints* yields advantages in energy efficiency as well as in information acquisition. These advantages have likely made "the trife of difference" for the organism to guarantee its survival. By presenting these examples side by side in this chapter one obtains a glimpse into the workshop of evolution. Physics is the hard reality hidden behind complex life activities. Three general results can be recognized: Physics enables, there are optima, physics sets limits.

Good physics allows animals to build their bodies economically, and perform efficiently the numerous functions in all walks of life and in all niches of the biosphere. This holds for strength and agility, camouflage, and the development of senses. The basic physics concepts and the biological boundary conditions are presented in Sect. 11.1. Opportunities beckon in all niches of the biosphere. One could say that the fitness landscape of life is molded by good physics. Often the same end can be reached with different body structures. This is co-evolution.

The laws of Physics impose stringent limits on some body parameters, and allow for certain optima, which are recapitulated in Sect. 11.2. Practically all parameters have lower and upper limits. The phase space of life is donut shaped. Every organism is an optimum solution to a many-parameter-problem of a particular niche in the biosphere. Shape or size, materials, and body functions may be optimized. In most situations too many parameters come into play to clearly see the optimization. However, in some cases the variation of a single parameter leads to an optimum with regards to energy use, or information collection. The fact that many animals found such optima must be considered a crowning achievement of the evolution.

When animals reach limits imposed by physical principles, quite often new physics tricks allow these restrictions to be circumvented. This is one aspect of evolution. Occasionally, the predator-pray interplay leads to a spiral of improved physics. Some thoughts on these topics are discussed in Sect. 11.3. The supreme command of physics by all living organisms and the steady search for other more suitable physics principles is part of the secrets of evolution. If energy can be saved, it will be saved. If information can be obtained readily it will be collected. If there are optima they will be found. If smaller size is sufficient, smaller size will be adopted. If there are limits that can be circumvented they will be circumvented ... by successful animals.

> Schläft ein Lied in allen Dingen   Sleeps a song in every being
> Die da träumen fort und fort,   That exists without effort
> Und die Welt hebt an zu singen   And creation lifts her secrets
> Triffs du nur das Zauberwort   If you find the magic word
>
> (translated by the author)

One gets a glimpse of these effects using the principal concepts of *Zoological Physics*: (1) capture the main events with a quantitative model that combines physics concepts and biological restriction, (2) look for optima, (3) find limits, and (4) separate primary causes from secondary effects. In Sect. 11.4 an attempt is made to describe the technique and the limitations of such physical modeling, and to suggest some additional topics that ought to be discussed in further studies. However, the reader should not take the derivations or conclusions as gospel, but rather treat the text as a set of methods and backup facts to interpret the actions and body designs of animals within the framework of physics laws. And thus it is hoped that the principal goals of this book shines through: to present physics models which give a deductive perspective that adds to the inductive method of observational biology, and to guide the reader towards developing his or her own zoological physics models.

11.1 Physics Enables Within the Framework of Biological Restrictions

The essence of living organisms is to assemble and continuously exchange with the environment the inorganic components of the cells, and to maintain (through metabolism) the essential parameters such as concentrations of molecules, ions, temperature, and pressure within narrowly defined limits. Life functions are enabled by controlling the fluxes associated with these parameters: food consumption to maintain the metabolic rate, charge fluxes to conduct signals in nerves, heat fluxes to maintain the body temperature, mass fluxes in body fluids to transfer materials throughout the body, and diffusion fluxes to feed every cell of the organism, thus enabling the activities of the body.

These processes obey the laws of physics and they occur within the limits set by biological restrictions. Physics laws help in the process of evolution, they mold the fitness landscape of organisms. The physics laws in question are the mathematical formulation of physical concepts dealing with energy and information. The biological restrictions have dissipative character. They are related to the metabolic rate, to the muscle efficiency, to the strength of materials, to transport coefficients, to molecular processes in the genes, and to information processing rates of senses, nerves, and brain. Some of the essence of physics content is conveyed by non-dimensional numbers. The power of physics principles is also illustrated by the process of co-evolution.

11.1.1 Physics Concepts

The physics concepts discussed here can be divided into energy related topics (metabolism, energy conversion, forces, flow, and the physics of getting around), and principles connected with the acquisition of information carried by light, sound, and electric fields.

In Chap. 1 we described concepts dealing with eating and heating. Animals consume food (with the chemical energy Δh J/kg) at a certain rate, which depends on the body mass M. This *metabolic rate* $\Gamma \propto M^{\frac{3}{4}}$ is the principal biological restriction on life functions. Pound by pound, the specific metabolic rate $\gamma = \Gamma/M$ decreases with body mass $\gamma = \Gamma/M \propto M^{-\frac{1}{4}}$. There are other *allometric* scaling laws for body parameters f that can be written in the form $f \propto M^{\alpha}$ such as the lifetime $\tau \propto M^{\frac{1}{4}}$. During its lifetime τ an animal consumes the energy $\Sigma E = \int \Gamma dt$. This energy is related to the increase of entropy in the universe caused by the animal, hence Γ places a time scale onto the *second law of thermodynamics*. Since the body temperature T is narrowly confined the specific lifetime entropy increase $\Delta s = \Delta S/M = \Sigma E/MT \approx \gamma \cdot \tau/T \approx \text{const } M^{-\frac{1}{4}} \cdot M^{\frac{1}{4}}/M$ is approximately the same for all animals.

Chapter 2 deals with energy and temperature. Heat ΔQ that is released in the metabolic process can be converted into work ΔW (force F times distance Δs) and internal energy ΔU, which increases the body temperature. The *conservation of energy* (first law of thermodynamics) demands $\Delta Q = \Delta W + \Delta U$. The biophysics of muscle tissue restricts the efficiency η at which chemical energy ΔH released as heat can be converted into mechanical work. The fraction $\Delta U = (1 - \eta)\Delta Q$ must appear as an increase of the internal energy. The internal energy U, depends on the temperature T, which must be controlled within narrow limits: for life in general between the freezing and the boiling point of water, and for warm blooded animals within a few degrees of 36 °C. This control requires careful management of conduction, convection, and radiation.

Chapter 3 deals with forces and structures. Newton's third law demands that forces occur in pairs. Body structure must withstand all force. There are static forces, like gravity and pressure that can be found in inanimate as well as in living matter. Biological restrictions on forces and structures arise from the magnitude of muscle forces (with typical values $f \approx 2 \cdot 10^5 \, \text{N/m}^2$), and the elastic response of materials. Some forces only appear when objects are in motion and are called dynamic forces. To stand up to various forces, all body dimensions scale with body mass in certain way.

Chapter 4 describes how fluids move through the body, by diffusion and convection. The *conservation of mass* requires that fluids move fast when they are squeezed through narrow vessels. Bernoulli's equation yields the pressures needed to to make the fluid flow when friction is negligible. Hagen Poiseulle's relation holds when the Reynolds number is small and friction becomes important. Material parameters like the diffusion of gasses through cells, viscosity of blood, and the oxygen-carrying capacity of red blood cells, are the biological restrictions of fluid flow.

Chapter 5 deals with motion. Motion may be analyzed with the help of the conservation of energy. However, one must realize that the conversion of energy from one form to another can only proceed at well-determined transfer times. The general form of motion analysis is based on Newton's second law, *the conservation of momentum*. Forces must be applied for an object to move. Many movements of body parts must be treated as relative motion: a foot is connected to the lower leg, which articulates about the knee, which in turn swings from the hip that is travelling at the speed of the whole body. Biological restrictions are imposed on motion processes by the strength of bones, tendons, and cartilage, by the speed of energy transfer, and by the strength and frequency at which muscled forces can be deployed.

Chapter 6 is concerned with locomotion, where arms, legs, and flippers swing periodically. There are dramatic savings in energy when the locomotion frequency is properly tuned to the appropriate internal *resonance*. Different gaits accomplish such tuning. Fish have acquired body aspect ratios where drag forces are small. Birds need a minimum flight velocity to stay aloft. Biological limits arise here due to the finite mechanical power that can be released from metabolism. Very big birds can never fly.

Chapter 7 describes the information fluxes that animals have learned to harvest. Light and sound are *wave phenomena* that have many common features, like the definition of intensity I [Watt/m^2], phase speed, refraction, diffraction, scatter-

Table 11.1. Physical concepts and biological restrictions

Chapter	biological restrictions	physics concepts
1. Food	metabolic rate $\Gamma = aM^{\alpha}$ Γ/T = entropy production rate	mechanical power of animals $P = \eta\Gamma$ entropy
2. Heat	thermal conductivity κ minimum of energy waste energy efficiency efficiency at maximum power	$\Delta Q = \Delta U + \Delta W$, food energy, heat ΔQ, internal energy ΔU, phases mechanical energy ΔW, *equation of state* $U = U(T)$, T temperature conduction, convection, radiation
3. Force	body shape, arrangement of bones, muscles, and tendons material strength Y, scaling specific muscle force $f_o = 2 \cdot 10^5$ N/m² efficiency η_{muscle} is a function of the muscle contraction speed	statics: $\Sigma F = 0$, $\Sigma\tau = 0$, free body diagrams, compression, tension various forces (static and dynamic)
4. Flow	haemoglobin: 1 cm³ air holds as much O_2 as 1 cm³ blood red blood cells $d \approx 10$ µm viscosity, diffusion constants O_2 concentration in air and water diffusion,	*conservation of mass* in fluid flow Reynolds number Re laminar, turbulent flow Bernoulli, Hagen Pouiseulle
5. Motion	strength of muscle forces, tendons, bones cartilage; protect body parts (brain) against excessive acceleration	*conservation of momentum* $\Sigma F = M s a$ relative motion *conservation of energy*, energy time scales
6. Loco motion	move body parts in resonance, metabolic power	fly, swim, walk/run, different gaits, tuning of resonance
7. Waves	phase velocity in materials, n attenuation of light and sound κ solar spectrum	waves: phase velocity v, intensity I, propagation direction; refraction, diffraction, scattering, absorption
8. Light	sensors for light and infrared biological mirrors, $n(\lambda)$ of bio-materials, bio-luminescence	photons, diffraction limit light collection and lenses
9. Sound	ears, detect pressure fluctuations, Fourier analyze, tissue impedance, voices, sonar, Sofar channel	intensity, sound level, particle oscillations, pressure fluctuations, diffraction limit, Doppler effect, impedance matching, $Z = \rho v$
10. E & M	cells, action potentials natural batteries, nerve conduction speed, diffusion limited production of biological E-fields	E-fields, circuits, resistors, capacitors B-fields

ing, and absorption. Restrictions beyond physics laws in this area are associated with refractive indices of biological materials for light and sound, and by environmentally modified absorption coefficients. There must also be a lower limit of capability for the senses, nerves, and brain to collect and evaluate information, which is likely related to the rate of change d/dt of the scaled information intensity $\Delta I/I$.

Chapter 8 discusses how various eyes collect and *focus light*, and how light waves are really torrents of *photons*. Each photon is associated with a specific *wave length*, which implies a *diffraction limit*. Special eyes are described, as well as tricks that animals use to become invisible with light. In response, predators have found ways to see the invisible ghosts with the help of polarizers in their eyes. The biophysics of photo detector properties, polarizing materials, reflectors and bioluminescence goes beyond the scope of this book.

Chapter 9 deals with ears and voices, and why the sound intensity is so drastically reduced when sound travels from the air into the inner ear. Concepts of *intensity* and *impedance* are explained. Numerous animals have learned that sound can be produced easily. The pitch of voices has to do with the size of the vocal organ. Voices are not only used for communication but also for sonar beams that replaces light thereby enabeling some animals to see in the darkness of the night or at the bottom of the ocean. Strong sonar beams even become lethal weapons. The impedance of tissue and bones, the functioning of the cells that convert minute pressure fluctuations into acoustic nerve signals of a wide frequency range, and voices, are the limiting biological parameters of acoustics.

Chapter 10 reminds the reader that all cells are charged batteries or *capacitors*, and nerve cells serve as conductors for traveling charge waves thus establishing communication inside the body. Ohms law, and the rules of charging, and discharging capacitors apply. Cells are charged by the metabolism which establishes the ion concentrations inside the cells. Electrical signals due to the flow of ions like Na^+, Ca^{++}, K^+, is therefore closely tied to the metabolism.

11.1.2 Fitness Landscape Molded by Physics

Live evolves through *successful errors*, where deviations from the old norm are reinforced by positive feedback from the environment. Every new form is a mistake compared to the body design of its parents. However, not always sticking to the script can be the secret of success. *Error is the chisel of evolution.* Body mass, energy, and information are the three cornerstones of life: information to shape the body mass, and energy to operate it. Information controls the actions and directs the construction of the body, but without energy no action can take place. Energy is the fuel of life, and it governs the ability to move, stay warm, and procreate.

Life has an extraordinary ability to maintain successful features. What worked once is not easily abandoned. At the same time, life possesses a vital originality and opportunistic inventiveness. The environment is subject to never ending changes. Each life form must adapt to it in its struggle to survive. Too much rigidity means losing the touch to reality, death may occur by calcification. Too much experimentation has the opposite dangers: most mutations are mistakes. The improved survival of the few is bought with the demise of the many unsuccessful variants. Between these poles of conservatism and experimentation *life* has managed to find for each environment a nearly optimum solution – over the dead bodies of countless unsuccessful mutations.

When one considers the multitude of combinations that can be stored in the genes the impression might arise that life could operate in uncountable ways, and that an infinite number of combinations of form, function, and design exist. However, the laws of macroscopic physics impose severe limitations on form and function. Organisms in general have evolved several evolution strategies.

1) Minimize the total energy consumption, make best use of the food supply.
2) Extend the living space to access new energy resources.
3) Modify the body form to generate the least mechanical losses.
4) Evolve material properties towards optimum values.
5) Increase the reaction speed by choosing the best operating temperature.
6) Increase the information from the surrounding by better senses and brain.
7) Find better ways to hide.
8) Become utterly undesirable for any potential enemy.
9) Increase information through language and education.
10) Use more appropriate physics.

These principles help in the selection of the survival of the fittest and thereby shape the path of evolution.

11.1.3 Examples of How Physical Principles Are Utilized

To illustrate how physics enables life functions, the gist of examples from every chapter is summarized.

Energy and Temperature: Homeotherm animals eat in order to generate mechanical energy and to stay warm. The body chemistry operates most efficiently at temperatures of about 37°C. Therefore, higher animals are warm-blooded. (1) The first law of thermodynamics shows that heat is generated as a byproduct when work is done. This heat can be used to warm the body: to stay warm in cold climates animals activate their muscles. (2) Animals make sure that they are well insulated with materials of low thermal conductivity. Air has the lowest values, so animals in cold climates surround themselves with layers of air, trapped within fur or feathers, two equally effective materials, invented independently by birds and mammals. (3) Sometimes it is advantageous to allow parts of the body to stay cold. Seabirds live with cold feet; tuna fish exist with cold hearts and warm muscles. Heat transfer by convection and counter-flow heat exchangers are employed by these critters. (4) Heterotherm animals must rely on the environment to reach the best working temperatures. Snakes and lizards wait for the sun to warm them up. They need less metabolic energy than warm-blooded species, so they can get by where food is scarce. However, they could easily fall prey to warm blooded carnivores. A yardstick for energy use is efficiency η, the ratio of work W performed to energy Q used in a process.

Features of **Body Forms and Forces** offer another glimpse into the benefit of using good physics. (1) Fish can control buoyancy by the use of a swim bladder.

More ancient species like sharks do not possess this organ. They must be in continuous motion. (2) The elasticity of tendons and ligaments plays a role in counteracting gravity forces and serve to temporarily store energy thus saving energy in locomotion. (3) The insignificant force of surface tension opens a niche for many species of small insects that walk on water. (4) All large structures must be composed of elements that can withstand tension (like skin and tendons), and other elements that can withstand compression (like bones, or shells). A yardstick for the properties of materials is Young's modulus Y, the ratio of stress F/A over strain $\Delta L/L$. (5) All larger animals developed bones or shells, which are cheap and effective. (6) Mobile animals adopted streamlined shapes that offer the least resistance in swimming, and flying. A yardstick for motion effective form is the drag coefficient C_D, the ratio of drag force F_D to Bernoulli force $0.5\,\rho A\, u^2$.

Fluid Flow in the Body mainly serves the exchange of material and heat. (1) Chemicals like oxygen and carbon dioxide, travel in the same vehicle; the hemoglobin molecule, which rides in red blood cells, is imbedded in the convective flow of the blood. (2) Laminar fluid flow is maintained where the fluid resistance is low. (3) Diffusion is employed in the terminals of the blood traffic lanes: the lung and the capillaries. The lung is a marvel of design with a huge surface area, pressed into a small volume. (4) The heart is a very efficient pump that achieves on average about three billion cycles, approximately 100 times more cycles than an automobile spark plug, or other technical devices. A yardstick of fluid flow is the Reynolds number Re, the ratio of inertia forces over viscose forces.

Many physics tricks are apparent in **Locomotion**. (1) Limbs are designed with small mass moments of inertia. Slender feet and thick muscles close to the hip allow quick motion of legs in birds as well as swift runners. (2) Periodic motion at resonance frequencies yield minimized energy expenses. Runners, flyers, and swimmers, make use of this technique. Such simple harmonic motion may involve potential, kinetic, and/or elastic energy. Motion frequencies should be tuned to internal resonance frequencies. (3) Landing with soft limbs, a trick shown for instance by cats, reduces the impact forces. It gives these animals the ability to come down safely in jumps from great highs. (4) Engineers were not the first to invent rocket propulsion. Animals, like squid and jellyfish, have practiced it for millions of years. (5) Swimming and flying at high speed with Reynolds numbers exceeding 100 is accomplished by vortex production. Vortices could impede the motion, presenting added mass, a burden on the animal, but careful timing of tails, wings, and flippers transforms the vortices into structures that help the propagation rather than hinder the motion. The yardstick of vortex shedding from an object of width D is the Strouhal number St, namely the ratio of vortex shedding frequency f to system frequency u/D.

Of the three necessities of life, matter, energy, and **information,** the last one bears the most profound assistance to the evolution of life on earth. The inherited information of the genes fixes what has been successful, but external information opens entirely new avenues for life. (1) Much of this comes as data encoded into light and sound waves, which tell the organism what is out there, where opportunities beckon, and where dangers lurk. (2) Information can travel in waves as slow or rapid intensity variation, or as frequency modulation. (3) The waves may be scattered, refracted, diffracted and attenuated. Scattering is useful for creating ambient light. Refraction helps to build imaging eyes, diffraction sets limits on the visible detail. (4) When these effects are properly applied, an *information landscape* of the surrounding can be constructed by the senses and the brain. The brain can then play out 'computer games' to decide which actions are most appropriate and yield the largest energetic cost benefit ratio.

Light is a steady stream of tiny energy particles: photons. These radiation bundles carry the minute energy $h\nu$. The light intensity can be measured as the number of photons of all frequencies ν that pass per second through a target area. (1) Most often the photon number is so huge that the light intensity may be treated as a continuum. Then the signals encoded in the light beam can be collected and read with pinhole eyes to form an image. (2) A small blob of transparent medium, shaped as a lens, increases the image resolution by many orders of magnitude, while at the same time increasing the collection power of the eye. (3) Infrared sensing eyes permit some animals to see in the dark. Lens eyes produce sharp images only for a certain distance. (4) To focus onto various distances, the lens must either be moved relative to the retina, or the focal length must be altered. Both principles are found in nature. (5) A yard stick for the light gathering power of lenses is the F number, the ratio of focal length to lens aperture. It can be optimized. (6) The image contrast may be improved by removing stray light. For this purpose animals have evolved ideal mirrors using the principle of interference. (7) Light travels freely through the air, arrives instantaneously since its speed is so large, but it is extinguished rapidly in water. Deep in the sea where no daylight penetrates, bioluminescence in various forms is found.

Sound is characterized by its speed v, its frequency, and its intensity I. (1) Sound helps to identify objects in the immediate surrounding. (2) Sound does not travel freely from air to condensed media, due to the difference of impedance Z in these materials. Therefore, ears must possess great amplification power. (3) Information travels at many sound frequencies, which ears can separate. (4) With sensitive ears animals can gather information about distance, direction, identity, and intentions of the source. (5) Sound is much easier to generate than light. It may be produced in many different ways facilitating communication and speech. (6) Sound can be used for active seeing. This is sonar. (7) Sound can become a weapon for scaring off attackers, or stunning prey. (8) The social fabric of higher animals and humanity is knitted together by sounds and language.

Electric Fields appear as byproducts of living cells. (1) They play a role in nerve conduction. (2) Some animals have evolved electrical senses. These yield information in situations where light and sound fail: in muddy, turbulent waters, at the bottom of streams, in the depth of the oceans. Even the weak **Magnetic Field** of the earth contains information that help animals to find their way, or detect the speed.

Thus every field of classical physics has enabled animals to exist, strive, and evolve.

11.1.4 Non-Dimensional Numbers and Scaling Relations

Physics provides much insight through ratios and non-dimensional numbers, and scaling relations. They are initially introduced to characterize structures, materials, and phenomena. But they have the added use to distinguish the important effects from the less important parameters. Such numbers have appeared in every chapter.

In Chap. 1 allometric scaling relations are given that allow sweeping conclusions about animal performance based solely on their body mass M. Most significant for the whole text is the metabolic rate Γ. The metabolic rate Γ controls the activities of the entire body. The life supporting energy mostly comes from the sun. Instead of letting the sun energy increase the background entropy immediately, organisms keep the energy for a while to support their life functions. Hence Γ is proportional to the rate at which an organisms contributes to the increase of entropy in the universe.

In Chap. 2 the efficiency η is encountered, which describes the ability to generate mechanical work out of heat. If an organism has a high efficiency η it is capable of generating mechanical power effectively.

In Chap. 3 Young's modulus Y, the ratio of stress over strain is discussed, which characterizes the tissue elasticity. Scaling relations for limb dimensions are given. Young's modulus of tension Y in spider webs is matched to the task at hand, soft for the orb, and elastic and strong for the drag lines. The Strouhal number $St = fD/u = D/\lambda$ is mentioned, namely the ratio of the lateral dimension D of an object and the wavelength of the vortex street that it causes in a flow of speed u. Strouhal numbers describe the size and frequency of the vortex production. Further discussed in this chapter is the ratio of the drag force to the Bernoulli force $\frac{1}{2} A \rho u^2$, called the drag coefficient C_D. The drag coefficient C_D can be altered by body aspect ratio. It has a minimum at a certain aspect ratio. Mechanical energy can be saved if the organism is built with this optimum aspect ration.

In Chap. 4 the Reynolds number Re is given, namely the ratio of inertia forces to viscous forces. At low Reynolds number only friction is important. At Re > 50 vortex shedding occurs, which is a principal lift mechanism in hovering flight.

In Chap. 6 more use is made of the Reynolds number and one encounters the Froude number Fr, defined as the ratio of centrifugal forces to gravitational forces.

In Chap. 7 the index of refraction $n = c_o/c$. A high index of refraction n helps to build lenses with high refractive power and short focal lengths.

For the formation of images, described in Chap. 8, the number of photon plays an important role, namely the intensity of a light beam divided by the photon energy. The number of image points for a lens eye is calculated as the ratio of image area divided by the diffraction limited pixel size. In this derivation the F-number f/D, appears as a factor. Low F-numbers characterize lenses with high collection efficiencies.

In Chap. 9 higher harmonics $n_i = f_i/f_o$ of the fundamental frequency in acoustic resonators are described. Higher harmonics n_i in resonance contribute to the richness of sound. The intensity I of sound is characterized by the sound level $\beta =$

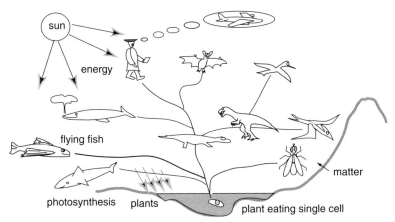

sun

energy

flying fish

photosynthesis   plants

matter

plant eating single cell

Fig. 11.1. Co-evolution of flippers and wings

$10 \log_{10}(I/I_0)$ and the impedance ratio Z_1/Z_2, which controls the transfer of sound from one medium to another. The sound level β reveals if sound may be used for communication or as a weapon.

In short the knowledge of the magnitude of such scaled quantities often is the first step in describing the performance and the actions of animals.

11.1.5 Co-Evolution

In the human sphere an inventor often acquires a patent to claim sole possession of a new idea. In the animal kingdom inventions often are *in the air* when the time is ripe. Suddenly many organisms seem to acquire similar or matched capabilities. Just as new and similar styles are adopted by fashion designers and automobile makers, as if they all had read the same script.

This principle of multiple and independent inventions is superbly demonstrated by co-evolution. There are numerous examples of new inventions that have been independently made by animals of different phyla: lens eyes of vertebrate and octopus, the flukes of whales and tails fins of fish, wings of birds and bats, Fig. 11.1.

Such similar forms appeared in the animal kingdom because the laws of physics called for nearly identical designs.

11.2 Optima and Limits

Animals have diverse requirements and many physical constraints. Important parameters are body mass, shape, temperature, position in space, speed, pressure, wave lengths of light, frequencies of sound, and intensities to name just a few.

Some of these parameters have stringent limits, others can be optimized in the attempt to save energy or optimize the information flux. Indeed, occasionally such optimization may even occur to satisfy conflicting demands by utilizing different ranges of an accessible parameter space [Davison and Shiner 2003]. But in the end evolution is a constant struggle to overcome limitations and to find ways to improve the optima just by the trifle of difference.

11.2.1 Size and Mass Ranges

The limits on body mass imposed by physics show up in many chapters. Often there is a minimum, and also a maximum mass for a particular body design, dictated by thermodynamics, statics, and dynamics, Fig. 11.3.

Big animals have less power per unit mass, and less surface heat losses. They can survive in cold climates but never have the agility of smaller animals. Warm-blooded animals need a minimum size, Fig. 11.2a. For birds it is about the size of a chickadee, Sect. 2.4.3. Insect bodies are too small to carry insulation. They can never be homeotherms. However, as a group (e.g. a beehive) they may well act like a warm blooded animal. Organisms larger than about 1 μm need an internal convection system for the transfer of nutrients and the removal of wastes and heat, Sect. 4.1.3. Eucariot cells can get by with diffusion, but insects and vertebrates must pump fluids through their body, Fig. 11.2b.

No organism heavier than a few grams can walk on water. The biggest water striders, which measure a few centimeters in length, tempt the limit of surface tension, Sect. 3.3.6. The legendary King Canute did not really walk on water. Big apes cannot swing through half arcs from branch to branch, Sect. 5.2.4, because centrifugal forces would pull their fingers from the handhold.

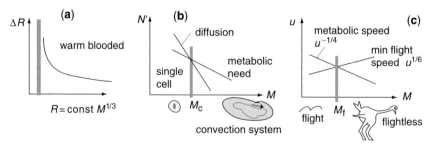

Fig. 11.2. (a) Insulation thickness ΔR of birds as function of body mass M. (b) Diffusion limit. (c) Flight limit

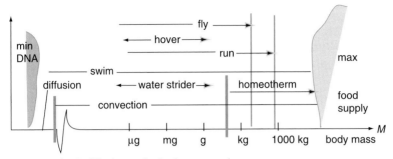

Fig. 11.3. Some physical limits on the body mass scale

Birds larger than about 15 kg cannot take to the air, Sect. 6.3.2. Condors and mute swans reach this limit, while the emu, the ostrich and the now extinct Moa belong to the group of flightless birds (Fig. 11.2c).

No animal has eyes to resolve structures of the size of a virus, which are much smaller that the wavelength λ of visible light. Light waves can only depict structures larger than λ, Sect. 8.2.6.

Eyes for focusing under water need large apertures A to collect low intensity light and small radii of curvature R, to compensate for the fact that the index of refraction of water is close to that of lenses. The limit of small radius of curvature, *and* large aperture for any lens, is the spherical form. Fish eyes have this design, Sect. 8.4.3.

The perception of sound is limited by the noise of thermal background motion. Ears of many animals have sensitivities reaching this limit, Sect. 9.3.3. Sounds are easy to make in a variety of ways, and sounds can carry signals.

Low frequency sounds, with large wavelength carry the farthest. That is why elephants and whales communicate with sound frequencies of a few Hertz, well below the threshold of human hearing, Sect. 7.3.4. They can make such low sounds because their bodies are large. Small bodies can only generate short wavelengths, or high frequencies. Hearing on land is complicated by the fact that only a very small fraction of sound energy is transmitted from the low-density air, into the water like body, Sect. 9.2.6.

Only after ears with large signal amplification had evolved could sound in air be perceived. However, once this hurdle was overcome, new opportunities arose in the evolution process of sound production and hearing. Animals evolved that possessed complete sonar systems, allowing their owners to *see with sound*, and even to determine the speed of their prey.

11.2.2 Living Space Ranges

On a macroscopic scale, the living space of all known organisms that we know is the earth. It is located within the *water shell* or the life belt centered about the sun, Fig. 11.4. It is the range where liquid water can exist, Sect. 2.4.1. On earth the water

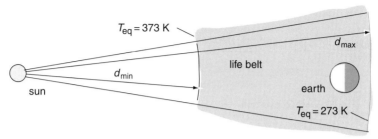

Fig. 11.4. The life belt around the sun

zone stretches from the Polar Regions and high mountain ranges to the waterless deserts. Animals in higher latitudes on the globe cope with the seasons by migration, or hibernation. Animals in arid regions use the daily temperature variations to persist in extreme locations.

Some ranges of the ocean are void of life for lack of light or nutrients. Yet once in a while an organism finds a way to cope with harsh living space conditions, and then strives due to absence of competition.

11.2.3 Information Limits

All higher animals collect information through senses to make them aware of their surrounding. Light, sound, and smell are the favorite data carriers.

Giant squid have pushed the limits of light collection to an extreme. Their huge eyeballs gather every photon at the bottom of the sea. They even have ink, an optical camouflage, to keep their soft bodies hidden from would be attackers. *Single photons* are the absolute lower limit of optical information bits. Even the human eye is capable of detecting single photons. Eyes reach the single photon sensitivity threshold; they could not be made more sensitive.

Sound is an organized vibration of the carrier medium. It does not travel through vacuum. Every medium is composed of atoms. Atoms are always agitated to some degree by thermal (Brownian) motion. The energy in this thermal noise is minute. It measures in $k_B T$, where k_B is the Boltzmann constant, and T is the temperature. Sound waves can only be distinguished if the kinetic energy in the vibrations is larger than the thermal noise. The ears on humans and other animals reach this *thermal noise threshold*. It is the physical limit of sound detection.

11.2.4 Energy Transfer Time Limits

The conservation of energy demands that no energy can ever disappear. Energy transfer is an effective way to reduce energy budgets. However, energy transfer takes time. Motion time scales must be tuned to energy transfer time scales.

In locomotion all animals generate and destroy kinetic energy, when they accelerate and decelerate their limbs. The energetic cost of transport can be substantially reduced, if part of the kinetic energy can be temporarily transferred into elastic or potential energy and reclaimed as kinetic energy in a subsequent motion phase. However, energy flows at certain fixed rates, see Sect. 5.1.7. It takes a certain time Δt to transfer a certain amount of energy ΔE. Therefore, intermediate energy storage only works if the motion phase available to the energy transfer is long enough. Alternately explained, the motion frequency f_L must be matched to the internal frequency f_0 associated with the energy storage mechanism. All periodic processes therefore occur in resonance. In fact, animals have found some ways to tune their internal resonances and adjust them to the desired locomotion frequencies.

11.2.5 Optima

Each individual body form seems to be an exquisitely shaped optimum to fit its niche in the biosphere. Looking closely, one can see that animals use the laws of physics ingeniously and often in a way to get by with a minimum of energy, and a maximum of information.

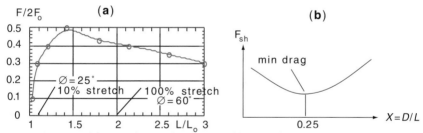

Fig. 11.5. (**a**) Lateral force F that can be supported by stretched spider silk. F_0 is the breaking strength of one strand. (**b**) Total drag force composed of skin friction and hydro drag as function of aspect ratio D/L

Examples of optima have been discussed in detail in the text. Some pertinent relations are sketched out here in a few examples.

Dragline spider silk, Sect. 3.8.4, stretches to $\Delta L/L \approx 40$, or $L/L_0 = \sqrt{2}$, see Fig. 11.5a. At this stretch the deflection angle is about 45°, and the lateral force that can be supported is a maximum at about 59% of the breaking strength $2F_0$ of two strands. For smaller deflection angles, the tension in the strands gets increasingly larger, for larger deflection angles the draglines are stretched so much that the cross section decreases significantly.

Fish can be described by their body height D and their length L. Fast fish have acquired quite similar body shapes, Sect. 3.4.4. Their aspect ratios $X=D/L$ are approximated as $X \approx 0.25$, with distinct tendency towards smaller X-values for larger masses. This aspect ratio yields a minimum for the total drag, Fig. 11.5b.

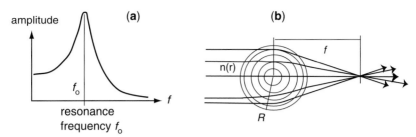

Fig. 11.6. (a) Resonance. (b) Fish eye lens

Of particular use in many different settings is the resonance of driven periodic motion, Sect. 5.3.5. If there is a free choice of frequency, the resonance frequency is always preferred because it yields the largest amplitudes for a given applied force. Resonance plays a role in walking, running, flying, voice production, hearing, and light detection, Fig. 11.6a.

The collection power by lenses is optimized when the lens is shrunk into a sphere. This is realized in fish eyes, Sect. 8.4.3, which have a gradually varying index of refraction $n=n(r)$ that generates a sharp image point, Fig. 11.6b.

Finally, the selection of the size $L \approx 50$ nm of the ferromagnetic Fe_3O_4 crystals in magnetotactic bacteria is also an optimization, Sect. 10.5.3. Smaller Fe_3O_4 crystals are no longer ferromagnetic. Larger crystals break into different Weiss domains with different orientation, which can only be lined up by external magnetization. In the unmagnetized state their average magnetic moment is smaller than that of the single domain 50nm crystals. A chain of single Fe_3O_4 crystals of $L \approx 50$ nm, which magnetic bacteria carry in their bodies, has the highest possible magnetic moment.

11.2.6 The Phase Space Arena of Organisms

Organisms and their life functions can be described with physical parameters such as temperature T, length L, mass M, lifetime t, pressure p, etc. Consider an n-dimensional phase space with these parameter axes. Then one can imagine that all life forms are suspended in a donut-shaped volume of this n-dimensional parameter space of organisms, marked in Fig. 11.7.

Physics principles dictate a range for every one of these parameters in which life can flourish. Examples were given throughout the text for limits on the parameter *body mass*. *Position in space* is also limited, each species has a certain living space on earth, which is closely connected to an acceptable *temperature and pressure range*. Organisms are generally restricted to the *temperature range* of liquid water, and to ranges of the ambient pressure. Ranges are also imposed on the *accessible information*. Physics determines the range of light wavelength and acoustic frequencies that can be detected. Physics also restricts the minimum size of information bits in the acoustic and the optical realm.

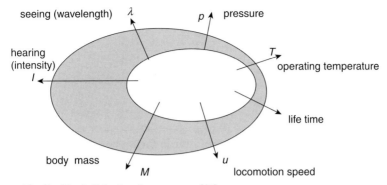

Fig. 11.7. The *livable* shell in the phase space of life

Many new achievements in the evolutionary chain push the phase space volume a bit further out, yet without fail, new limits are set by the laws of physics and the magnitude of material constants.

Consider for instance the temperature range in which life can flourish. Organisms are found in the entire interval between the freezing and boiling temperature of water. The edges are not *sharp* because some critters with antifreeze in their body can live on ice below 0°C. Certain bacteria are found in hot thermal vents at the bottom of the ocean at temperatures of 106°C. Within this range higher animals prefer to maintain their bodies at 37°C where the life functions are optimized. But other conditions may come into play. For instance the phytoplankton in the oceans, which is the bottom of the food chain nourishing all higher animals, grows most rapidly in the frigid waters of the Arctic and Antarctic during the summer months. Here daylight is much longer than in the tropics, and the cold water suspends more oxygen than warmer waters, so that these microscopic plants have all what they need to grow rapidly: O_2 and sunlight.

Life has experimented with myriads of design plans that often went to the limits of this phase space, models that could not meet the boundary conditions of the constantly changing geological world were discarded, leaving untried a probably even larger number of embodiments. Of course today's animals look much different than the first critters that roamed the earth. But the fossil records reveal that significant steps of evolution were often enabled by innovative uses of physics that had so far not been employed. As if the reward of mastering a new physical principle was a superior edition of an old design. Already at the time of the T-rex, this year's edition made last years model T look like an old dog.

11.3 Overcoming Limitations: The Workshop of Evolution

The fourth major point of Zoological Physics research – that new physics tricks become stepping stones for evolution – is only indirectly documented here.

One cannot fail to notice that physics rules supreme and governs the design and the actions of all animals. From this perspective physics becomes the grammar in the book of life that explains and links the zoological pictures. It seems that physical inventions are made quite independently at various levels of evolution. There is a continuous struggle between aggression and defense, where the stakes are raised in terms of physics sophistication. The workshop of evolution shows the continuous attempt to overcome limitations. Critters do not like to take *no* for an answer. Co-invention can be seen in such organs as wings or eyes. From the arms race of senses, one can read the arrow of time in the sophistication of the acquired physics. This observation also suggests that biospheres, which might exist on other planets in the universe, should likely evolve similar models of body and organs, like eyes, flippers, and wings, as we have here on earth. The laws of physics rule the inanimate world from atoms to stars. They also govern the body design of animals, and influence the actions of all living beings.

11.3.1 Measures and Countermeasures

Much of evolution hinges on increased sophistication of the use of physics laws. However physics is free to be used by anyone. The aggressor-defender arms race never stops. Some physics tricks can be annulled by physics countermeasures.

For example, bats learned to use sonar to detect prey that was previously hidden in the darkness of the night. However, eventually the moths that bats like to eat invented acoustic countermeasures, namely noises that jam the bat sonar.

Another example is the visibility of small and slow animals in the ocean. It is not easy to hide in the vastness of the sea. One possible trick is to become transparent – invisible. Jellyfish do it. They try to transmit and scatter light just like the medium water in which they live. Transparent as they are, they still scatter more light than the surrounding water. If sunlight is scattered at a right angle, it becomes polarized. Then the eyes of clever predators of jellyfish have adapted to recognize polarized light. Now the disguise of the jellyfish disappears and they become visible again to the better-equipped predators.

11.3.2 New Territory

Animals are fiercely territorial. Living space is the first condition for survival. The strongest individua claim the best traditional space right at the center. The weaker ones move as much as possible into previously unoccupied areas and into untested waters. Eventually every acceptable spot is taken. But what is acceptable?

New opportunities always beckon at the edges. Change your taste, change your habits, and change your body ever so slightly in response to new circumstances. Suddenly a desolate spot becomes hospitable for the modified species. Of course there are limits: there shall be light, and water, and ambient temperatures between 0 and 100 °C. But wait, what about the critters near the thermal vents at the bottom of the ocean? There is no light. Instead the organisms at the bottom of the food chain extract energy from the heat of the thermal vents, and not by capturing photons. What about the Galapagos finches, who live on islands without potable water? They replenish their body fluid by picking at the head of chicks of seabirds nesting there. What about small insects living on the ice, why does their body fluid not freeze up, rupturing body tissue with ice crystals? Antifreeze of course! They have chemicals in their body that lowers the ice point, below the ambient temperature.

Darwin's message that the merest trifle of advantage would drive evolution was cited at the beginning of this chapter. Since Darwin we have learned that the acquisition of new physics yields such advantages. Time and again the new models outperformed and displaced the *old dogs*. While we can clearly see the direction of evolution, and with some fantasy might even predict what else might come, provided mankind does not untimely destroy the habitat on our planet, the length of time steps of the evolution are harder to predict.

The animal kingdom is full of surprises. When one begins to look with the eyes of a physicist at the embodiments and the actions of animals, there arise many opportunities to ply the trade of a physicist: either to interpret what is there, or to speculate what could be there, and then look for the critter that has already done it, or just to suggest new experiments to the natural observer of animals.

11.4 The Open Door

This book was initially written as a teaching exercise to show how Physics complements the understanding of zoological problems. However, quite a number of the examples turned into research activities. The most important general lesson is that Physics restricts, or enables life functions, and that life functions and choices of organisms in turn limit the range of physical options.

The strength of Zoology is observing and reporting in great detail the design and behavior of animals. The strength of Physics is to make accurate predictions about grossly simplified *model* structures. This is an open door behind which loom new opportunities for gaining further insight. In writing this book a few rules for generating such zoological physics models became apparent. Most models are lower order approximations to more complicated problems, like tangents representing the local slope of curves. Different models may disagree when they represent different points of view, like the slope of the tangent on a circle taken at 12 o'clock and at 3 o'clock differ by 90°. Models are always less than the reality but they give quantitative results and they open the mind to advance the understanding of recalcitrant problems.

The opportunities for further research in *Zoological Physics* begin with inquiries about body design or functions of animals. Some obvious or provocative questions are posed in the last section of this chapter.

11.4.1 Elements of Zoological Physics Modeling

In order to analyze a body design or the function of an organ from the physics perspective, one must first have a clear picture of the biological function of the subject, and one must establish the important physical principles. A good Zoological Physics model *translates and simplifies* the requirements of biological activities into one or more equations between physical quantities. Then this set of equations is reduced into a single relation for a particular biological parameter (for instance the speed needed for a bird to stay aloft). The construction and testing of a model involves the following steps:

- Make two lists, one of the *biological functions* and another list of the *physics concepts* that might come into play.
- Write down pertinent physics equations for the concepts found in (1), identify biological material parameters that appear in these equations, and express the biological constraints (minimum biological conditions that must be satisfied). Most biological functions have equivalent physics relations. Recognize that if there are more unknowns than equations the system permits choices. Choice is an essential attribute of all living creatures (e.g. fight or flight).
- From this set of equations eliminate as many parameters as possible, and express a particular biological parameter f as function of the remaining physics parameters $p_1, p_2, p_3, \ldots$ This is your model. The remaining parameters can be freely chosen. (for instance an animal may chose to have a certain body mass, a certain length of limb, a certain lung displacement, or a certain preferred locomotion speed.)
- Test the model by choosing values for the parameters p_i and calculate the model quantity f with the help of your model equation. Compare this model prediction to empirical values. Make a list of the limitations of the model based on the assumptions, and attempt an error analysis.
- Reassess the model, especially the assumptions. What other body functions are affected by the biological system that was modeled? For example the big eyes of eagles, fixed in their head necessitates mobility of the neck. Look for limits imposed by physics laws onto the design or the activities of the animal.

Enjoy what you have been doing, but don't take your results too seriously. Your model likely is a one sided view of a complex life form. It could also well be that your results duplicate or contradict details that have long been known to the specialists in the field. Remember that your model may have errors in the assumptions or execution. A physical model is just a tool to interpret zoological phenomena: a viewing glass to look at particular aspects of a complex problem. Along the way to unravel the mysteries of the physics of animals there is ample room for a physicist to ply his or her trade.

11.4.2 Limitations of Modeling

A model is not an absolute truth. There may well be other ways to derive similar or even identical results. For instance, the refraction of light was explained by Huygens as a wave property, and by Newton as a particle phenomenon. As another example the pressure drop with height in the atmosphere, Sect. 3.3.1, can be derived from mechanics, from the kinetic theory of gases, as a thermodynamic cycle, and from the theory of diffusion [see Becker 1950].

This co-evolution of physics models is somewhat akin to the co evolution of animal organs. Since the starting points of these derivations are completely different assumptions one might conclude that all the results should be mutually exclusive and therefore wrong. The opposite is the case. Most likely all of them carry a self-consistent truth. All of them are acceptable partial explanations of the same phenomenon. In this vein physics models of animal functions or animal activities, constructed with quite different starting assumptions, may converge to the same truth. Yet there may also be alternate starting assumptions for describing the same animal functions that lead to different solutions altogether. Alternate solutions are likely to bring to the fore different aspects of the same phenomenon, and thereby advance the state of the art.

Finally some body functions must provide for multiple uses – like the air flow in the lung that serves breathing as well as speech. Then a model that tries to optimize the flow rate for the metabolic needs may be grossly inadequate for speech production.

11.4.3 Some Open Questions

When looking at such examples of the use of physics by animals, it becomes obvious that the same principles have been invented independently by entirely different species. Flying surfaces have evolved from folds of skin in bats and from scales growing into feathers of birds. Noise making apparatus evolved from the legs of crickets and from the vocal cords of mammals. This is evidence that what is possible will be achieved eventually. However, one can see that these applications evolve in small steps, never by a stroke of genius that created a fully developed organ in a giant leap of novelty. Generally only small deviations of the previous design are likely to be tolerated by any organisms. While error is the chisel of evolution, too large deviations mean certain death.

Out of these observation arise some questions that could be pursued in the future with the methods of *Zoological Physics* demonstrated here.

It should be possible to trace all *allometric relations* back to physics principles, and one should investigate the body mass range where each allometric relations is valid. By combining the metabolic rate $\Gamma \propto M^{3/4}$ and the allometric relation for the life time $T \propto M^{1/4}$ one can derive the total lifetime energy consumption e_L per unit mass as $e_L = \text{const } M^{3/4} \cdot M^{1/4}/M = C_L$. What is the significance of the life constant C_L?

Life means action and change. Change requires the expenditure of energy and change is associated with time steps Δt or frequency f. But energy can only flow at a certain rate from any given potential form. One must revisit results and conclusions that are only based on the conservation of energy to see if the resulting actions conform to the inherent energy transfer rates. This problem is particularly significant for locomotion where energy can be saved by resonance. When a resonance is identified one should look for mechanisms to *tune the natural frequencies* on.

What is the real cost of information to an organism. It should be calculated as an energy conserving exercise. Include must be the energy costs of growing and maintaining the sensors and the analyzing machinery.

When looking at the many forms of co-evolution one may wonder *is evolution inevitable*? Evolution is the deviation from the previous form, and error compared to the old design. After observing the steady progress that animals in every group have made as they evolved to higher forms, one might be tempted to suggest that there is a systematic sequence of these error steps, some optimization principle by which a) the direction, and b) the speed of evolution can be evaluated. There might be a way to assess the chance of life on other planets in the universe, where living conditions exist. In fact, if there is life on other planets somewhere in the universe, then creatures with organs such as lens eyes, wings, ears, bio-luminescence, and sonar would eventually evolve.

Is there a physics principles that by necessity advances the evolution of life? The genes do not drive the evolution, genes throw the dice, but physics decides which throw should count. Genes are the faithful scribes that record the laws for the next generation, using the genetic script as convenient alphabet. Since life only flourishes if energy sources are available it is likely that successful living means an optimization of the total available energy. Squander energy (increase entropy) *at the least possible rate*? This function would include as gains all the energy absorbed from food and sunlight, and contain as losses the total operating expense of the organism (idling power, locomotion costs, and expenses related to body growth and reproduction) as well as the expenses associate with the collection of food, and the acquisition of information. In such studies one might want to take into account that some early construction decisions might exclude the later adoption of certain techniques. For instance, since fish do not breathe air they have apparently foregone the ability to create acoustic, communication education and sonar.

It is an ambitious but very challenging research plan to look for such principles. However it is quite unlikely that we will ever quantify in physical terms the marvel of emotions associated with love, hate, friendship, and curiosity.

Problems and Essays

P 11.1 The Archer Fish Texotes Sagittarius Trepans

The archer fish $M \approx 0.05$ kg catches insects that perch on branches above the water, by spitting a mouthful of water at them (water bullet, $\Delta m = 0.0007$ kg) causing them to fall into the water. One of these fish, lets call him "Archie", cruises at a depth of 0.06 m underneath a large leaf that has fallen onto the water surface, Fig. 11.8. As he emerges from under the leaf Archie suddenly spots a juicy fly (measuring 11 mm from toe to wing-tip) sitting on a branch 0.30 m above the water. The edge of the leaf is 0.22 m away from the leaf, as shown.

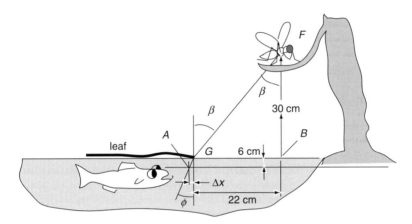

Fig. 11.8. The archer fish

a) At which distance Δx from the edge of the leaf can Archie first see his potential meal? b) Archie's eye ball has a diameter of $D=1.5$ mm. His spherical fish-eye lens has the F-number $f/D=1.28$. What is the focal length f, and what is the size of the image on Archie's retina? c) If Archie would stealthily swim to point B and shoot, what is the minimum energy that Archie has to impart to the water bullet to reach the fly at the height $h=0.30$ m? d) If the water bullet must have a velocity of 2.1 m/s when it reaches the prey in order to knock the fly off its footing, how large would the required energy of the water bullet have to be? e) What is the velocity at which Archie has to spit out the water in part d)? (Assume that the fish sticks its mouth just out of the water surface.) f) The water bullet is ejected by an over pressure, Δp generated in Archie's mouth. Determine Δp. (Give your answer in Pa, 1 Pa $=1$ N/m².) g) Each meal yields $\Delta m = 0.01$ g of protein. If Archie's metabolism could be described as a point on the mouse to elephant curve, how many flies would Archie need to consume in one week? (Assume $\Delta h = 30$ kJ/g of protein.) h) Consider the energetics of this hunt. Suppose Archie hits, and eats his victims in one out of 12 attempts. Is the exercise worth the energetic effort? Take into account that Archie has to swim around to spot the flies in the first place.

P 11.2 Size and Sound of the Past

The paleontologist Dr. Saurophil junior finds the partially preserved remains of *Parasaurolophus Clavacauda Curzonii*, Fig. 11.9. Dr. Saurophil records some data in his sketchbook. A copy is reproduced below. This animal had small claw-like front legs and very large hind legs on which it balanced its massive body. The head was supported by light neck bones 2.2 m long as shown and held in the extended position by strong tendons. From attachment points on the bones one can determine that the tendons had a cross section of 6.0 cm². The neck bones are anchored at a point about 1.2 m below the tendon attachment point. The head is only partially preserved, but it shows a crest tube which obviously served as an organ pipe like resonator that was closed at one end. The tube is angled, defining a 0.45 m long open front section, and the 1.45 m long closed tube end. The slender tail ended in a massive maze like ball of mass $M_c = 40$ kg, carrying two vicious spikes to fend of attackers. Most of the tailbones are lost, but the tendon attachment point indicates a cross section of 4.5 cm² and the tailbones are linked from a point 0.5 m below the tendon attachment.

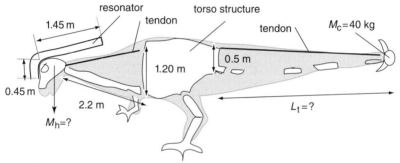

Fig. 11.9. Parasaurolophulus clavicaudae curzonii

Some questions bother Dr. Saurophil: How massive was the head? How long was the tail? What where the frequencies at which this animal would call its friends? a) Determine the mass of the head, neglecting the weight of the neck. Assume that the tendons are stretched to carry 1/8 of the maximum tensile strength of tendon material of animals living today. b) Determine the length of the tail, again neglecting the weight of the bones and muscles in the tail. c) Find the resonant frequencies of the resonance tube.

P 11.3 A Dogs Life

Consider a dog, called Casey, of body mass $M = 30$ kg. Latent heat of evaporation $L_v = 2257$ kJ/kg, viscosity of blood $v_{bl} = 4 \cdot 10^{-6}$ m²/s. a) What is Casey's basic metabolic rate in Watt? Express this number in Joule/day. How much dog food (specific enthalpy $\Delta h = 25$ kJ/g) does Casey have to eat just to support her basic metabolic rate? b) If Casey runs up a flight of stairs 2.8 m high in 2 s, how much mechanical power does her leg muscles generate? c) If the dog's leg muscles have an efficiency of $\eta = 20\%$, how much total power is generated by the dog as she runs up?

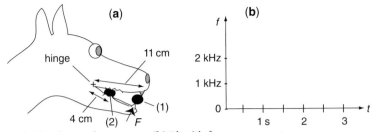

Fig. 11.10. (a) Dog's mouth geometry. (b) Plot it's frequency spectrum

d) On a hot summer day Casey keeps cool by sticking out her tongue. When she is lying on her favorite spot, not doing any work (basic metabolic rate) how much water must she evaporate in an hour in order to get rid of the heat? e) Casey often cracks nuts with her back teeth, which are as close to the jaw joint as the chewing muscles, Fig. 11.10a. If it takes a force of $F=300$ N to break a nut, how much force can she exert at location (1) with her incisors? f) How fast is the blood flowing in her aorta? Assume that the Reynolds number of the flowing blood in the aorta is just below the laminar – turbulent transition. g) Case's eye is somewhat like your own eye just smaller (let's assume the eye is exactly ½ the size of the human eye, but the rods and cones have the same lateral width $\Delta=1/400$ mm). Suppose she looks at a tennis ball (diameter 65 mm) at an object distance of $o=3$ m. What is the diameter of the image on the retina? h) How many rods and cones are covered by the image of the ball? i) Sound level: Casey can hear very faint sounds. She will raise her head when she hears a whisper of sound level $\beta=12$ dB. What is the pressure amplitude of this sound? j) Sound spectrum: Casey sometimes bangs her tail at a door in order to get in, and sometimes she howls with a clear tone (say of 880 Hz for about two seconds. Sketch the frequency – time traces of these two sounds into the frequency – time diagram, Fig. 11.10b.

P 11.4 Shared Meal
A true story, observed by the author in February 2001 in Vancouver at the university of British Columbia, Fig. 11.11. On a frosty winter day a crow ($M=\rho V=0.40$ kg), sitting on the roof of the Student Union Building (SUB), watches a student, $M=63$ kg sitting on a nearby the bench eating his SUB-standard 6″ turkey sandwich prepared at the Pendulum Restaurant for \$5.99, (rated as $\Delta H=1100$ kJ). Solar constant $S=1.37$ kW/m². After 5 minutes the student has eaten ½ of his meal, while the crow just absorbed the mild winter sun (intensity only 20% of its full power).

a) How much energy has the student ingested, and how long could he survive from this alone if his metabolic activity rate is a factor $b=3.5$ above his basic metabolic rate Γ_o. b) How much sun energy has the bird absorbed during this time? (Assume she is huddled into a sphere with feathers ruffled. Assume an absorption coefficient $\alpha\approx0.8$.) c) Suddenly the student becomes aware of a pretty girl nearby, and does not watch his sandwich. The crow sees her chance, and swoops down to steal the last half of the sandwich. How long was the student's attention side-

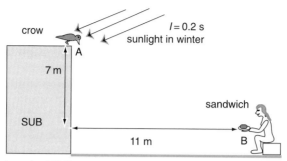

Fig. 11.11. Shared meal at UBC

tracked? (Calculate the time it took the bird to get from A to B. Find the crow's flight speed from the great flight diagram.) d) The crow flies back to her initial perch, and shows off her catch. What was the height h of the image of the $d=6''$ sandwich on the retina of the crow when she was eying her potential meal from the roof? (Treat the crow's eye as a miniature version of the human eye with $f=i=$ 6 mm.) e) Being University-bred, the crow croaks a polite *craw crow* (meaning *thank you, Sir* in Crowcanese) which the student (who unfortunately did not take Crowcanese 101) perceives only as a noise of $\beta=55$ dB. How large is the intensity I of this sound at the location of the student, and what it the acoustic power P emitted by the crow.

Essays

E 11.1 Size
What is the benefit of being small or large, and what limits the size? Describe some physical effects which small animals can utilize in their niche, and what physical effects give a lower limit to the size. Describe some physical effects, which big animals use to their advantage, and what physical effects may limit the maximum size of animals.

E 11.2 Choice of Senses
a) Is there a relation between the size of the animal and the range of distance of its dominant distance sense? b) Do land animals rely more on sight and sea animals more on sound? If so why? c) How can the senses be confused? d) How are the senses affected by daily or seasonal changes in the environment? e) Why do some animals mainly rely on their ears, other on their eyes, and some others on their sense of smell?

E 11.3 Help from Physics in Special Niches of the Biosphere
Describe the physical principles which one of the animals below uses to survive in its niche: alligator, bat, crow, dolphin, honey bee, monkey, octopus, pit snake, shark, or wolf.

E 11.4 Milestones of Evolution

From basic metabolism to language: describe some physical effects that evolution has" discovered" on the way from a single cell organism to *Homo sapiens*.

E 11.5 Survivors

Describe a physical principle which at one time in the history of life appeared to be a disadvantage but which some critter eventually has turned to its advantage.

E 11.6 Co-Evolution

Describe organs derived from different tissue that evolved into similar final forms.

Sample Solutions

S 11.1 The Archer Fish

a) The ray FG is refracted upon entering the water. Apply Snell's law; $\sin\beta/\sin\phi = n_w = 1.333$, where $\beta = \tan^{-1}(22/30) = 36.3°$. Then $\phi = \sin^{-1}(\sin 36.3/1.33) = 27.3°$. $\tan\phi = \Delta x/6$ cm, or $\Delta x = 6.0$ cm $\cdot \tan 26.4° = 3.10$ cm. b) $D/f = 1/1.28$, or $f = 1.28 \cdot D$. For pupil diameter $D = 1.5$ mm, $f = 1.5 \cdot 1.28 = 1.92$ mm. c) Minimum energy $E_p = mgh = 7 \cdot 10^{-4}$ kg $\cdot 9.8$ m/s$^2 \cdot 0.3$ m $= 2.1 \cdot 10^{-3}$ J. d) $E_{tot} = E_p + \frac{1}{2}mv^2 = 2.1 \cdot 10^{-3}$ J $+ 0.5 \cdot 7 \cdot 10^{-4}$ kg $\cdot$ (2.1 m/ s)$^2 = 3.64 \cdot 10^{-3}$J. e) $E_{tot} = \frac{1}{2}mv_0^2$. Hence $v_0^2 = 2 E_{tot}/m$, or $v_0 = 3.22$ m/s. f) Use the Bernoulli equation: Inside the mouth the water bullet is at rest, but it is exposed to the over pressure Δp. Outside the mouth the over pressure disappears, but the water has the kinetic pressure $\Delta p = \frac{1}{2}\rho v_0^2$, $\rho = 10^3$ is the density of water. Hence $\Delta p = 500 \cdot 3.22^2 = 5.18 \cdot 10^3$N/m^2. g) The metabolic need $\Gamma = 3.6 \cdot (0.05$ kg$)^{3/4} = 0.38$ W must be satisfied by N flies of energy content $E_f = 0.01$ g $\cdot 3 \cdot 10^4$J/g $= 300$ J/fly in the time interval one week: $\Delta t = 3600 \cdot 24 \cdot 7 = 6.05 \cdot 10^5$ s. Then 0.38 W $= N \cdot$ 300 J/fly/6.05 $\cdot 10^5$ s, or $N = 0.38 \cdot 6.05^5/300 = 766$ flies/week. h) Spitting effort: $E_{spit} = 12 \cdot E_{tot} = 12 \cdot 3.64 \cdot 10^{-3} = 4.36 \cdot 10^{-2}$ J. This is a very small fraction of the energy $E_f = 300$ J. The energy acquisition is well worth the effort.

S 11.4 Shared Meal

a) One half of the meal is $\Delta H = 0.5 \cdot 1.1 \cdot 10^6 = 5.5 \cdot 10^5$ J. The metabolic rate of the student is $\Gamma = b \cdot \Gamma_0 = 3.5 \cdot 3.6 \cdot 63^{3/4} = 282$ W. From $\Gamma = \Delta H/\Delta t$ find the time interval Δt during which the digested food could support the metabolism. $\Delta t = \Delta H/\Gamma = 5.5 \cdot 10^5$ J/282 W $= 1950$ s $= 32.5$ min. b) Solar energy absorbed by the bird: treat the bird as a sphere of radius R, and body mass $M = 0.4$ kg $= \rho \cdot (4\pi/3)R^3$. Solve for the radius $R = (3M/4\pi\rho)^{1/3} = (3 \cdot 0.4/4\pi \cdot 1000)^{1/3} = 4.57 \cdot 10^{-2}$ m. The bird (modeled as a sphere) absorbs the sun light from an area the size of its shadow. The shadow of a sphere is a circular disc of area $A = \pi R^2 = \pi \cdot (4.57 \cdot 10^{-2}m)^2 = 6.56 \cdot 10^{-3}$ m^2. Assume an absorption coefficient $\alpha = 0.8$ If the winter sun has an intensity of $I = 20\%$ of the full sun light ($S = 1.37$ kW/m^2). Thus $I = 0.2 \cdot 1.37 \cdot 10^3$ W/m$^2 = 274$ W/m^2, and the bird intercepts with its body the solar power $P = A \cdot \alpha \cdot I = 6.57 \cdot 10^{-3}$ m$^2 \cdot 0.8 \cdot 274$W/m$^2 = 1.44$ W. During $\Delta t = 5$ min $= 300$ s the crow absorbs the solar energy $E = P \cdot \Delta t =$

1.44 W $\cdot$ 300 s $=432$ J. c) First find the flight time of the crow. The crow must cover the distance $\Delta s = v\,(7\ m^2 + 11\ m^2) = 13.03$ m. Assume that the crow flies at speed from the great flight diagram, according to $U = 15\,M^{1/6} = 15 \cdot 0.4^{1/6} = 12.88$ m/s $= \Delta s/\Delta t_f$. Then the crow will cover the distance Δs in the time $\Delta t_f = \Delta s/U = 13.03\,m/12.88\,m/s = 1.01$ s. d) The image size I of a sandwich with the object height $O = 6$ inch $= 6$ inch $\cdot$ 25.4 mm/inch $= 0.152$ m on the retina of a standard eye ($\frac{1}{4}$ the size of a human eye with $i = f_{human} = 23$ mm $= 2.3 \cdot 10^{-3}$m) can be found if one knows the object distance $o = 13.03$ m. From similar triangles: $I/O = i/o = 0.25 \cdot 0.023\,m/13.03\ m = 4.4 \cdot 10^{-4}$, thus $I = i \cdot 4.4 \cdot 10^{-4} = 6.72 \cdot 10^{-5}$m. e) The sound intensity at the location of the student is $\beta = 55$ dB $= 10\,\log_{10}(I/10^{-12})$. Solve for the intensity I: first divide the equation by 10 so that $\beta/10 = 5.5 = \log_{10}(I/10^{-12})$, then raise the equation to the power 10, namely $10^{5.5} = 3.162 \cdot 10^5 = I/10^{-12}$. Thus $I = 3.162 \cdot 10^5 \cdot 10^{-12} = 3.162 \cdot 10^{-7}$. This intensity comes from a source at a distance $\Delta s = R = 13.03$m $I = P_{acoustic}/4pR^2$. Solve for $P = I \cdot 4pR^2 = 6.74 \cdot 10^{-4}$W.

Epilog

The topics of this book have been collected from many sources, and many other subjects could have been included. In the attempt to lay out the text and develop individual sections from first physics principles, often in ignorance of ongoing work in specialized areas, these pages may contain derivations and conclusions that differ from the accepted views. Some of the derivations may duplicate results and conclusions that are – without the authors knowledge – well known to the specialists in the field. Also, with certainty, these pages will contain *errors* and misrepresentations. They are hard to avoid when entering a new field. For these shortcomings I apologize.

Some of the data reported in tables or graphs do not have proper references. Over the years I had collected these date like common knowledge without recording the sources and now I don't know where they came from, and I am sorry not to be able to give proper credit. However, while acknowledged authorship is a badge of honor, an even higher achievement is to see one's work accepted permanently into the body of common knowledge.

> *Just as a tree knows not which sources nourished it to bloom*
> *So I don't know who added indirectly to this work.*
> *Thus, I extend my thanks to all my teachers,*
> *To colleagues and to students for their help.*

Boye Ahlborn

List of Tables

Energy Related Quantities

Information Related Quantities

References

Ahlborn, B.: "Thermodynamic Limits of Body Dimensions of Warm Blooded Animals", J. Non-Equilib. Thermodyn. **25**, 87–102 (2000)

Ahlborn, B. and Blake, R.W.: "Walking and Running at Resonance", Zoology **105**, 165–174 (2002b)

Ahlborn, B. and Blake, R.W.: "Why Birds Cannot be Smaller than Bees", Can. J. Zool. **79**, 1724–1726 (2001a)

Ahlborn, B. and Blake, R.W.: "Lower Size Limit of Aquatic Mammals", Am. J. Phys. **67**, 920–922 (1999)

Ahlborn, B., Chapman, Stafford, R.S., Blake, R.W., Harper, D.G.: "Experimental Simulation of the Thrust Phases of Fast-Start Swimming Fish", J. Exp. Biology **200**, 2301–2312 (1997)

Ahlborn, B., Harper, D.G., Blake, R.W., Ahlborn, D., Cam, M.: "Fish without Footprints", J. Theor. Biol. **148**, 521–533 (1991)

Ahlborn, B., Lefrançoise, M., King, H.D.: "The Clockwork of Vortex Shedding", Physics Essays **11**, 144–154 (1998a)

Ahlborn, B.: *How Animals Make Use of Physics*, Lecture Notes on Zoological Physics Edition 2002 (University of British Columbia, Vancouver, 2002a)

Ahlborn, B., Blake, R.W., Megill, W.: "Minimum Drag and Profile Thickness", J. Morphology **248**, 201 (2001b)

Ahlborn, B., Seto, M., Noack, B.: "On Drag, Strouhal Number, and Vortex Street Structure", Fluid Dynamics Research **30**, 379–399 (2002)

Ahlborn, B., Blake, R.W., Jacobs, S.: "The Hemoglobin Connection" in *Adsorption in Porous Solids* (VDI Reihe 3, Nr. 555, ed. Staudt, R., VDI Verlag, Düsseldorf, 1998b), p. 171–179

Andresen, B., Shiner, J., Uehlinger, D.E.: "Allometric Scaling and Maximum Efficiency in Physiological Eigen Time", Proc. Natl. Acad. Sci. USA **99**, 5822–5824 (2002)

Astrand, P.O. and Rodahl, K.: *Textbook of Work Physiology* (Mc Graw Hill, New York, 1970), pp. 210, 681

Balmer, R.T. and Strobusch, A.D.: "Critical Size of Newborn Homeotherms", J. Appl. Physiol.: Respirat. Environ. Exercise **42**, 571–577 (1977)

Bastian, J.: "Electrosensory Organisms", Physics Today **2**, 30–37 (1994)

Batchelor, G.K.: *An Introduction to Fluid Dynamics* (Cambridge University Press, 1967), p. 460–467

Becker, R.: *Vorstufe zur Theoretischen Physik* (Springer Verlag, Berlin, Goettingen, Heidelberg, 1950), p. 137–143

Bejan, A.: *Shape and Structure* (Cambridge Univ. Press, 2000), p. 260

Bejan, A.: *Advanced Engineering Thermodynamics* (John Wiley, New York, 1997), p. 631

Bennet–Clark, H.C.: "How Cicadas Make their Noise", Scientific American **5**, 338 (1998)

Bennet-Clark, H.C.: "Scale Effects in Jumping Animals" in *Scale Effects in Animal Locomotion* (ed. Pedley, T.J., Academic Press, London), p. 185–201

420   References

Blakemore, R. and Frankel, R.B.: "Magnetic Navigation in Bacteria", Scientific American **6**, 373 (1981)

Blevin, R.: *Applied Fluid Dynamics Handbook* (von Norstrand Reinhold & Company, 1984), p. 365

Bloom, M. and Mouritsen, O.G.: "Evolution of Membranes", *Handbook for Biological Physics*, Chap. 2 (Elsevier Sciences, 1995)

Busch, W.: Illustration from "Plitsch und Plum" (1882), p. 271

Busch, W.: Illustration from "Balduin Bählam" (1883), p. 31

Busch, W.: Illustration from "Maler Klecksel" (1884), p. 129

Clark, A.J.: *Comparative Physiology of the Heart* (Mac Millan, New York, 1927), pp. 157, 144

Clarkson (1968): cited in *Trilobites a Photographic Atlas* (eds. Riccardo Levi Setti, Univ. Chicago Press, 1975), p. 279

Conniff, R.: *So Sweet, so Mean* (Smithonean, Sept. 2000), p. 72–82

Curzon, F.L. and Ahlborn, B.: "Efficiency of Carnot Engine at Maximum Power Output", Am. J. Phys. **43**, 22–24 (1975)

Darveau, C.A., Suarez, R.K., Andrews, R., Hochatchaka, P.W.: "Allometric Cascade as Unifying Principle of Body Mass Effects on Metabolism", Nature **417**, 166–170 (2002)

Darwin, Ch. (Sept. 18, 1861): in a letter to a friend, see Shermer, M.: Scientific American **4**, 38 (2001)

Davidovits, P.: *Physics in Biology and Medicine* (Harcourt Academic Press, 2001), p. 176

Davison, M. and Shiner, J.S.: "Optimization in the Face of Contradictory Criteria ", J. Non-Equilibr. Thermody. **27**, 205–216 (2003)

Denny, M.: *Air and Water* (Princeton University Press, Princeton, 1993), pp. 89, 232

Denton, E.: "Reflectors in Fishes", Scientific American **6, 97–105** (1971)

Dickinson, M.: "Solving the Mystery of Insect Flight", Scientific American **6**, 49–57 (2001)

Ditfurth, H.v.: *Der Geist fiel nicht vom Himmel* (Deutscher Taschenbuch Verlag, 1980), pp. 116 ff.

Dumont, A. and Waltham, C.: "Walking", The Physics Teacher **35**, 372–376 (1997)

Farley, C.T., Glasheen, J., McMahon, T.A.: "Running, Springs, Speed, and Animal Size", J. Exp. Biology **185**, 71–86 (1993)

Ferris, D.P., Louie, M., Farley, C.T.: "Running in the Real World: Adjusting Leg Stiffness for Different Surfaces", Proc. R. Soc. London **B265**, 989–994 (1998)

Fritzsch, B. and Manley: 6th International Conference on Vertebrate Morphology, Jena 2001, Journal of Morphology **248**, 232 (2001)

Gamow, R.T. and Harris, J.F.: "Infrared Senses of Snakes", Scientific American **5, 66–73** (1973)

Gordon, J.M.: "Spherical Gradient Index Lenses as Perfect Imaging and Maximum Power Transfer Devices", Applied Optics **39**, 3825–32 (2000)

Goos, F. and Hänchen, H.: "Ein neuer und fundamentaler Versuch zur Totalreflexion", Annalen der Physik **1**, 333 (1947)

Gray, J. and Hancock, G.J.: "The Propulsion of the Sea Urchin Spermazoa", J. Exp. Biol., 802–814 (1955)

Handbook of Optics, 2nd Ed., Vol. 1 (1995), p. 43.28

Harris, J.E. and Crofton, H.D.: "Structure and Function in Nermatodes: Internal Pressure and Cuticular Structure in Ascaris", J. Exp. Biol. **34**, 116–130 (1957)

Hartmann, B.: "How we Localize Sound", Physics Today **11**, 24–29 (1999)

Hemilia, S. Nummela, Reuter, T.: "A Model for the Odsontocete Middle Ear", Hearing Research **133**, 82–97 (1999)

Hochachka, P.W. and Somero, G.N.: *Biochemical Adaptation – Mechanisms and Processes in Physiological Evolution* (Oxford University Press, New York, 2002)

Huggenberger, S., Rauschmann, M.A., Vogel, T.J., Oelschlaeger, H.H.A.: Abstracts 6th Int. Con. Vertebrate Morphogy, Jena 2001, Journal of Morphology **248**, 243 (2001)

Jacob, K.: "Brüchiger Schild", Bild der Wissenschaften **9**, 38–43 (2002)

Johnson, S.: "Transparent Animals", Scientific American **2**, 80 (2000)

Keeton, W.T.: "The Mystery of the Homing Pigeons", Scientific American **12**, 375 (1974)

Kleiber, M.: "Body Size and Metabolism", Hilgardia **6**, 315–353 (1932), see also Kleiber, M.: *The Fire of Life* (John Wiley & Sons, New York, 1961), p. 10

Konishi, M.: "Listening with two Ears", Scientific American **4**, 66–73 (1993)

Koretz, J.K. and Handelman, G.H.: "How the Human Eye Focusses", Scientific American **7**, 92–99 (1988)

Kryzka, J.: "Walking on a Treadmill", Zoological Physics P438 Project Report, UBC Vancouver (1999)

Leung, V.: "The Range of Pitch in Voice and Smoke", P438 Report, The University of British Columbia (April 1999)

Loewen, S., Ahlborn, B., Filuk, A.B.: "Statistics of Surface Flow Structures on Decaying Grid Turbulence", Phys. Fluids **29**, 2388–2397 (1986)

Luneburg, R.K.: *Mathematical Theory of Optics* (Brown Univ. Press, Providence, R.I., 1944)

Lutz, H.: *Ultraschallfiebel* (Springer-Verlag, Berlin, Heidelberg, New York, 1999)

Matthiessen, L.: "Über die Beziehungen zwischen dem Brechungsindex und Dimensionen des Auges", Pflüger's Arch. **27**, 510–528 (1886)

McKay, A.: UBC, private communication (2002)

McMahon, T.A. and Bonner, J.T.: *On Size and Life* (Scientific American Library, W.H. Freeman and Company, New York, 1983), p. 188

Megill, W.: private communication (2000)

Megill, W.: "The Tunable Oscillator of the Yelly Fish Polyorcas", PhD Thesis (University of Bristish Columbia, Vancouver, 2002)

Miller, D.T.: "Retinal Images an Vision at the Frontier of Adaptive Optics", Physics Today **1**, 31 (2000)

Mochon, S. and McMahon, T.A.: "Ballistic Walking: An Improved Model", Math. Bioscience **52**, 241–260 (1981)

Motani, R.: "Rulers of the Jurassic Seas", Scientific American **12**, 51–59 (2000)

Narins, P.M.: "Frog Communication", Scientific American **8**, 76–83 (1995)

Numella, S., Reurter, T., Hemillä, S., Holmberg, P., Paukku, P.: "Anatomy of the Killer Whale Middle Ear", Hearing Research **133**, 61–70 (1999)

Numella, S., Wägar, T., Hemilä, S., Reuter, T.: "Scaling of the Cetacean Middle Ear", Hearing Research **133**, 71–80 (1999)

Padian, K.: "Breathing Life into T-Rex", Scientific American **9**, 119 (1999)

Pettigrew, D., Manger, P.R., Fine, S.L.B.: "The Sensory World of Platybus", Phil. Trans. Roy. Soc. London, B. Biol. Sci. 1998, 353, (1372), 1199–1210

Pflegl Nickols, E.: "Nerve Conduction" in Tipler: *Physics for Scientists and Engineers* (Worth, 1991), p. 740 ff.

Philbrink, N.: *In the Heart of the Sea* (Penguin Books, New York, 2001)

Pickett, J.P.: Scientific American **9**, 104 (2001)

Preuschoft, H.: private communication (1999)

Preuschoft, H. et al.: "Size Influence on Primate Locomotion and Body Shape", Folia Primatol. **66**, 93–112 (1996)

Rossing, T.D.: *The Science of Sound* (Addison Wesley, New York, 1990)

Sataloff, R.T.: "The Human Voice", Scientific American **12**, 108–115 (1992)

Schleich, H.: *Biophysics of Electroreception, Biophysics* (Springer Verlag, Berlin, Heidelberg, 1973)

Schleich, H. et al.: "Electroreception and Electrolocation in Platybus", Nature **319**, 401–2 (1986)

Schmidt-Nielsen, K.: "Counter Current Systems in Animals", Scientific American **5** (1981)

Schmidt-Nielsen, K.: "Locomotion, Energy Cost of Swimming, Flying, and Running", Science **177**, 222–226 (1972)

Schmidt-Nielsen, K.: *Scaling: Why is Animals Size so Important?* (Cambridge University Press, Cambridge, 1984, 1993), p. 123

Shermer, M.: Scientific American **4**, 38 (2001)

Spatz, H.C.: "Circulation, Metabolic Rate, and Body Size in Mammals", J. Comp. Physiology B **161**, 231–236 (1991)

Strick, T. et al.: "The Manipulation of Single Bio-Molecules", Physics Today **10**, 46 (2001)

Strong, J.: *Concepts of Classical Optics* (Freemann and Company, San Francisco, London, 1958), p. 313

Suga, N.: "Bio-Sonar and Neural Computation in Bats", Scientific American **6**, 60–68 (1990)

Tennekes, H.: *The Simple Science of Flight* (MIT Press, 1997), p. 12

Tenny, S.M. and Remmer, J.E.: "Comparative Quantitative Morphology of the Mammalian Lung: Diffusing Area", Nature **197**, 54–56 (1963)

Tomatis, A.A.: *Der Klang des Lebens* (Rowohlt Verlag, Reinbeck bei Hamburg, 1987), p. 16

Turner, J.S. and Schroter, R.C.: "Why are Small Homeotherms Born Naked? Insulation and the Critical Radius Concept", J. Therm. Biol. **10**, 233–238 (1985)

Versluis, M., Heydt, A.v.d., Schmitz, B., Lohse, D.: "Snapping Shrimp", Science **289**, 2114 (2000)

Vogel, S.: *Life's Devices* (Princeton Univ. Press, 1988), p. 49

Wakeling, J.M. and Johnstone, J.A.: "Muscle Power Output Limt its Fast Start Performance of Fish", J. Exp. Biol. **201**, 1506–1526 (1998)

Walcott, C., Gould, J.L., Kirschvink, J.L.: "Pigeons Have Magnets", Science **205**, 1027–1028 (1979)

Walker, M.M, Diebel, C.E., Green, C.R.: "Structure, Sense and Use of Magnetic Sense in Animals", J. Appl. Physics **87**, 4653–4658 (2000)

West, G.B., Brown, J.H., Enquist, B.J.: "A General Model for the Origin of Allometric Scaling Laws in Biology", Science **276**, 122–126 (1997)

White, F.M.: *Heart Transfer* (Addison-Wesley, Reading MA, 1984)

Wilke, T.: "Die Tödliche Farbe des Mäuseurins", Bild der Wissenschaften **6**, 40–44 (2002)

Winklhofer, M., Holtkamp-Rotzler, E., Hanslik, M., Fleissner, G., Petersen, N.: "Clusters of Superparamagnetic Magnetite Particles in the Londoupper-beak Skin of Homing Pigeons: Evidence of a Magnetoreceptor?", Eur. J. Mineral. **13**(4), 659–669 (2001)

Yeung, M.Y.: "The Body Aspect Ratio of Fast Swimming Animals", Zoological Physics P**438**, Project Report, University of BC, Vancouver (2001)

Young, D. and Bennet-Clark, H.C.: "The Role of Tymbal in Cicada Sound Production", J. Exp. Biol. **198**, 1001–1019 (1995)

Zimmer, C.: "The Electric Mole", Discover **16**, 8 (1993)

Zoeger, J., Dunn, J.R., Fuller, M.: "Magnetic Material in the Head of the Common Pacific Dolphin", Science **213**, 892–894 (1981)

Index

Printing: Saladruck, Berlin
Binding: Stein+Lehmann, Berlin